New Frontiers in Oceanography

New Frontiers in Oceanography

Editor: Theodore Roa

www.callistoreference.com

Callisto Reference,
118-35 Queens Blvd., Suite 400,
Forest Hills, NY 11375, USA

Visit us on the World Wide Web at:
www.callistoreference.com

ISBN: 978-1-64116-114-5 (Hardback)

Cataloging-in-Publication Data

New frontiers in oceanography / edited by Theodore Roa.
p. cm.
Includes bibliographical references and index.
ISBN 978-1-64116-114-5
1. Oceanography. 2. Ocean. 3. Marine sciences. 4. Earth sciences. I. Roa, Theodore.
GC11.2 .N49 2019
551.46--dc23

Table of Contents

Preface

I am honored to present to you this unique book which encompasses the most up-to-date data in the field. I was extremely pleased to get this opportunity of editing the work of experts from across the globe. I have also written papers in this field and researched the various aspects revolving around the progress of the discipline. I have tried to unify my knowledge along with that of stalwarts from every corner of the world, to produce a text which not only benefits the readers but also facilitates the growth of the field.

The science of oceanography is the study of the physical and biological aspects of the oceans. It also covers the study of ocean currents, plate tectonics, seafloor geology and ecosystem dynamics. Oceanography helps in understanding diverse phenomena in the fields of meteorology, climatology, geology, hydrology, etc. It helps in developing a complete understanding of the world's oceans and the varied marine ecosystems. It branches out into marine biology, ocean chemistry, marine geology and marine physics. Modern oceanography also delves into ocean acidification, ocean heat content, ocean currents and temperatures, climatic phenomena of global warming and receding ice shelves. There has been a change in ocean chemistry leading to concerns over adaptability of marine life to changing ecosystem. This book attempts to understand the multiple branches that fall under the discipline of oceanography and how such concepts have practical applications. It compiles valuable contributions from researchers across the world, which have significance in the development of this field. Students, oceanographers and marine biologists will be immensely benefited from the interdisciplinary approach of this book.

Finally, I would like to thank all the contributing authors for their valuable time and contributions. This book would not have been possible without their efforts. I would also like to thank my friends and family for their constant support.

Editor

1

Extreme events in lake ecosystem time series

Ryan D. Batt,[1,a*] *Stephen R. Carpenter*,[1] *Anthony R. Ives*[2]
[1]Center for Limnology, University of Wisconsin – Madison, Madison, Wisconsin; [2]Department of Zoology, University of Wisconsin – Madison, Madison, Wisconsin

Scientific Significance Statement

Extreme events are large and potentially surprising. Although they are often studied in environmental variables like precipitation, they are far less often investigated in biological variables like population size. Are massive events common in biology? What types of dynamics tend to produce extremes? To answer these questions, we analyzed hundreds of decades-long time series collected from 11 lakes. We found that biological variables had the strongest tendency to produce record-breaking annual extremes, greater than the meteorological, physical, and chemical variables that, in part, drive the biological variables. Furthermore, the biological extremes were driven by within-year dynamics. The results imply that we should expect to be surprised by extreme fluctuations, particularly biological ones.

Abstract

Climate change has generated growing interest in extreme events, and extremes are known to have important consequences for ecosystems. Theoretical mechanisms generating frequent extremes apply to both environmental and biological processes, yet past studies of ecological extremes have focused primarily on the abiotic environment. The rarity or commonness of extremes in biological time series is unknown. We evaluated the statistical tendency to produce extreme events in 595 biological, chemical, physical, and meteorological time series taken from 11 lakes. We found that extreme events were much more likely to occur for biological variables than for other categories. Additional analysis revealed that the tendency to produce extremes was driven primarily by within-year dynamics, suggesting that processes occurring at short time scales underlie the high frequency of extremes in biological variables. These results should lead us to expect surprises in long-term observations of biological populations.

Extreme but uncommon events, such as floods and pest outbreaks, pose significant risks to the environment and society. The frequency and substantial ecosystem impacts of extreme events in environmental variables like temperature and precipitation are well studied (Gaines and Denny 1993; Peters et al. 2004; Denny et al. 2009; Fey et al. 2014). However, the prevalence of extremes in ecosystem variables is unclear because they have not been surveyed systematically. Furthermore, the historical tendency for ecosystem variables to show extremes may depend on whether they amplify or dampen sources of exogenous variability like weather extremes, which are increasing in frequency in many places (Mitchell et al. 2006; IPCC 2012). Thus, comparing the historical frequency of extreme events across components of ecosystems will further our basic understanding of ecosystem dynamics, particularly in the face of global change.

Extreme value theory (Coles 2001) is commonly employed to gauge the probabilities of large perturbations in

*Correspondence: battrd@gmail.com

[a]Present address: Department of Ecology, Evolution, and Natural Resources, Rutgers University, New Brunswick, New Jersey.

Author Contribution Statement: RDB and SRC conceived the study. ARI designed the mechanistic model and simulations. RDB and ARI wrote the R code and analyzed the data; analysis was guided by feed-back from SRC. RDB, SRC, and ARI wrote the paper.

time series (Palutikof et al. 1999; Katz et al. 2002, 2005). These large events can be minima or maxima, but henceforth we restrict our discussion to maxima because ecosystem variables are usually defined over the interval $[0, \infty)$. The "tailedness" of a distribution refers to the shape of the tail of a probability distribution, affecting its ability to produce frequent extremes and to generate record-breaking events. We measured tailedness by fitting the generalized extreme value distribution (GEV) to observations. The GEV has three parameters, location (μ), scale (σ), and shape (ξ). ξ measures tailedness: negative ξ indicates a bounded upper tail, ξ near 0 indicates a thin (exponential) tail such as demonstrated by a normal distribution, and positive ξ indicates that the maxima are fat-tailed.

Fat tails have been found in many types of data, although there is little empirical description of the tailedness of time series of most ecosystem variables, particularly biological variables. Anecdotally, ecologists often recall large events having been observed in their study systems, but rigorous quantification of the size and frequency of these events is lacking. Tailedness has been evaluated for environmental time series like flood stage (Villarini and Smith 2010), stream discharge and precipitation (Katz et al. 2002), sediment deposition (Katz et al. 2005), water temperature (Gaines and Denny 1993), wind speed (Gaines and Denny 1993; Palutikof et al. 1999), and wave force (Gaines and Denny 1993; Denny et al. 2009). Biological time series, by contrast, are rarely analyzed for tailedness. In the case of Segura et al. (2013), tailedness was calculated for the *changes* in population size of microbes, but not for population size itself. Schmitt et al. (2008) calculated the tailedness of the size of a single population of marine zooplankton, finding the population to be fat-tailed. We could not find more examples quantifying tailedness in biological time series. However, extreme events produced by fat-tailed time series are known to have significant impacts in other settings, making it important to identify their prevalence in biological variables.

Despite the paucity of studies of tailedness in biological variables, it is plausible that biological time series could be fat-tailed. One way for a biological time series to be fat-tailed is for a population to amplify the variability of the surrounding environment. As a heuristic example for how such enhancement may occur, consider a population governed by birth and survival processes that are in turn influenced by environmental (stochastic) drivers (Supporting Information: Extremes in a model population). Birth and death processes often act multiplicatively (Pielou 1977; Gurney and Nisbet 1998), which causes the population to have thicker tails than the environment (Fig. 1). The ubiquity of multiplicative processes in biological systems makes this mechanism for amplifying tailedness potentially widespread, although other mechanisms can generate fat tails such as some types of nonlinear dynamics (Sornette 2006; e.g., chapter 14).

Here, we document the fat-tailedness and time series characteristics of biological and environmental variables to

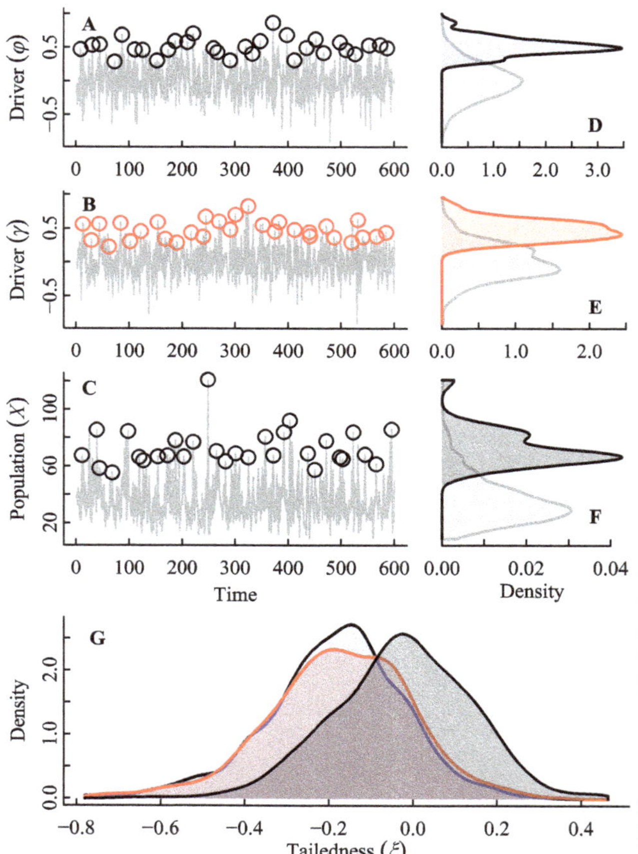

Fig. 1. The relationship between the tailedness of a population and its environmental drivers in the simple illustrative model presented in Supporting Information: Extremes in a model population. (**A**) Time series of a hypothetical environmental variable φ that influences birth rates, with annual maxima indicated by blue circles. Data were simulated for 30 yr with 20 measurements per year to represent the typical time series in our empirical analyses. (**B**) A hypothetical environmental variable γ influencing survival rates, with red circles for annual maxima. (**C**) Population size (X), with black circles for annual maxima. (**D–F**) Empirical density plots for the parent distributions in **A–C** (gray lines) and for annual maxima (blue, red, and black lines). (**G**) The blue, red, and black lines show density plots of ξ estimated from 500 simulations of the process exhibited in **A–C**. From theory, as the number of years becomes large, the estimates of ξ_X will always exceed ξ_φ and ξ_γ, although this is not always the case for the simulated 30 yr of data.

better understand the frequency and origin of extreme events in an ecosystem context. Our goal is to fill the knowledge gap about how often biological variables may be fat-tailed, and how the tailedness of biological variables compares to the tailedness of their environment. We surveyed 595 long-term lake time series to quantify the relative frequency of extremes in biological, chemical, physical, and meteorological variables. These variables were measured within the same ecosystems, providing a direct comparison between biological tailedness and the tailedness of environmental variables, many of which are expected to affect variation in the biological variables. We then analyzed these time series to determine if they showed evidence for generating fat tails at time scales shorter or longer than the sampling interval. This comparison constitutes a direct empirical assessment of the tailedness of a large number of biological time series, and may provide a clue as to if and how biology amplifies variability in exogenous drivers to produce extreme events.

Materials and methods

Data collection

Raw data were downloaded from the online databases of the North Temperate Lakes Long-Term Ecological Research (NTL-LTER) program (lter.limnology.wisc.edu/; Supporting Information). Data were downloaded on 18 April 2013, except wind data, which were downloaded on 30 March 2014. Most time series from the northern Wisconsin lakes begin in 1981. Most time series from the southern lakes begin in 1996, except fish time series that begin in 1995, zooplankton that begin in 1976, and three ice cover time series that begin in 1851, 1852, and 1877. Most time series end between 2010 and 2012. Measurement frequencies varied among time series; meteorological time series were typically daily, fish time series were annual, and other lake time series were quarterly or every 2–6 weeks, depending on the study site and time of year. Additionally, some lake time series were only sampled once during the winter, depending on ice conditions. Empirically, observation frequency did not have an influence on tailedness (Supporting Information Fig. S1). Northern Wisconsin lakes include Allequash Lake, Big Muskellunge Lake, Crystal Bog, Crystal Lake, Little Rock Lake, Sparkling Lake, Trout Bog, and Trout Lake. The only time series from Little Rock Lake was ice cover. Southern Wisconsin lakes include Fish Lake, Lake Mendota, Lake Monona, and Lake Wingra.

Meteorological variables included air temperature (daily minimum, maximum, range), sum daily precipitation, snow depth, and daily average wind speed. Physical variables included dissolved oxygen concentration, dissolved oxygen percent saturation, water temperature, the fraction of light at depth relative to light at surface, light extinction coefficient, Secchi depth, lake level, duration of the ice-free season, and the depth of the epilimnion. Chemical variables included alkalinity, bicarbonate-reactive silica (unfiltered), calcium, chloride, conductivity, dissolved inorganic carbon, dissolved organic carbon, dissolved reactive silica (filtered), iron, potassium, magnesium, manganese, sodium, ammonium, nitrate plus nitrite, pH, sulfate, total inorganic carbon, total organic carbon, total nitrogen (unfiltered), total phosphorus (unfiltered), and total particulate matter. Biological variables included the abundance of fishes and zooplankton, and epilimnetic chlorophyll concentration. Each genus of fish or zooplankton comprised its own time series.

Distribution fitting and waiting times

We fit a generalized extreme value distribution (GEV) to the time series of annual maxima for each variable; this is the theoretical limiting distribution of maxima that is approached for many statistical processes with large enough sample sizes, much like the normal distribution is the limiting distribution of sums of random variables (Coles 2001). The cumulative distribution function of the GEV is

$$F(x;\ \mu,\ \sigma,\ \xi)=\begin{cases}\exp\left\{-[1+\xi(x-\mu)/\sigma]^{-1/\xi}\right\} & \xi\neq 0\\ \exp\{-\exp[-(x-\mu)/\sigma]\} & \xi=0\end{cases}$$

where the case of $\xi\neq 0$ is subject to the constraint that $1+\xi(x-\mu)/\sigma>0$. Variables that follow very different distributions can nonetheless have their tails compared by the GEV because it is the limiting distribution for maxima. The extremes to which we fit the GEV were defined via the block-maximum method (Palutikof et al. 1999) with blocks corresponding to years to yield annual maxima. Each time series contained between 15 and 158 annual maxima (median = 29) and was categorized as biological ($n = 283$ lake time series, including chlorophyll, 28 fish genera, and 34 zooplankton genera), chemical ($n = 220$, e.g., ion concentration), physical ($n = 80$, e.g., water temperature), or meteorological (12 regional time series, e.g., mean daily wind speed).

We used the GEV to assess tailedness and estimate waiting time between record-breaking extremes. The GEV parameters were fit using maximum likelihood after removing a linear temporal trend from time series of maxima (Katz et al. 2005). We calculated the unconditional expected waiting time as the inverse of the probability p (calculated from the cumulative distribution function) of observing a value ≥ 10% over the current record, multiplied by the observation frequency (F, yr^{-1}) of maxima in the time series (F was usually 1); thus, waiting time = $1/p^*F$. This calculation does not account for possible autocorrelation in the annual maxima in the time series; if there is autocorrelation, then the expected conditional waiting time will depend on the most recent value of the time series. Statistical analyses were performed in the statistical programming language R (R Core Team 2014). In some cases (79 of the 595 time series; 0 biological, 44 chemical, 32 physical, 3 meteorological), waiting times were undefined because the value 10% over the record (x) was outside the support of the generalized extreme value distribution (GEV):

$$\begin{aligned} &x\in[\mu-\sigma/\xi,+\infty) && \text{when } \xi>0,\\ &x\in(-\infty,+\infty) && \text{when } \xi=0,\\ &x\in(-\infty,\mu-\sigma/\xi] && \text{when } \xi<0\end{aligned}$$

For example, the return level could be outside the support if the estimate of ξ is less than zero, and the return level is greater than the limit of the upper tail (μ–σ/ξ). Undefined waiting times were excluded from comparisons of waiting time between variable types. Note that removing the 79 time series with undefined (infinite) waiting times makes Fig. 2 conservative with respect to showing that biological time series have shorter waiting times than the other categories.

ARMA models

We used autoregressive moving average (ARMA) models to assess the relative contributions of processes occurring at time scales shorter (within-sample) or longer (between-sample) than a sampling interval (Supporting Information: Using ARMA to assess the time scale of sources of fat tails). ARMA models are a generic class of time series model capable of approximating many processes by predicting the current value of a variable from its past values and with past values of error terms. The parameter values of an ARMA model can be used to make general inferences about a variable's temporal dynamics, such as if it has long-term "memory," which is the autocorrelation of current values with values from many time steps ago. Furthermore, the residuals from the fitted ARMA models give information about the dynamics after memory has been removed. We fit ARMA models to parent time series, with the number of AR (p) and MA (q) parameters being chosen by AIC, subject to the constraint that $1 \leq p \leq 3$ and $q < p$ (Supporting Information: ARMA models). If the set of block maxima for a time series is fat-tailed and that for the ARMA residuals is not, this indicates that between-sample dynamics are responsible for generating the fat tail. Conversely, if both the time series and the ARMA residuals are fat-tailed, then within-sample dynamics are responsible at least in part for the generating the fat tail (Ives et al. 2003, 2010). ARMA is only an approximation of these between-sample processes, and could fail in some cases. Nonetheless, this approach can be informative and provides a common analysis for comparisons across our large set of diverse time series.

Results

Frequency of extremes

We found a wide range of tailedness (ξ) in each variable category, but on average the biological variables had higher values of ξ than other categories of variables (Fig. 2A; Supporting Information Table S1), even after controlling for length of time series and the identity of the lake (pairwise comparisons with biological time series: chemical, $z = -10.1$, $p \ll 0.0001$; physical, $z = -11.7$, $p \ll 0.0001$; meteorological, $z = -2.1$, $p = 0.0234$ (Supporting Information Table S2). For waiting times to break a current record, the pattern across categories was similar to that for ξ: biological variables were likely to break their current records sooner on average than other categories of variables (Fig. 2B). These results indicate that the biological variables tended to produce extremes more frequently than other categories of ecosystem variables.

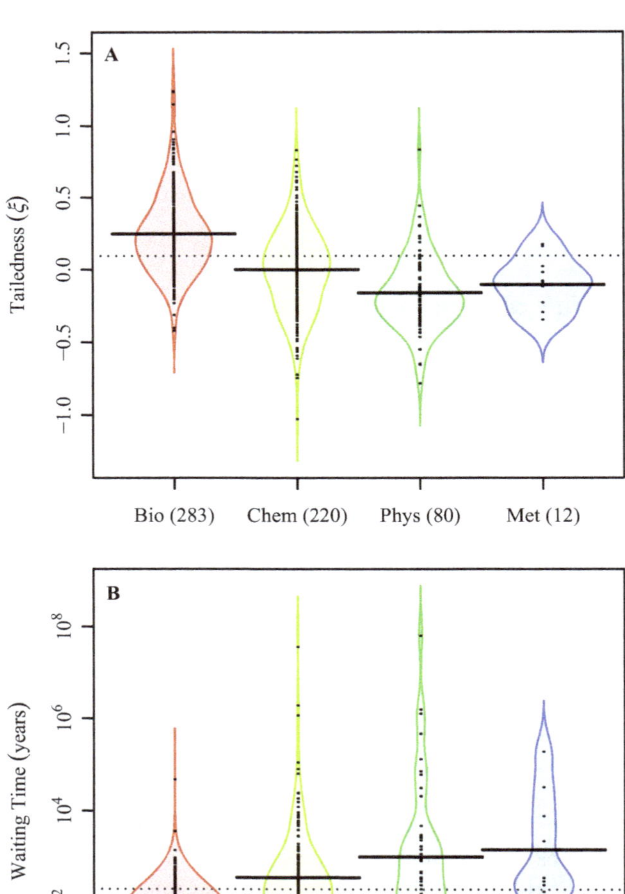

Fig. 2. (**A**) Bean plots of tailedness (ξ) estimated from the GEV, and (**B**) waiting times to the next record-breaking event estimated from the GEV fits. Waiting times are the time until the maximum observation in a time series is broken by at least 10%. Each "bean" consists of a mirror image of the density plot of ξ or of the waiting times in each category (red = biological, yellow = chemical, green = physical, blue = meteorological). The values for individual time series are indicated by the small black dashes inside each bean, and the number of time series in each category is in parentheses. Horizontal dotted line is the mean across all categories. Solid horizontal lines are the means within each category.

To further compare the different categories while accounting for uncertainty in the estimates of ξ, we tested whether the time series were fat-tailed by, for each time series, comparing the null hypothesis that $\xi = 0$ against the alternative $\xi > 0$ for positive estimates or $\xi < 0$ for negative estimates (1-tailed test with $\alpha = 0.05$, not correcting for multiple comparisons). We present results that are not corrected for multiple comparisons, and subsequently those with p-values that are corrected to maintain the false discovery rate (Benjamini and Hochberg 1995). Of the 283 biological time series, 38%

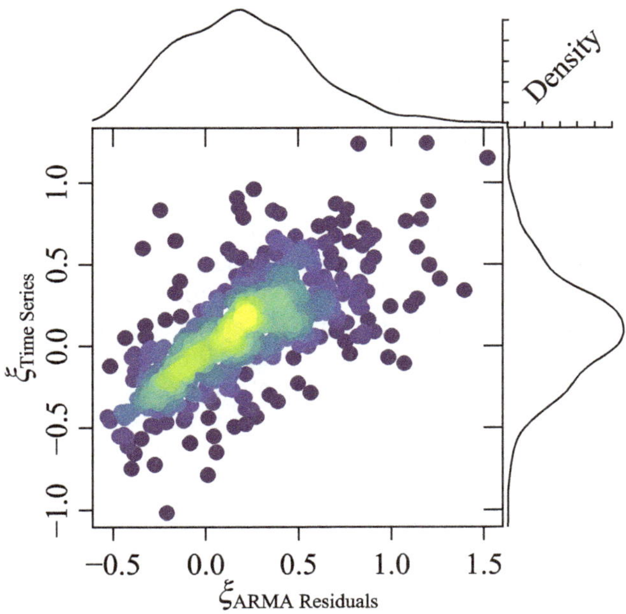

Fig. 3. Scatter plot of time series ξ vs. ARMA residual ξ. Density plots in the margins reflect the distribution of the parallel axis. The color of the points in the scatter plot is proportional to the number of points in that coordinate region, with yellow colors in regions overlain with many points.

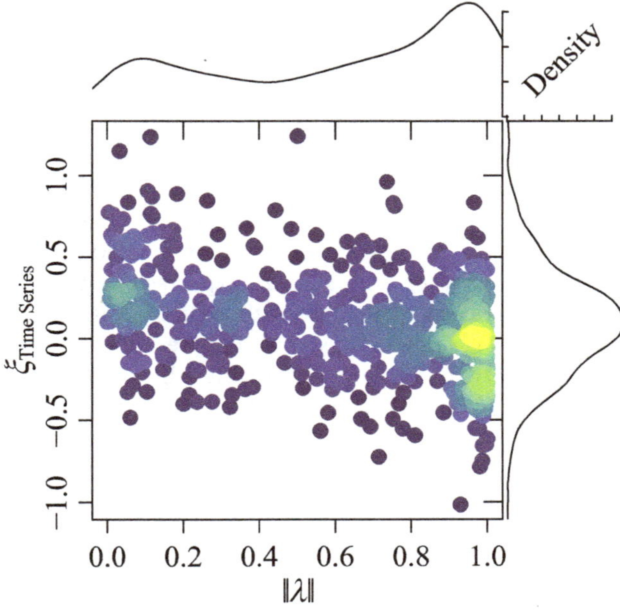

Fig. 4. Scatter plot of time series ξ vs. the dominant eigenvalue ($||\lambda||$) of the autoregressive parameters from the ARMA fits. Marginal plots and colors are as in Fig. 3.

had an estimate of ξ statistically significantly greater than 0 (corrected = 12%), by far the highest frequency of statistically significant fat-tailedness among categories (13% for chemical, 3.8% for physical, 0% for meteorological; corrected = 1.8%, 0%, and 0%, respectively). Using the same procedure to investigate distributions with bounded upper limits ($\xi < 0$), only 0.7% of biological time series had statistically significant estimates of $\xi < 0$ (corrected = 0.4%), which was the lowest frequency among variable categories (22% for chemical, 51% for physical, 42% for meteorological; corrected = 8.2%, 25%, 25%, respectively). Among the biological variables, the fish, zooplankton and chlorophyll time series with the largest estimates of ξ were, respectively, *Lepomis* spp. in Lake Mendota ($\xi = 1.2$; 90% CI: $0.7 < \xi < 1.8$), *Kellicottia* spp. in Crystal Bog ($\xi = 1.2$; 90% CI: $0.7 < \xi < 1.7$), and chlorophyll in Crystal Lake ($\xi = 0.5$; 90% CI: $0.2 < \xi < 0.8$).

ARMA dynamics

If extreme events were generated largely by long-term dynamics created by correlations across multiple time steps in the parent time series, then estimates of ξ for the original time series should be uncorrelated with estimates of ξ for the ARMA residuals. Additionally, the magnitude of correlation across time steps, measured by the eigenvalue of the ARMA process, should be positively correlated with the tailedness. Both of these expectations were false. Estimates of ξ for the time series were positively related with the estimates of ξ calculated for the ARMA residuals (slope = 0.40, $p \ll 0.0001$; Supporting Information Fig. S2, Fig. 3). Moreover, estimates of ξ for the times series were negatively related with the eigenvalues (slope = −0.23, $p \ll 0.0001$; Fig. 4). These correlations are consistent with the inference that extreme events were generated primarily by processes that occur within a time step.

Discussion

Biological time series had distributions with fatter tails than time series of other ecosystem variables. As tailedness (ξ) increases, the tail probabilities become large or "fat" relative to distributions like the normal, implying that extreme events will be more common. A fatter tail also decreases the time that passes until an event of arbitrarily large size is observed, increases the largest value observed after an arbitrary amount of time, and makes the chances of observing "massive" and "big" events more similar. Thus, in our analyses biological variables have larger and more frequent extremes. As values of ξ increase for a distribution, the statistical moments of the distribution $\geq 1/\xi$ become undefined (Coles 2001); for example, at $\xi \geq 1$ the mean (first moment) is infinite, and at $\xi \geq 0.5$ the variance (second moment) is infinite. By contrast, biological variables are often characterized by distributions that have defined means and variances (e.g., lognormal). However, the estimates of ξ for some of the fish and zooplankton time series were greater than 1, implying that their annual maxima will not converge to a long-term mean as more data are collected. Therefore, even distributions like the lognormal would underestimate the frequency and intensity of extremes in these fat-tailed biological variables. This result underscores the challenge that fat-tailed distributions pose for management: prevalent extreme events can make it impossible statistically to characterize even the mean abundance of a species.

Compared to meteorological, physical, and chemical time series, the biological time series we analyzed had larger values for tailedness (ξ) and shorter expected waiting times until record-breaking events, even though the variables were measured in the same ecosystems. This result provides evidence to suggest that biological variables should be expected to produce larger and more frequent surprises than the other categories of variables. Overall, these analyses provide evidence that biological variables are more likely to be fat-tailed than other types of variables in the same ecosystems.

Mechanisms

Mechanisms for promoting or diminishing fat tails are common in biological processes (Sornette 2006). Negative density dependence can lead to thinner tails and is common in population dynamics (May 1976; Ziebarth et al. 2010), giving reason to expect less fat-tailedness in biological variables relative to the environmental variables that drive them. However, nonlinear models fit to time series of disease (Sugihara et al. 1990), flour beetles (Dennis et al. 2001), spruce budworms (Ludwig et al. 1978), and phytoplankton (Benincà et al. 2008) can produce fat tails under certain parameter combinations, even deterministically. Nonlinear processes are common in biology (Hsieh et al. 2005), suggesting that they may produce fat tails in biology just as they are known to do in other settings (Barabási 2005; Newman 2005).

Our ARMA analysis suggests that mechanisms within a time step ($\leq$ 1 year) often generate observed fat tails, and that increasingly long memory cannot explain the observed fat tails. A simple process for generating such fat tails is for a population to amplify the variability of its environment. Multiplicative birth and survival processes influenced by stochastic environmental drivers are one way that fat tails could be generated within time steps (Fig. 1 and Supporting Information: Extremes in a model population). Comparing alternative mechanisms using models tailored for specific biological systems is needed if we are to better understand how fat tails are generated.

Concluding remarks

Our analysis included all of the variables typically collected by limnologists (Magnuson et al. 2006), and because our results might have differed if we had investigated different subsets of the variables, we analyzed all of the data that were available in a comprehensive analysis. Estimating a statistic based on rare events requires long time series (Palutikof et al. 1999; Katz et al. 2005), and estimates of ξ will improve with continued data collection. In fact, this need for long time series may serve as an explanation for the relative paucity of empirical assessments of biological tailedness—time series like those used in this study have only recently matured to a length suitable for extreme value analysis, whereas long environmental time series are more common. Therefore, continuing long-term monitoring of ecological variables and characterizing the tailedness of additional time series are critical for understanding ecological extremes.

Because fat tails dictate the frequency and intensity of massive events produced by an ecosystem, biological fat tails stand to greatly impact the management of populations that are critical to human well-being, such as fish stocks, invasive species, pests and diseases, or phytoplankton that form harmful blooms. Although there are several classes of general causal mechanisms for fat tails (literature cited above), more research is needed to identify causes of fat tails for specific populations. Our study focused on 11 lakes, but similar analyses of additional ecosystems are needed to test the generality of the patterns found here. Tests in additional ecosystems are also crucial for determining whether certain taxa consistently have higher tailedness, or if certain ecosystem characteristics can affect the tailedness of its denizens. Such patterns could help make predictions of extremes for ecosystems without long-term records, and would provide further insight into the causes of biological tailedness, its impacts on ecosystem functioning, and its implications for human use and management of ecosystems.

It is dangerous to consider the future as a set of norms from the past (Oppenheimer et al. 2007; Burgman et al. 2012), and forecasting future events, especially extreme events, is difficult. Our empirical results indicate that biological time series present frequent surprises. If this pattern was caused by directional environmental change in climate, nutrient inputs, habitat loss, or other factors, then extremes may become more common in ecosystem time series.

References

Barabási, A.-L. 2005. The origin of bursts and heavy tails in human dynamics. Nature **435**: 207–211. doi:10.1038/nature03459

Benincà, E., J. Huisman, R. Heerkloss, K. D. Jöhnk, P. Branco, E. H. Van Nes, M. Scheffer, and S. P. Ellner. 2008. Chaos in a long-term experiment with a plankton community. Nature **451**: 822–825. doi:10.1038/nature06512

Benjamini, Y., and Y. Hochberg. 1995. Controlling the false discovery rate: A practical and powerful approach to multiple testing. J. R. Stat. Soc. Ser. B **57**: 289–300.

Burgman, M., J. Franklin, K. R. Hayes, G. R. Hosack, G. W. Peters, and S. A. Sisson. 2012. Modeling extreme risks in ecology. Risk Anal. **32**: 1956–1966. doi:10.1111/j.1539-6924.2012.01871.x

Coles, S. 2001. An introduction to statistical modeling of extreme values. Springer.

Dennis, B., R. A. Desharnais, J. M. Cushing, S. M. Henson, and R. F. Costantino. 2001. Estimating chaos and complex dynamics in an insect population. Ecol. Monogr. **71**: 277–303. doi:10.1890/0012-9615(2001)071[0277:ECACDI]2.0.CO;2

Denny, M. W., L. J. H. Hunt, L. P. Miller, and C. D. G. Harley. 2009. On the prediction of extreme ecological events. Ecol. Monogr. **79**: 397–421. doi:10.1890/08-0579.1

Fey, S. B., and others. 2014. Recent shifts in the occurrence, cause, and magnitude of animal mass mortality events. Proc. Natl. Acad. Sci. USA **112**: 1083–1088. doi:10.1073/pnas.1414894112

Gaines, S. D., and M. W. Denny. 1993. The largest, smallest, highest, lowest, longest, and shortest: Extremes in ecology. Ecology **74**: 1677–1692. doi:10.2307/1939926

Gurney, W. S. C., and R. Nisbet. 1998. Ecological dynamics. Oxford Univ. Press.

Hsieh, C., S. M. Glaser, A. J. Lucas, and G. Sugihara. 2005. Distinguishing random environmental fluctuations from ecological catastrophes for the North Pacific Ocean. Nature **435**: 336–340. doi:10.1038/nature03553

IPCC. 2012. Changes in climate extremes and their impacts on the natural physical environment. In C. B. Field and others [eds.], Managing the risks of extreme events and disasters to advance climate change adaptation. Cambridge Univ. Press.

Ives, A. R., B. Dennis, K. L. Cottingham, and S. R. Carpenter. 2003. Estimating community stability and ecological interactions from time-series data. Ecol. Monogr. **73**: 301–330. doi:10.1890/0012-9615(2003)073[0301:ECSAEI] 2.0.CO;2

Ives, A. R., K. C. Abbott, and N. L. Ziebarth. 2010. Analysis of ecological time series with ARMA(p,q) models. Ecology **91**: 858–871. doi:10.1890/09-0442.1

Katz, R. W., M. B. Parlange, and P. Naveau. 2002. Statistics of extremes in hydrology. Adv. Water Resour. **25**: 1287–1304. doi:10.1016/S0309-1708(02)00056-8

Katz, R. W., G. S. Brush, and M. B. Parlange. 2005. Statistics of extremes: Modeling ecological disturbances. Ecology **86**: 1124–1134. doi:10.1890/04-0606

Ludwig, D., D. Jones, and C. Holling. 1978. Qualitative analysis of insect outbreak systems: The spruce budworm and forest. J. Anim. Ecol. **47**: 315–332. doi:10.2307/3939

Magnuson, J. J., T. K. Kratz, and B. J. Benson [eds.]. 2006. Long-term dynamics of lakes in the landscape: Long-term ecological research on North Temperate lakes. Oxford Univ. Press.

May, R. M. 1976. Simple mathematical models with very complicated dynamics. Nature **261**: 459–467. doi:10.1038/261459a0

Mitchell, J. F. B., J. Lowe, R. A. Wood, and M. Vellinga. 2006. Extreme events due to human-induced climate change. Philos. Trans. R. Soc. A **364**: 2117–2133. doi:10.1098/rsta.2006.1816

Newman, M. 2005. Power laws, Pareto distributions and Zipf's law. Contemp. Phys. **46**: 323–351. doi:10.1080/00107510500052444

Oppenheimer, M., B. C. O'Neill, M. Webster, and S. Agrawala. 2007. The limits of consensus. Science **317**: 1505–1506. doi:10.1126/science.1144831

Palutikof, J. P., B. B. Brabson, D. H. Lister, and S. T. Adcock. 1999. A review of methods to calculate extreme wind speeds. Meteorol. Appl. **6**: 119–132. doi:10.1017/S1350482799001103

Peters, D. P. C., R. A. Pielke, B. T. Bestelmeyer, C. D. Allen, S. Munson-McGee, and K. M. Havstad. 2004. Cross-scale interactions, nonlinearities, and forecasting catastrophic events. Proc. Natl. Acad. Sci. USA **101**: 15130–15135. doi:10.1073/pnas.0403822101

Pielou, E. C. 1977. Mathematical ecology, 2nd. John Wiley & Sons.

R Core Team. 2014. R: A language and environment for statistical computing. R Foundation for Statistical Computing.

Schmitt, F. G., J. C. Molinero, and S. Z. Brizard. 2008. Nonlinear dynamics and intermittency in a long-term copepod time series. Commun. Nonlinear Sci. Numer. Simul. **13**: 407–415. doi:10.1016/j.cnsns.2006.04.005

Segura, A. M., D. Calliari, H. Fort, and B. L. Lan. 2013. Fat tails in marine microbial population fluctuations. Oikos **122**: 1739–1745. doi:10.1111/j.1600-0706.2013.00493.x

Sornette, D. 2006. Critical phenomena in natural sciences: Chaos, fractals, selforganization and disorder: Concepts and tools. Taylor & Francis US.

Sugihara, G., B. T. Grenfell, and R. M. May. 1990. Distinguishing error from chaos in ecological time series. Philos. Trans. Biol. Sci. **330**: 235–251. doi:10.1098/rstb.1990.0195

Villarini, G., and J. A. Smith. 2010. Flood peak distributions for the eastern United States. Water Resour. Res. **46**: 1–17. doi:10.1029/2009WR008395

Ziebarth, N. L., K. C. Abbott, and A. R. Ives. 2010. Weak population regulation in ecological time series. Ecol. Lett. **13**: 21–31. doi:10.1111/j.1461-0248.2009.01393.x

Acknowledgments

We thank M. Pinsky, J. Gaeta, C. Herren, J. Kurtzweil, A. Latzka, J. Magnuson, S. Oliver, M. Turner, L. Winslow, and students in Zoology 955/6 (Spring 2013) and 995 (Spring 2014) at UW–Madison for discussions and comments on the manuscript. We thank the North Temperate Lakes Long-Term Ecological Research program for providing the data. RDB was supported by a postdoctoral fellowship from the Institute of Marine and Coastal Sciences at Rutgers University. Additional financial support was provided by NSF grants DEB-1440297 (NTL-LTER), DEB-1144683 (SRC), and DEB-Dimensions-1240804 (ARI).

2

The root of the problem: Direct influence of riparian vegetation on estimation of stream ecosystem metabolic rates

Walter K. Dodds,[1] Flavia Tromboni,[2] Wesley Aparecido Saltarelli,[3] Davi Gasparini Fernandes Cunha[3]*

[1]Kansas State University, Manhattan, Kansas; [2]Departamento de Ecologia, IBRAG, Universidade do Estado do Rio de Janeiro, Rio de Janerio, Brazil; [3]Departamento de Hidráulica e Saneamento, Escola de Engenharia de São Carlos, Universidade de São Paulo, São Paulo, Brazil

Scientific Significance Statement

Whole-system estimates of ecosystem respiration and gross primary production of aquatic systems are generally thought to reflect biological processes occurring in the water. However, living roots of plants growing next to the system might alter metabolism, and little is known of how important they are. We show that respiration by roots of riparian vegetation can substantially influence dissolved oxygen dynamics in a small stream, and do so with a diurnal pattern that mimics those caused by in-stream photosynthesis. We show that the intimate relationship between riparian vegetation and streams goes beyond leaf inputs, shading, and leaching of dissolved materials from terrestrial to aquatic habitats.

Abstract

Abundant living roots can be found in some streams and other shallow marine and freshwater habitats. A reach of a small Brazilian forested stream had 28% cover by live roots and exhibited diurnal trends in dissolved oxygen that could be attributed to gross primary production, but we hypothesized that activity of riparian tree roots in the channel caused this pattern. During sunny periods, trees transpire deoxygenated water from roots to the canopy but not in the dark, resulting in diurnal cycles of dissolved oxygen. Whole-stream shading experiments showed that photosynthesis in the stream is not responsible for the pattern. Sealed chamber measurements showed living roots of riparian vegetation had substantial respiratory activity and ammonium and nitrate uptake, and rates per unit area were greater than sand and less than silt (the other two dominant substrata), indicating roots can substantially alter in-stream biogeochemistry.

Whole-system metabolism (ecosystem respiration, ER, and gross primary production, GPP) is a central aquatic ecosystem function and is often used as an index of stream ecosystem health (Fellows et al. 2006; Correa-González et al. 2014) and the trophic state (Dodds 2006, 2007). Metabolism is affected by natural features of the aquatic and terrestrial ecosystems (e.g., biofilms, light availability, canopy cover and rainfall) as well as by the impacts from anthropogenic activities (Wang et al. 2003; Frankforter et al. 2010). However, we know of little work on an additional biological component in some streams, living tree roots in the open channel.

Streams and other aquatic habitats (e.g., lake shores, estuarine environments) can have exposed living roots associated with nearby vegetation. These roots and their associated biofilms might be an important component of benthic metabolism but their importance to whole systems is not well

*Correspondence: wkdodds@ksu.edu

Author Contribution Statement: WKD, DGFC, and FT obtained funding for the project. WKD initiated the idea for the project, participated in the field work, analyzed the data, and wrote the manuscript. FT constructed chambers, participated in field work, laboratory work, data analyses, and manuscript editing. DGFC and WAS participated in field work, laboratory work, and edited the manuscript.

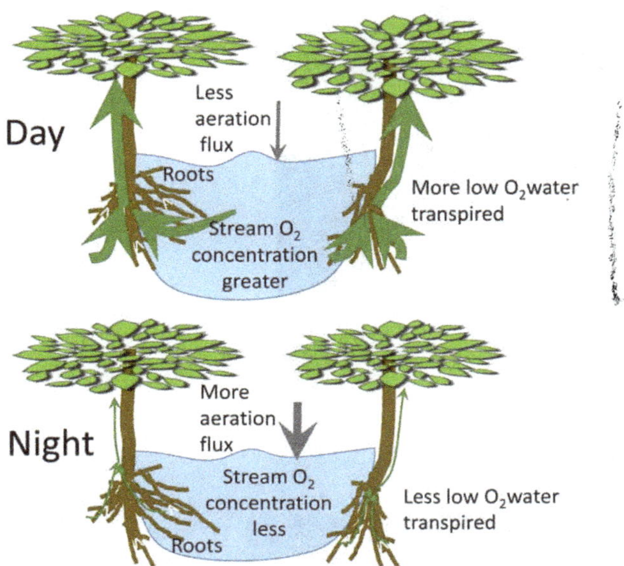

Fig. 1. Conceptual diagram of how diurnal changes in riparian vegetative evapotranspiration and water transport could influence dissolved oxygen concentrations in streams.

known. We found a diurnal swing in O_2 in a small tropical stream that could be GPP-related. However, little light reached the stream bottom and there was a very high mass of live tree roots lining the edges of the channel and streaming into the main flow. We suspected the roots might be involved in the diurnal O_2 swings. Evapotranspiration of plants is greater during the day than during the night, moving more water from the roots into the leaves during the day (Lambers et al. 1998). Thus, we hypothesized stream O_2 was related to evapotranspiration through roots (Fig. 1), and that tree roots were significant players in biogeochemical processing.

In the day, evapotranspiration rates increase, moving water from the roots into the aboveground plant tissues and at night evapotranspiration decreases and there should be less water movement. Roots continuously respire and consume O_2. During the day the O_2-depleted intercellular water in the roots tends to move up into the plant, lessening apparent O_2 demand in the stream. At night, all the O_2 demand needs to be met from O_2 diffusing into the stream from the atmosphere. Additionally, these roots could be very important to whole-stream metabolism and biogeochemical cycling. We tested these hypotheses by: (1) measuring light and dark metabolism of the roots and other substrata from two streams in sealed recirculating chambers to rule out photosynthesis, (2) we covered 100 m of a stream in dark plastic for 24 h to more convincingly rule out photosynthesis using whole-stream measurements, and (3) we tested N uptake rates of substrata in chambers using ^{15}N-ammonium or ^{15}N-nitrate to assess activity of tree roots and other substrata commonly observed in a stream.

Methods

We used two sites located in relatively pristine watersheds dominated by Cerrado vegetation and dense and diverse tropical forest streamside vegetation in São Paulo State, Brazil. Point transects within the studied reaches ($n = 30$) indicated sand, silt, and roots were the dominant cover types in these streams. The primary site is Espraiado Stream (S 21° 58.825′, W 47° 52.42′). This narrow and deep stream (average depth 24 cm, average width 67 cm) had a discharge of 14 L s^{-1} during our experiments. We made secondary measurements of root metabolism at a similar stream, Broa (S 22° 11.569′, W 47° 53.879′) about 20 km from the Espraiado Stream. This stream had a discharge of 35 L s^{-1}, with channel average depth and average width of 28 cm and 84 cm, respectively.

Metabolism and N isotope uptake measurements were made in recirculating chambers (Rüegg et al. 2015). The root measurements were made with a modified top to the chambers and other substrata with the original chamber top. For roots, we added an acrylic stand tube about 8 cm diameter and 5 cm high just upstream of the working area of the chamber (Fig. 2). Living roots were sealed into the chamber from the atmosphere using a latex glove with some fingers chopped out and rubber bands to tighten the glove to the roots and the wrist of the glove put around the stand tube.

Silt and sand were taken with minimally disturbed cores of 160 cm^2, to a depth of 1 cm and placed gently in trays in the chambers. Roots were left attached to the plants. Two incubations at Espraiado were made with a common streamside understory riparian plant (the nonnative *Hedychium coronarium*) that could easily be separated from the bank with the root systems intact. The remaining tree roots (species could not be determined given the forest diversity and intermingling of roots) were measured in submerged chambers.

Once substrata were sealed in the chambers, they were darkened and O_2 was logged every 30 s with an YSI ProODO O_2 meter (Yellow Springs Instruments, Yellow Springs, Ohio) with an optical sensor (± 0.1 mg L^{-1} accuracy). After at least a 0.03 mg O_2 L^{-1} decrease or 20 min, dark plastic was removed and chambers were incubated an equal amount of time in the light. All measurements were made during the daytime. The probe reads to 0.01 mg O_2 L^{-1} and a drop of only about 0.03–0.05 mg O_2 L^{-1} allows a reliable linear estimate of metabolic activity in many substrata.

Following incubations, substrata were removed and returned to the laboratory. Sand and silt samples were shaken with water and poured off to leave behind the large inorganic particles. The samples were then suspended and subsampled for dry mass, ash-free dry mass (Eaton and Franson 2005), and N isotope composition. Roots were dried and weighed.

Nitrogen uptake rates of individual substrata were measured in light incubations. We used stock solutions of ^{15}N

Fig. 2. Stream before plastic (**A**), after covering by plastic (**B**), chamber in stream with roots (**C**), and chamber outside of stream with roots still attached and going into the stream at the left behind the chamber (**D**).

$NaNO_3$ (98+% ^{15}N Isotec Company) of 108.6 mg L^{-1} of 0.1 N HCl (to preserve solutions in the field) and ^{15}N NH_4Cl (98+% ^{15}N, Aldrich) of 21.6 mg L^{-1} of 0.1 N HCL. We added 50 μL stock ^{15}N ammonium or nitrate stock solution per chamber with a 1.2- to 1.5-fold increase in ^{15}N for nitrate and an 11- to 40-fold increase for ammonium. The acid addition was too small to substantially alter the pH of the $\sim$ 10 L chambers. We sampled water chemistry before and after each incubation; samples were filtered in the field (Whatman GF/F) into acid-washed bottles, returned to the laboratory and frozen for analyses of nitrate (Eaton and Franson 2005) and ammonium (Mackereth et al. 1978). Incubations were terminated by gently rinsing substrata in label-free water in the field and filtering (sand and silt samples, see above) or drying (roots) immediately upon return to the laboratory. Additional samples were also collected for natural abundance background levels. Dried samples were ground and filters were analyzed directly for ^{15}N content and % N at the SIMSL facility at Kansas State University.

The flux of ^{15}N per unit area was calculated from the change in ^{15}N content of substrata over time and the mass of N per unit area. The total uptake of N was then scaled by the ratio of the $^{15}N/^{14}N$ in the chamber water. This ratio was calculated from the mass of ^{15}N added, the measured ammonium or nitrate concentration in the chamber, and the volume of the chamber. Equations for these calculations were taken from Dodds et al. (2000).

We covered a reach of 100 m of stream with black plastic (> 99.9% absorption) to darken the channel and directly establish that in-stream GPP was not responsible for the observed diurnal O_2 trends. Following 2 weeks of sunny and dry conditions, we monitored the uncovered stream for 24 h (starting 23 April 2016), placed 100 m of plastic covering over the stream for the next 24 h, and then removed the plastic for the last day. Plastic was not touching the water surface (Fig. 2), to avoid altering aeration rates. We monitored light above and at the bottom of the stream, water O_2 and temperature at the upstream end and at the bottom of the plastic-covered reach (two MiniDO_2T above and one Onset logger and one MiniDO_2T below), and depth (Hobo depth logger). Atmospheric pressure at surface was taken from a nearby monitoring station to correct depth readings.

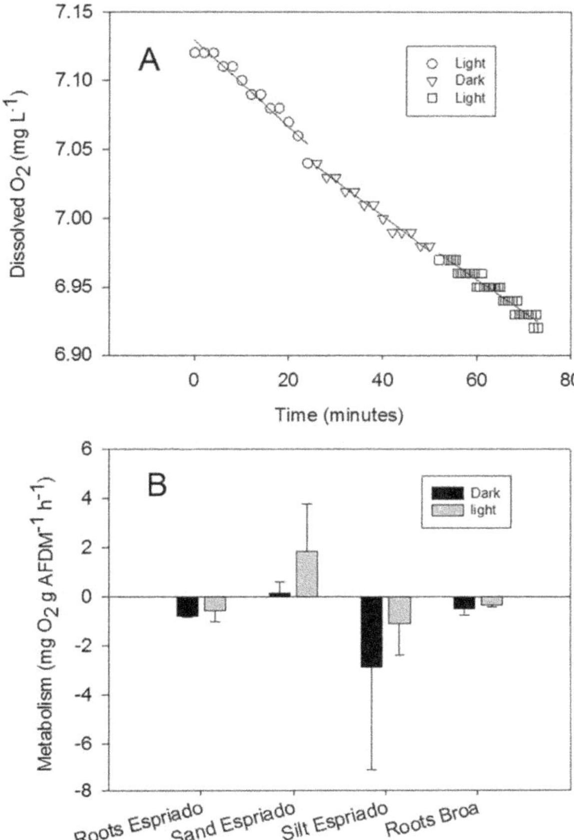

Fig. 3. Dissolved oxygen trace from a single root experiment alternating from light to dark and back to light from Broa stream, with regression slopes not significantly different (**A**) and metabolic rates of stream compartments from Espraiado and Broa normalized to ash free dry mass, in the light and in the dark (**B**). Error bars are one standard deviation.

Probes were calibrated against air-saturated water before and after deployment to account for drift during the 3-day deployment. We estimated canopy cover with a spherical densiometer (Jennings et al. 1999).

We measured aeration and discharge in Espraiado for modeling of metabolism and estimation of the "footprint" of each O_2 probe (Hall et al. 2016). Aeration was measured by releasing a solution of 238 g NaCl in 1 L of H_2O with 60 mL of SF_6 that had been equilibrated overnight, at 22 mL min^{-1} with a metering pump (FMI, New York). Conductivity was monitored at the bottom of the reach until plateau was reached. Gas was sampled along the reach and analyzed by gas chromatography (Hall and Tank 2005). Stream widths were taken at 20 transects across the reach. Discharge was measured with a pulsed addition of NaCl (750 g NaCL in 5 L H_2O) (Kilpatrick and Cobb 1985) metered with a logging conductivity meter at the bottom station (Hanna Instruments, Limena, Italy). Metabolism in the experimental reach was modeled for the day before, the day of, and the day after the plastic was placed on the stream with a nonlinear parameter estimation program (Riley and Dodds 2012).

Analysis of variance (ANOVA) was used to test for differences in metabolism and ^{15}N enrichment in Espraiado chamber experiments (Satistica 10.0, Statsoft, Tulsa, Oklahoma)

Results

Ambient light in the stream or streamside had little influence on metabolic rates (GPP or ER) of roots or other compartments in the stream as measured in recirculating chambers (Fig. 3). Canopy cover at Espraiado was 66% and at Broa was 72%. A light probe placed in the bottom of the Espraiado channel had about 6% of the light on average (when the stream was not covered with plastic) as compared to a probe above the water but below the canopy (Fig. 4). A single chamber run in the Broa stream that started in the light, was covered, and then reopened had no difference in regression slopes of O_2 vs. time ($p > 0.05$) suggesting no effect of light on net production and therefore no measurable GPP (Fig. 3A). Similarly, ash free dry mass-specific rates of ER (Fig. 3B) were not significantly influenced by light. Two-way ANOVA on Espraiado rates indicated that substrata had a significant effect on rates ($p = 0.023$) but light did not ($p = 0.139$).

In Espraiado Stream, transects put sand (31%), roots (28%), and silt (21%) as dominant cover types. However, the mass of organic material was dominated by silt followed by roots (Table 1).

When we scaled rates of respiration to the entire ecosystem, they were dominated by silt followed by roots (Table 1). This is because more organic material was associated with fine sediment and the rate of respiration per mass of dry sediment was greatest (Fig. 3). Still, respiration rates associated with roots and their attached biofilms were measurable and made up a substantial portion of respiration by the dominant substrata.

For the ^{15}N-ammonium and nitrate additions two-way ANOVA indicated that the amount of label differed significantly by substratum ($p = 0.040$) and treatment ($p = 0.042$), with a marginal interaction between substrata and treatment ($p = 0.054$). Uptake rates of nitrate and ammonium by organic material in sand did not differ significantly from zero (Table 1). Roots were more active in nitrate uptake than silt or sand, but uptake rates of ammonium were greater for silt than roots.

The darkening of the stream channel did not influence the daily oscillation of O_2 in Espraiado Stream (Fig. 4) and reduced light in the stream channel below detectable limits. While O_2 was always below saturation, there were consistent increases in O_2 during lighted periods. The peak O_2 concentration preceded the peak light, which preceded peak temperature.

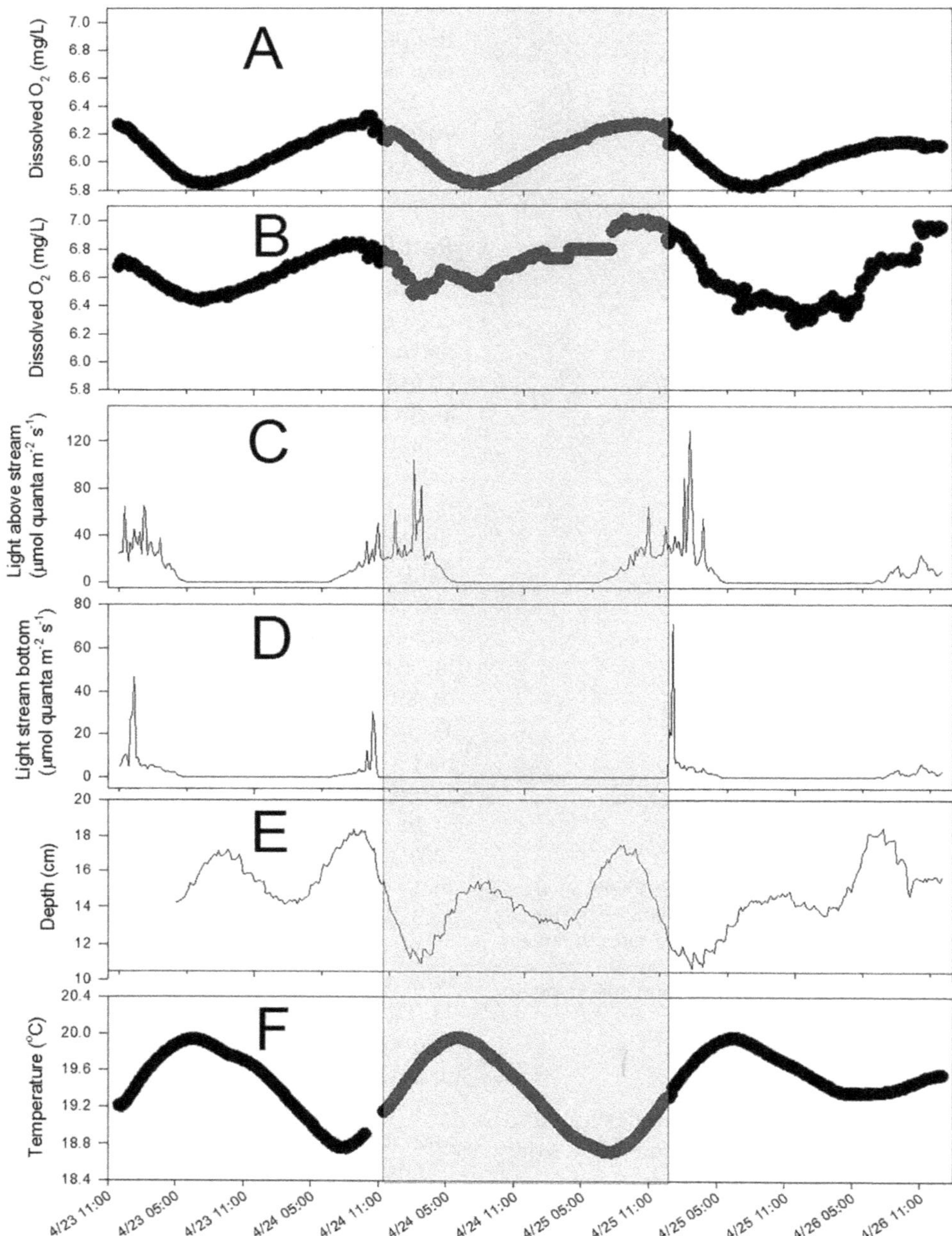

Fig. 4. Environmental conditions in Espraiado Stream before, during and after 24 h of stream being covered with black plastic. The time period with the plastic cover is indicated by the gray shading. (**A**) O_2 at bottom of treatment reach, (**B**) O_2 at top of treatment reach, (**C**) light above stream, (**D**) light at bottom of stream (note expanded scale relative to **C**, **E**) stream depth, and (**F**) stream temperature.

The ~ 7% difference in concentrations at peak O_2 and minimum O_2 during the shaded portion of the experiment suggests roughly 7% of the discharge would need to be transpired (assuming anoxic water is removed through the roots) to account for the concentration difference. The maximum diurnal effect on depth was a 42% decrease, meaning the water loss from the channel could more than account for the concentration difference. The differences between the upstream and the downstream probe during the shaded portion of the experiment were 0.4 and 0.7 mg L^{-1} O_2 during day and night, respectively. The diurnal difference of 0.3 mg L^{-1} O_2 that developed over the 100 m stretch of stream each day is greater than the accuracy of the O_2 probes suggesting the shaded region was long enough for our experiment.

The first 2 days of the experiment were done under clear skies. The third day was partly cloudy with somewhat lower

Table 1. Mass of material per unit area and rates of respiration, nitrate and uptake per unit mass and per unit stream area, and incubation nitrogen concentrations for dominant compartments (by area) in the Espraiado Stream. Where negative values for uptake are reported, they are essentially below detection as our method did not allow determination if mineralization exceeded uptake, likewise with a positive value for respiration.

		Roots		Silt		Sand	
Measurement	**Units**	**Mean**	**SD**	**Mean**	**SD**	**Mean**	**SD**
Total mass	g m^{-2}	103	57	119	106	47	20
Ash free dry mass	g m^{-2}	68	37	44	39	17	7
Respiration	g O_2 m^{-2} d^{-1}	−1.25	0.05	−2.51	1.80	0.03	0.18
NO_3^--N conc.	μg L^{-1}	38.1	18.4	−1.6	2.6	−0.2	0.1
NO_3^--N areal uptake	mg m^{-2} h^{-1}	38	18	−236	382	−97	79
NO_3^--N mass-specific uptake	μg g^{-1} min^{-1}	6.2	3.0	−33.1	53.5	−34.4	27.9
NH_4^+-N conc.	μg L^{-1}	184	92	1970	1818	−536	379
NH_4^+-N areal uptake	mg m^{-2} h^{-1}	30	15	276	255	−190	134
NH_4^+-N mass-specific uptake	μg g^{-1} min^{-1}	38	18	−236	382	−97	79

SD, standard deviation.

light at the stream surface (Fig. 4) and had a lower O_2 peak. This is consistent with the fact that diurnal oscillation in O_2 is not evident under completely cloudy and rainy conditions at other times of year (data not shown).

The plastic did not influence temperature or oscillations of depth in any obvious fashion. The difference in temperature from upstream and downstream did not change with or without plastic on the stream (data not shown).

Estimated GPP rates (as modeled by Riley and Dodds 2012) using the 2-station method across the experimental reach were 0.43 g O_2 m^{-2} d^{-1}, 0.72 g O_2 m^{-2} d^{-1}, and 0 g O_2 m^{-2} d^{-1} on days 1–3, respectively. The highest GPP estimate was on the day the plastic was covering the reach and the lowest on the cloudiest day where evapotranspiration would be expected to be lowest. Therefore, the model incorrectly indicated measurable GPP in a completely dark reach. The rates of ER for these 3 days were 15.7 g O_2 m^{-2} d^{-1}, 17.3 g O_2 m^{-2} d^{-1}, and 15.3 g O_2 m^{-2} d^{-1} on days 1–3, respectively, exceeding the weighted and scaled up sum of the rates from chambers by about fivefold.

Discussion

Our data show that living riparian roots that are in the stream channel can have a substantial impact on N uptake and O_2 dynamics in a small tropical stream. They were the second most important site of respiration and N uptake. We know of no other published measurements of metabolism or N uptake of roots and their associated biofilms while they are still attached and in the stream. We assume this lack of measurement is because of the technical difficulties that we solved with our modified chamber design. While tree roots in lotic environments can be important animal habitat and alter hydrology (e.g., Fritz et al. 2004), their biogeochemical role is less well defined.

There are several published examples of N uptake by roots from streams, or the nearby hyporheic. For example, natural abundance isotopic data shows that marine-derived nitrogen from anadromous salmon enters riparian vegetation (e.g., Mathewson et al. 2003). Also, a stream ^{15}N addition to a desert stream indicated that N moved from the stream into riparian vegetation (Schade et al. 2005). In-stream measurements of N uptake into different stream compartments based on ^{15}N addition experiments (e.g., Dodds et al. 2000) have not assessed the importance of living tree roots to N uptake to our knowledge. Our experiments suggest that roots can be an important N sink in stream nutrient budgets. We do not know if biofilms on the roots were responsible for the uptake, or the roots themselves. Macrophytes in European streams can form important substrates for biofilms that dominate N uptake (Levi et al. 2015), and biofilms on roots could be important as well.

Ammonium uptake rates were greater than nitrate uptake rates per unit biomass for roots and silt. Nitrate uptake can provide the bulk of the uptake in temperate forest trees (Nadelhoffer et al. 1984), but this might not be true for roots in our stream. If the nutrient taken up by the roots enters the trees, it might not be directly available for other stream compartments (but we did not separately estimate uptake by epiphytic biofilms). Tree leaves can fall back into the stream channel, and tree roots can be broken down by the microbial community in tropical and subtropical streams (Bloomfield et al. 1993; Fritz et al. 2006) and presumably enter the stream food web, so all the N entering the roots might not be lost to aboveground vegetation.

We find no other explanation for diurnal changes in stream flow and O_2 concentrations other than O_2-depleted water from within and around the roots moving up into the riparian vegetation. The main water sources for plant transpiration and for streamflow have been assessed for several ecosystems (e.g., Penna et al. 2013). Evapotranspiration by streamside trees can significantly influence discharge of streams, as mediated by groundwater and stream water depletion (Constantz et al. 1994) although riparian vegetation does necessarily use water from the stream (Dawson and Ehleringer 1991). Curiously, we observed two peaks and troughs of water depth per day and cannot explain this pattern, but the data indicate that diurnal cycles do drive changes in discharge. Tree roots can also facilitate O_2 transport to sediments through the vascular system of the plant and even pressurized transport can occur through the plant stem's intercellular vascular system (Grosse et al. 1992), but we suspect these processes are too slow to explain the observed O_2 patterns.

Chamber and shading experiments clearly ruled out photosynthesis in the stream channel as the causative factor of O_2 increases during the day. Only the sand exhibited positive (but nonsignificant) O_2 net production (Fig. 3). Given the low light in the stream channel, and that stream shading did not alter O_2 trends, or 2-station estimates of GPP, primary producers in the stream almost certainly did not cause the observed O_2 patterns.

Measured aeration rates can be used to estimate zone of upstream influence (Hall et al. 2016). Using this approach, the 80% O_2 turnover distance was 1.2 km, indicating that the diurnal trends in O_2 probably occurred over relatively long pieces of stream channel. However, O_2 dynamics can be influenced by processes relatively close to the point of measurement (e.g., Dodds et al. 2013). If measured water velocity characterizes upstream velocities, the expected travel time for this distance is 4.7 h. This distance would lead to substantial observed lags in the response of photosynthesis to light and tend to push the observed O_2 peak later than solar noon if in-stream photosynthesis caused the peak.

Our data suggest that increased evapotranspiration by the trees during sunny days move more O_2-poor water from the roots up into the plant than during the dark, leading to an observed increase in O_2 concentration in the stream channel. Temperature increases should decrease O_2 as saturation concentration decreases and respiration rates increase. We observed the opposite trend.

Some of the strongest evidence for our evapotranspiration hypothesis comes from the observation that the O_2 peak preceded peak sunlight, not the expected delayed peak that occurs after solar noon (Chapra and Di Toro 1991). The earlier peak is consistent with an explanation linked to plant evapotranspiration. The seasonally dry tropical tree *Simarouba glauca* had leaf hydraulic conductance that peaked mid-morning but was depressed mid-day to conserve water (Brodribb and Holbrook 2004). While we do not have data on tree species in the Cerrado area where our streams are located, our results are consistent with this general explanation. Saturation of photosynthesis is not a likely cause of the peak in O_2 preceding the peak in sunlight as the onset of saturation of photosynthesis for periphyton in streams is generally at substantially greater irradiances (if it occurs at all) than observed in the stream channel (Dodds et al. 1999).

Our results open the broader question of how important are roots in other aquatic habitats? Buried root activity is probably mostly ascribed to sediment respiration, and diffusive processes should average out the effect over time leading to less diurnal O_2 fluctuation. Exposed roots tend to occur where water movement precludes sedimentation, although some cases of exposed roots (e.g., floating vegetation islands or vines from trees hanging over still water) may occur in more quiescent waters. However, shallow waters in streams, wetlands, and perhaps estuarine areas can all have exposed roots. When living roots are in the water we expect the strongest effects occur when water depth is relatively shallow, high aeration does not flatten the diurnal curve (e.g., high energy streams), and the relative areal cover of living roots is fairly high.

In conclusion, tree roots can play a large role in biogeochemical processing in streams, overriding the temperature influence on O_2 solubility and respiration rates. Nighttime rates of ER should most strongly reflect total respiration occurring in the stream channel, as little water is moving into the trees in the dark. This particular stream had a very high biomass of active roots in the stream channel, so probably represents an extreme in the continuum. We know of no other measures of activities of roots and their associated biofilms in streams while they are still attached to living plants and they can be important in stream ecosystem function.

References

Bloomfield, J., K. A. Vogt, and D. J. Vogt. 1993. Decay rate and substrate quality of fine roots and foliage of two tropical tree species in the Luquillo Experimental Forest, Puerto Rico. Plant Soil **150**: 233–245. doi:10.1007/BF00013020

Brodribb, T., and N. M. Holbrook. 2004. Diurnal depression of leaf hydraulic conductance in a tropical tree species. Plant Cell Environ. **27**: 820–827. doi:10.1111/j.1365-3040.2004.01188.x

Chapra, S. C., and D. M. Di Toro. 1991. Delta method for estimating primary production, respiration, and reaeration in streams. J. Environ. Eng. **117**: 640–655. doi:10.1061/(ASCE)0733-9372(1991)117:5(640)

Constantz, J., C. L. Thomas, and G. Zellweger. 1994. Influence of diurnal variations in stream temperature on streamflow loss and groundwater recharge. Water Resour. Res. **30**: 3253–3264. doi:10.1029/94WR01968

Correa-González, J. C., M. del Carmen Chávez-Parga, J. A. Cortés, and R. M. Pérez-Munguía. 2014. Photosynthesis, respiration and reaeration in a stream with complex dissolved oxygen pattern and temperature dependence. Ecol. Modell. **273**: 220–227. doi:10.1016/j.ecolmodel.2013.11.018

Dawson, T. E., and J. R. Ehleringer. 1991. Streamside trees that do not use stream water. Nature **350**: 335–337. doi:10.1038/350335a0

Dodds, W. K. 2006. Eutrophication and trophic state in rivers and streams. Limnol. Oceanogr. **51**: 671–680. doi:10.4319/lo.2006.51.1_part_2.0671

Dodds, W. K. 2007. Trophic state, eutrophication and nutrient criteria in streams. Trends Ecol. Evol. **22**: 669–676. doi:10.1016/j.tree.2007.07.010

Dodds, W. K., B. J. F. Biggs, and R. L. Lowe. 1999. Photosynthesis-irradiance patterns in benthic microalgae: Variations as a function of assemblage thickness and community structure. J. Phycol. **35**: 42–53. doi:10.1046/j.1529-8817.1999.3510042.x

Dodds, W. K., and others. 2000. Quantification of the nitrogen cycle in a prairie stream. Ecosystems **3**: 574–589. doi:10.1007/s100210000050

Dodds, W. K., A. M. Veach, C. M. Ruffing, D. M. Larson, J. L. Fischer, and K. H. Costigan. 2013. Abiotic controls and temporal variability of river metabolism: Multiyear analyses of Mississippi and Chattahoochee River data. Freshwater Sci. **32**: 1073–1087. doi:10.1899/13-018.1

Eaton, A. D., and M. A. H. Franson. 2005. Standard methods for the examination of water & wastewater. American Public Health Association.

Fellows, C., J. Clapcott, J. Udy, S. Bunn, B. Harch, M. Smith, and P. Davies. 2006. Benthic metabolism as an indicator of stream ecosystem health. Hydrobiologia. **572**:71–87.

Frankforter, J. D., H. S. Weyers, J. D. Bales, P. W. Moran, and D. L. Calhoun. 2010. The relative influence of nutrients and habitat on stream metabolism in agricultural streams. Environ. Monit. Assess. **168**: 461–479. doi:10.1007/s10661-009-1127-y

Fritz, K. M., M. M. Gangloff, and J. W. Feminella. 2004. Habitat modification by the stream macrophyte *Justicia americana* and its effects on biota. Oecologia **140**: 388–397. doi:10.1007/s00442-004-1594-3

Fritz, K., J. Feminella, C. Colson, B. Lockaby, R. Governo, and R. Rummer. 2006. Biomass and decay rates of roots and detritus in sediments of intermittent coastal plain streams. **556**: 265–277. doi:10.1007/s10750-005-1154-9

Grosse, W., J. Frye, and S. Lattermann. 1992. Root aeration in wetland trees by pressurized gas transport. Tree Physiol. **10**: 285–295. doi:10.1093/treephys/10.3.285

Hall, R. O. J., and J. L. Tank. 2005. Correcting whole-stream estimates of metabolism for groundwater input. Limnol. Oceanogr. Methods **3**: 222–229. doi:10.4319/lom.2005.3.222

Hall, R. O., J. L. Tank, M. A. Baker, E. J. Rosi-Marshall, and E. R. Hotchkiss. 2016. Metabolism, gas exchange, and carbon spiraling in rivers. Ecosystems **19**: 73–86. doi:10.1007/s10021-015-9918-1

Jennings, S., N. Brown, and D. Sheil. 1999. Assessing forest canopies and understorey illumination: Canopy closure, canopy cover and other measures. Forestry **72**: 59–74. doi:10.1093/forestry/72.1.59

Kilpatrick, F. A., and E. D. Cobb. 1985. Measurement of discharge using tracers. Department of the Interior, US Geological Survey.

Lambers, H., F. Chapin, III, and T. Pons. 1998. Plant physiological ecology. Springer-Verlag.

Levi, P. S., T. Riis, A. B. Alnøe, M. Peipoch, K. Maetzke, C. Bruus, and A. Baattrup-Pedersen. 2015. Macrophyte complexity controls nutrient uptake in lowland streams. Ecosystems **18**: 914–931. doi:10.1007/s10021-015-9872-y

Mackereth, F. J. H., J. Heron, J. F. Talling, and Freshwater Biological Association. 1978. Water analysis: Some revised methods for limnologists. Freshwater Biological Association Scientific Publication, 36. Titus Wilson & Sons Ltd, Kendal, 117 p.

Mathewson, D., M. Hocking, and T. Reimchen. 2003. Nitrogen uptake in riparian plant communities across a sharp ecological boundary of salmon density. BMC Ecol. **3**: 1. doi:10.1186/1472-6785-3-4

Nadelhoffer, K. J., J. D. Aber, and J. M. Melillo. 1984. Seasonal patterns of ammonium and nitrate uptake in nine temperate forest ecosystems. Plant Soil **80**: 321–335. doi:10.1007/BF02140039

Penna, D., and others. 2013. Tracing the water sources of trees and streams: Isotopic analysis in a small pre-alpine catchment. Procedia Environ. Sci. **19**: 106–112.doi:10.1016/j.proenv.2013.06.012

Riley, A. J., and W. K. Dodds. 2012. Whole-stream metabolism: Strategies for measuring and modeling diel trends of dissolved oxygen. Freshwater Sci. **32**: 56–69. doi:10.1899/12-058.1

Rüegg, J., J. D. Brant, D. M. Larson, M. T. Trentman, and W. K. Dodds. 2015. A portable, modular, self-contained recirculating chamber to measure benthic processes under controlled water velocity. Freshwater Sci. **34**: 831–844. doi:10.1086/682328

Schade, J. D., J. R. Welter, E. Martí, and N. B. Grimm. 2005. Hydrologic exchange and N uptake by riparian vegetation in an arid-land stream. J. North Am. Benthol. Soc. **24**: 19–28. doi:10.1899/0887-3593(2005)024<0019:HEANUB>2.0.CO;2

Wang, H., M. Hondzo, C. Xu, V. Poole, and A. Spacie. 2003. Dissolved oxygen dynamics of streams draining an urbanized and an agricultural catchment. Ecol. Modell. **160**: 145–161. doi:10.1016/S0304-3800(02)00324-1

Acknowledgments

We thank FAPERJ (Fundação de Amparo à Pesquisa do Estado do Rio de Janeiro) Process Number E-26/100.018/2015 and CAPES (Coordenação de Aperfeiçoamento de Pessoal de Nível Superior) for the financial support to build the chambers, for isotope analyses, and scholarship support for F. Tromboni. We thank CAPES program Science Without Borders for the financial support (Process Number 88881.068045/2014-01) to W. K. Dodds as a visiting researcher in Brazil. W. A. Saltarelli thanks CNPq (Conselho Nacional de Desenvolvimento Científico e Tecnológico) for the scholarship and DGF Cunha thanks FAPESP (Fundação de Amparo à Pesquisa do Estado de São Paulo) for the financial support (Process Number 2014/02088-5). Page charges, some chamber construction, and salary for W. K. D. were funded by U.S. NSF Macrosystems grant #EF1065255. We thank Matt Whiles for urging us to figure this out and the reviewers and editors for helpful comments. This is contribution no. 17-157-J from the Kansas Agricultural Experiment Station.

3

Ice formation and the risk of chloride toxicity in shallow wetlands and lakes

Hilary A. Dugan,[1]* *Greta Helmueller*,[1,2] *John J. Magnuson*[1]
[1]Center for Limnology, University of Wisconsin – Madison, Madison, Wisconsin; [2]University of South Florida, College of Marine Science, St. Petersburg, Florida

Scientific Significance Statement

In urban environments across Midwest and Northeast North America, chloride concentrations are increasing in freshwater wetlands and lakes as a result of runoff from road salt (deicer) application. While it is known that ice formation on lakes can lead to a buildup of ions beneath an ice cover, no limnological research has linked ion exclusion owing to ice formation to chloride concentration under ice. Our study examines how the process of ice formation on shallow waterbodies may act to elevate chloride concentrations in already impacted systems, and increase chloride concentrations above toxicity thresholds.

Abstract

The process of ice formation in shallow waterbodies can increase chloride to toxic levels in waterbodies already impacted by chloride loading from road salt (NaCl deicer) application. Chloride concentrations were measured bi-weekly in a shallow, urban wetland in Madison, Wisconsin. We found that in this shallow waterbody, ice thickening doubled chloride concentrations from ion exclusion as the water froze. To understand the role of ice formation and ion exclusion, we constructed a numerical model to predict chloride concentrations beneath the ice resulting from ion exclusion. Where chloride levels already are elevated above background and flushing rates are low, ice thickening can push concentrations well above toxicity thresholds for much of the winter. The compounding effects of road salt runoff and ice formation should be considered in the management of water quality and ecosystem health in shallow urban water bodies or waterbodies receiving road salt runoff from nearby roadways.

In north temperate ecosystems, a major threat to freshwater quality is chloride runoff from road salt (specifically sodium chloride, NaCl) application (Findlay and Kelly 2011). Road salt runoff impacts wetlands (Richburg et al. 2001; Hill and Sadowski 2016), streams (Thunqvist 2004; Kaushal et al. 2005; Corsi et al. 2010), lakes (Novotny et al. 2008; Chapra et al. 2009; Dugan et al. 2017), and groundwater (Williams et al. 2000; Panno et al. 2006), and affects most lakes surrounded by impervious surfaces in areas of the U.S. Midwest and Northeast (Dugan et al. 2017).

Long-term increases in chloride concentrations in lakes has been shown across North America (Dugan et al. 2017), but these long-term trends are found in deeper lakes with medium to long residence times. The residence time of a lake is considered the mean time water or a dissolved substance (like chloride) spends in the lake before being flushed

*Correspondence: hdugan@wisc.edu

Author Contribution Statement: JJM designed the study; JJM and GH collected all data; HAD led the data analysis and modeling efforts. All authors contributed to developing and writing the manuscript.

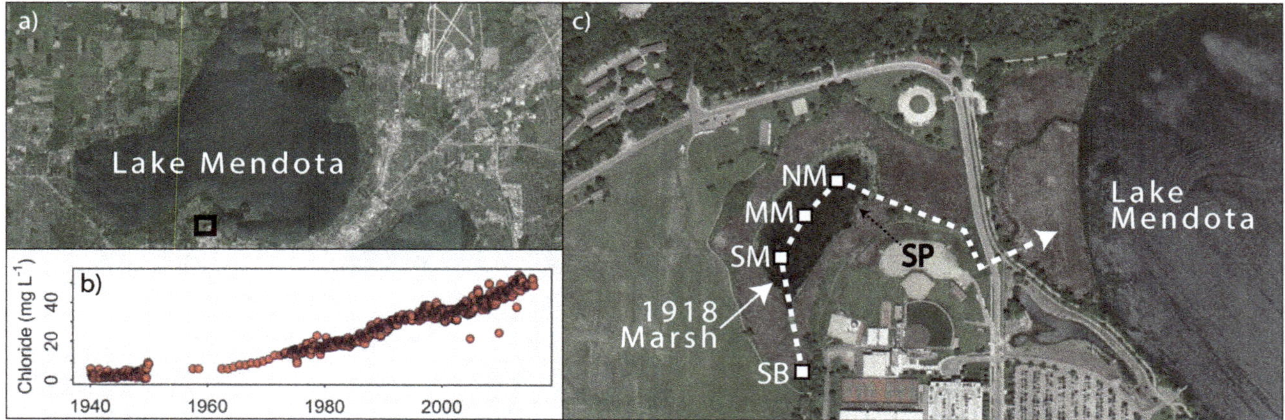

Fig. 1. (**a**) Lake Mendota (39.6 km^2) borders the city of Madison, Wisconsin, U.S.A. Outlined by the black box is the 1918 Marsh, a small marsh (0.074 km^2) on the south side of Lake Mendota. (**b**) Red circles represent chloride concentrations in Lake Mendota. Chloride has been increasing steadily since measurements began in the 1940s. (**c**) Water flows through the 1918 Marsh from South to North and into Lake Mendota. Sampling sites included South Bridge (SB), South Marsh (SM), Mid Marsh (MM), and North Marsh (NM). The campus snow storage pile (SP) is located immediately east of the marsh. Water drains from the snow pile toward North Marsh.

out. Longer residence times allow chloride concentrations to build up in a lake. For example, Lake Mendota in Madison, Wisconsin, has a mean depth of 12.8 m and a residence time of 4 yr. Chloride concentrations have risen from background levels of < 3 mg L^{-1} in 1940 to 50 mg L^{-1} in 2016 (Fig. 1) (Center for Limnology; NTL LTER 2012). In recent years, this equates to an increase of over 700 tons of chloride per year in Lake Mendota. Streams and shallow lakes may not experience long term increases in chloride concentrations owing to the short residence time of the system, but they are much more susceptible than larger lakes to rapid increases (and decreases) in chloride concentrations and the influence of chloride contaminated groundwater (Panno et al. 1999; Williams et al. 2000; Marsalek 2003). Shallow wetlands and lakes may be more vulnerable to disturbances, akin to river systems, than lake systems with longer water residence times (Resh et al. 1988).

Elevated chloride levels in wetlands and lakes can harm biological function (Corsi et al. 2010; Van Meter and Swan 2014; Hintz et al. 2017), and lead to decreased survival and impaired reproduction in a range of aquatic species and inhibition of plant growth (EPA 1988). We have used 230 mg L^{-1} as a conservative concentration below which there is likely to be relatively low chloride toxicity based on published criteria by the U.S. Environmental Protection Agency (EPA) and the Wisconsin Department of Natural Resources (WI DNR). The chronic and acute chloride criteria are set at 230 mg L^{-1} and 860 mg L^{-1} by the EPA (EPA 1988), and 395 mg L^{-1} and 757 mg L^{-1} by the WI DNR (Wisconsin DNR 2008). The "chronic" threshold is defined as the concentration that should not be exceeded by a 4-d average chloride concentration more than once every 3 yr. The "acute" threshold should not be exceeded by a 1-h average concentration more than once every 3 yr (EPA 1988). These criteria are recommended guidelines, and, in reality, specific aquatic plants and animals display a range of tolerances to chloride loading.

In northern climates, waterbodies (including lakes, ponds, and wetlands) are susceptible to high chloride loading from anthropogenic road salt runoff. In shallow waterbodies, this loading can be exacerbated by chloride exclusion during ice cover thickening. When water begins to crystallize into ice, dissolved ions are rejected into the surrounding water, and not incorporated into the ice lattice (Notz and Worster 2009). We hypothesize that ice cover formation in shallow water bodies with high chloride concentrations from anthropogenic inputs can raise chloride levels to toxic levels during winter. In cold climates, ice formation can freeze half the volume of a shallow waterbody, thereby potentially doubling ion concentrations. The process of ion exclusion from freshwater ice has only been documented in a few lakes (see Table 1).

Table 1. Ion and solute exclusion rates from lake ice.

Site	Solute	Exclusion rate (%)	Citation
Wisconsin	Chloride	60, 89, 99	This study
Laboratory	Conductivity	95	Bluteau et al. (2017)
Canada	Conductivity	87–99	Pieters and Lawrence (2009)
Canada	Conductivity	80–97*	She et al. (2016)
China	TDS	80	Zhang et al. (2012)
Canada	Ions	63–99†	Belzile et al. (2002)
Canada	CDOM	29–99†	Belzile et al. (2002)

* Calculated from pond and ice core concentrations.
† Converted from exclusion factor (*ef*: water column concentration : ice cover concentration) by 1–(1/*ef*). Biased rates, as pre-ice water concentrations are unknown.

Here we focus on the role of ice formation and thickening in elevating chloride levels in a shallow, open-water marsh already impacted by road salt runoff in Madison, Wisconsin. We use observational data and a numerical model to assess changes in chloride concentration based on chloride exclusion and dilution from changes in ice thickness.

Research site

The Class of 1918 Marsh is part of the Lakeshore Nature Preserve on the campus of the University of Wisconsin - Madison (Fig. 1). Surface water sources include runoff from parking lots, building roofs, sidewalks, a large area of undeveloped playing fields, and a snow storage pile for snow removed from campus streets, sidewalks, and parking lots (Fig. 1). Water plumbed from the roof of the hospital was added to the south-inlet storm sewer to dilute contaminants in the runoff from streets and parking lots. Presently the outlet waters are pumped into a storm sewer that passes under University Bay Drive and enters Lake Mendota's University Bay. Shallow groundwater likely contributes some inflow; however, both shallow ground- and surface water flows are reduced significantly owing to freezing conditions during the winter. The marsh in 2015 had a total area including the cattails of ~ 74,000 m^2, and open water area of 23,400 m^2. On 12 Dec 2012, water depths at 18 locations across the open water averaged 0.2 m and ranged from 0.1 m to 0.44 m.

Methods

Routine sampling sites (Fig. 1) included the south-inlet storm sewer (South Bridge) and three open water locations (South Marsh, Mid Marsh, and North Marsh). Water flows from South Marsh, to Mid Marsh, to North Marsh and eventually into Lake Mendota. Sampling was conducted every 2 weeks from Nov 2014 through the summer of 2015. Samples could not be obtained on 14 March 2015 because ice conditions were too dangerous for access. In 2014–2015, total winter precipitation (Dec–Mar) was 50% less than the climatological average (1960–2015) (Lawrimore et al. 2011).

All samples were collected at the surface of the marsh. If ice was present, a hole was first drilled with an ice drill or ice chisel (spud) or spud. Water was collected via syringes, and filtered through a 25 mm 0.45 μm GMF filter into plastic scintillation vials in the field. All samples were stored at 4°C and analyzed at the University of Wisconsin's Center for Limnology on an ion chromatograph (Dionex ICS 2100) using an electro-chemical suppressor. For each sampling event, water temperature, water depth, ice thickness, and snow depth were recorded. Water depth was measured from the water surface to the marsh bottom. Of the three open-water sites, no site was consistently the deepest or shallowest. The full range of water depths was between 0.12 m and 0.65 m; however, we suspect the shallower measurements were prone to error owing to soft bottom sediments and benthic macrophytes as well as by missing the narrow deep site at the South Marsh site. Given missing data and the uncertainty in water depths, we ran the numeric model using individual water depth measurements as well as minimum, mean, median, and maximum water depths across the three sites. Using maximum water depths across the three sites returned the lowest root mean square error (RMSE) in model fit.

Numerical model

We wrote a numerical model to predict the concentration of chloride in 1918 Marsh resulting from ice formation at three sites (South Marsh, Mid Marsh, and North Marsh) using measurements of water depth, ice thickness, and inflowing chloride concentrations. Because we had no flow data, our model does not include outflow. Therefore, it is only applicable for a system with a residence time greater than the model time step, otherwise any ion exclusion from ice cover formation would be flushed quickly downstream. During winter, the assumptions of no outflow and no evaporation/sublimation are acceptable, as freezing temperatures and short days minimize these fluxes.

Thus, we assumed that increased water depth in a two-week period was the result of inflow from the upstream area. The proportion of water depth attributed to inflowing water (**Pinflow**) was calculated as:

$$\mathbf{Pinflow} = \frac{\mathbf{water\ depth}_t - \mathbf{water\ depth}_{t-1}}{\mathbf{water\ depth}_t} \tag{1}$$

The inflow concentration was taken as the mean of the chloride concentration measured at time $\boldsymbol{t}$ and $\boldsymbol{t_{-1}}$. The proportion of the water column beneath the ice attributed to new ice formation $\boldsymbol{P\Delta}\mathbf{Ice}$ was calculated as the change in ice thickness divided by the average water depth:

$$\boldsymbol{P}\,\Delta\mathbf{Ice} = \frac{0.9 \times (\mathbf{Icethickness}_t - \mathbf{Icethickness}_{t-1})}{0.5 \times (\mathbf{Waterdepth}_t + \mathbf{Waterdepth}_{t-1})} \tag{2}$$

Ice thickness was converted to water equivalents by multiplying by 0.9, the normal density ratio of freshwater ice to water.

Chloride concentration was calculated differently based on whether ice was thickening or melting:

Ice thickening

Chloride concentration at time $\boldsymbol{t}$ ($\mathbf{Marsh[Cl]}_{\boldsymbol{t}}$) was calculated as the weighted average of two water parcels, (i) new water flowing into the marsh ($\mathbf{Inflow[Cl]} \times \mathbf{Pinflow}$) and (ii) water already present in the marsh ($(1 - \mathbf{Pinflow}) \times \mathbf{Marsh[Cl]}_{\boldsymbol{t}-1}$). The concentration of the latter increases from additional chloride due to ion exclusion $[(1 - \mathbf{Pinflow}) \times \mathbf{Marsh[Cl]}_{\boldsymbol{t}-1} \times \boldsymbol{P}\Delta\mathbf{Ice} \times \boldsymbol{Pexc}]$. The new concentration of chloride is divided by $(1 - \boldsymbol{P}\Delta\mathbf{Ice})$ to reflect the smaller volume of water present beneath the ice cover. $\boldsymbol{Pexc}$ represents the proportion of ions excluded from the ice, as some chloride ions are likely trapped between ice crystals. Since **Pexc** is not well constrained for freshwater systems, we ran the model 100 times for each site using ***Pexc*** values between 0% and 100%. Goodness of fit was evaluated based on the RMSE of the residuals.

In entirety, chloride at time ***t*** is modeled:

$$\mathbf{Marsh}[\mathbf{Cl}]_t = (\mathbf{Pinflow} \times \mathbf{Inflow}[\mathbf{Cl}]) + (1 - \mathbf{Pin\,flow}) \times \frac{[\mathbf{Marsh}[\mathbf{Cl}]_{t-1} \times (1 - \boldsymbol{P\Delta}\mathbf{Ice})] + [\mathbf{Marsh}[\mathbf{Cl}]_{t-1} \times \boldsymbol{P\Delta Ice} \times \boldsymbol{Pexc}]}{1 - \boldsymbol{P\Delta}\mathbf{Ice}} \quad (3)$$

Decreases in water depth were either the result of outflow or evaporation/sublimation, and were not taken into account in our model. Therefore, when **Pinflow** = 0, and Eq. 3 is simplified to:

$$\mathbf{Marsh}[\mathbf{Cl}]_t = \frac{[\mathbf{Marsh}[\mathbf{Cl}]_{t-1} \times (1 - \boldsymbol{P\Delta}\mathbf{Ice})] + [\mathbf{Marsh}[\mathbf{Cl}]_{t-1} \times \boldsymbol{P\Delta Ice} \times \boldsymbol{Pexc}]}{1 - \boldsymbol{P\Delta}\mathbf{Ice}} \quad (4)$$

Ice melt

When ice melt leads to a decrease in ice thickness, ***PΔIce*** is negative. ***Marsh*[*Cl*]*ₜ*** is the weighted average of inflowing water, water already present, and ice melt:

$$\mathbf{Marsh}[\mathbf{Cl}]_t = (\mathbf{Pinflow} \times \mathbf{Inflow}[\boldsymbol{Cl}]) + (1 - \mathbf{Pinflow}) \times \{[\mathbf{Marsh}[\mathbf{Cl}]_{t-1} \times (1 + \boldsymbol{P\Delta}\mathbf{Ice})]\} + [\mathbf{Marsh}[\mathbf{Cl}]_{t-1} \times - \boldsymbol{P\Delta}\mathbf{Ice} \times (1 - \boldsymbol{P}\mathbf{exc})] \quad (5)$$

When **Pinflow** = 0, and Eq. 5 is simplified to

$$\mathbf{Marsh}[\mathbf{Cl}]_t = [\mathbf{Marsh}[\mathbf{Cl}]_{t-1} \times (1 + \boldsymbol{P\Delta}\mathbf{Ice})] + [\mathbf{Marsh}[\mathbf{Cl}]_{t-1} \times - \boldsymbol{P\Delta}\mathbf{Ice} \times (1 - \boldsymbol{P}\mathbf{exc})] \quad (6)$$

The assumptions of this model, includes:

- A constant ion exclusion rate from ice cover thickening.
- The chloride concentration in the water remains the same during the time step.
- Surface and groundwater discharge over the time step is negligible.

For the 1918 Marsh, we ran the model at two-week time steps based on the availability of observational data. The inflow chloride concentrations ($\mathbf{Marsh}[\mathbf{Cl}]_{t=1}$) to South Marsh were taken from the South Bridge sampling site. South Marsh was then considered the inflow to the Mid Marsh site, and likewise Mid Marsh was taken as the inflow to the North Marsh. Observation data were used for initial chloride concentrations ($\mathbf{Marsh}[\mathbf{Cl}]_{t=1}$).

Alternatively, the model can be run independently for each time step, by using observational data for $\mathbf{Marsh}[\mathbf{Cl}]_{t-1}$ at each time step (as opposed to only setting initials conditions) and allowing the exclusion rate to change. At each time step, Pexc was chosen as the lowest calculated RMSE.

Results

Chloride concentrations in the 1918 Marsh increased from ~ 240 mg L^{-1} on 26 Oct 2014 to a maximum of 1250 mg L^{-1} on 1 March 2015. Following ice melt, concentrations in the marsh decreased and declined slowly from April to December, 2015 (Fig. 2a,b, Helmueller et al. pers. comm.). Overall, the mean concentration at the three open-water sites of 307 mg L^{-1} is approximately 6 times higher than the 2015 concentration of neighboring Lake Mendota, and 100 times above likely background concentrations (< 3 mg L^{-1}) based on early Lake Mendota values (Fig. 1b).

Chloride concentrations were similar at all open-water sites, except for 1 Mar 2015 when concentrations at North Marsh spiked to 1251 mg L^{-1}, compared to 894 mg L^{-1} at Mid Marsh, and 619 mg L^{-1} at South Marsh (Fig. 2a). Maximum chloride concentrations coincided with the maximum ice thickness (0.41 m) recorded on the marsh on 1 Mar 2015. At all open-water sites (South-, Mid-, and North Marsh), the change in chloride over a 2-week time step was correlated positively with the change in ice thickness ($r^2 = 0.69$, Supporting Information Fig. S1).

The model simulations of the 1918 Marsh were run from 26 Oct 2014 to 11 Apr 2015. The best model fits (lowest RMSE) at South-, Mid-, and North Marsh, were for exclusion rates of 60% (RMSE = 47, bias = 4%), 89% (RMSE = 97, bias = −9%), and 99% (RMSE = 179, bias = −16%) (Fig. 3). When using ion exclusion rates greater than 60%, the resulting water concentrations were markedly higher than those using an ion exclusion rate of 0% (Fig. 3) and were similar in magnitude and patterns to measured chloride over the ice cover period. The numerical model slightly underestimated chloride concentrations at the Mid- and North Marsh, and overestimated chloride during the winter at the South Marsh site (Fig. 3). Model fits were compared to observational data by calculating the RMSE of residuals. Because the model only considers the observational chloride concentration at the first time-step, the predictions are not "reset" following a poor estimate. Specifically looking at the South Marsh site, when **Pexc** > 0.6, the model overestimates chloride on 7

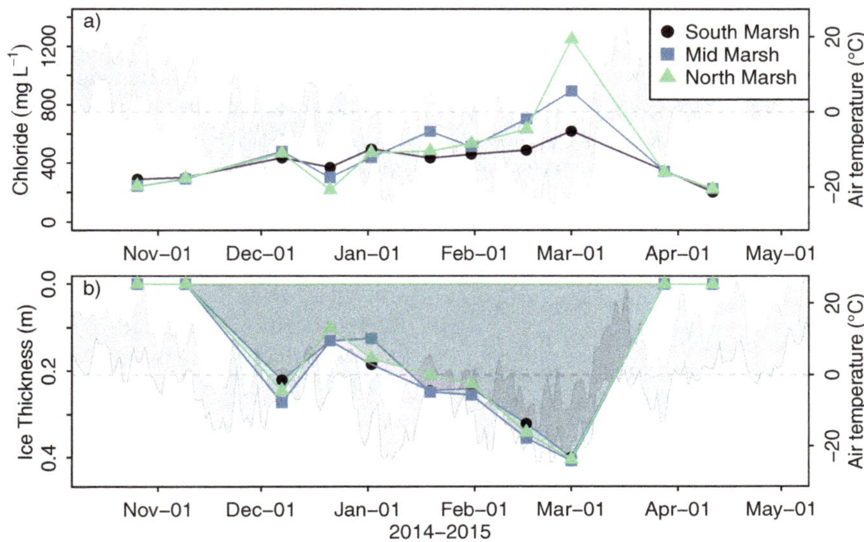

Fig. 2. (**a**) Chloride concentrations in South-(circles), Mid-(squares), and North Marsh (triangles) from Oct 2014 to May 2015. The gray timeseries represents the range in daily air temperature. (**b**) Symbols denote the measured ice thickness at the three sites. The lines and shaded area represent the seasonal ice thickness.

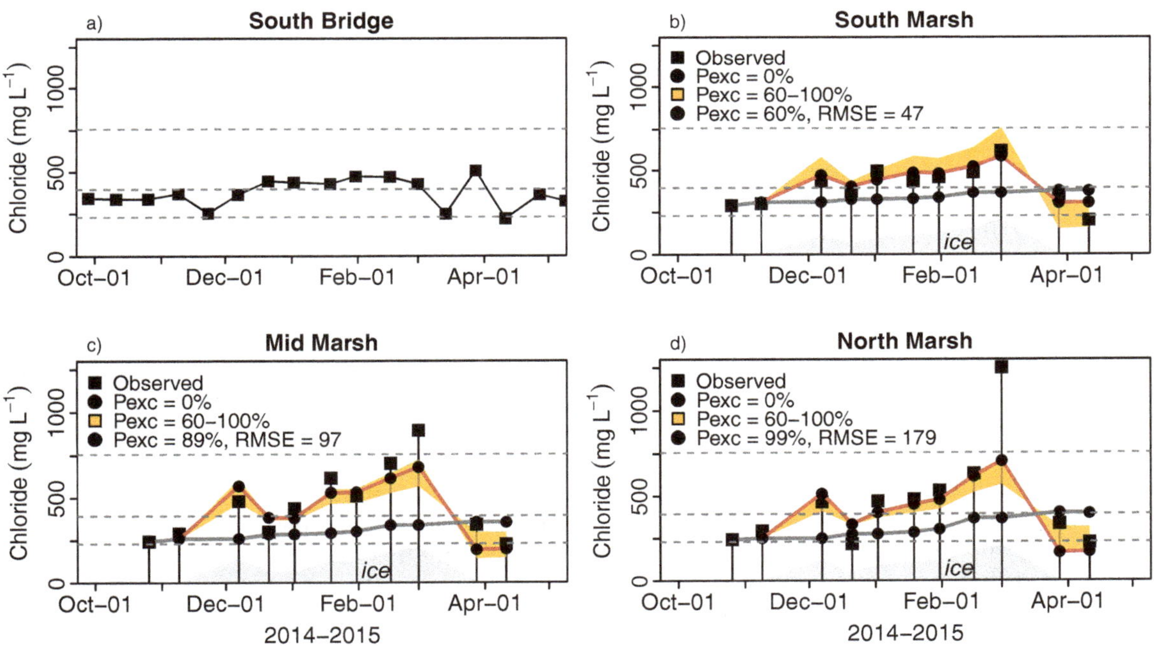

Fig. 3. (**a**) Chloride concentrations at South Bridge inlet. (**b–d**) Chloride concentrations at South-, Mid-, and North Marsh from late Oct 2014 to May 2015. Droplines are provided to show sampling dates. Black squares denote observed values while gray circles/line represent modeled chloride concentrations given 0% ion exclusion ($P_{exc} = 0\%$) during ice thickening. The yellow area represents the range in modeled chloride concentrations given rejection rates between 60% and 100%. Red circles/lines represent the best model fit (lowest RMSE). Dashed horizontal lines denote the WI DNR's acute (757 mg L^{-1}) and chronic (395 mg L^{-1}) chloride water quality criterion, and the EPA's chronic water criteria (230 mg L^{-1}). Ice thickness is represented by a gray polygon (not to scale).

Dec and all subsequent predictions are too high. However, the predictions match the pattern among dates. The model predictions were more precise for the Mid- and North Marsh sites, but at no time did the model predict chloride concentrations above 750 mg L^{-1}, less than the maximum recorded concentration of 1251 mg L^{-1}.

The model was also run independently for each time step, by using observational data for $\mathbf{Marsh[Cl]}_{t-1}$ at each time step and allowing the exclusion rate to change. This method improved the overall model fits at Mid Marsh (RMSE = 76), and North Marsh (RMSE = 176), but provided a poorer fit at South Marsh (RMSE = 53) (Supporting Information Fig. S2).

Discussion

At all three sites in the Marsh, chloride concentrations rose above the chronic chloride water quality criterion of both the EPA (230 mg L^{-1}) and the WIDNR (395 mg L^{-1}) during the winter months. Following ice melt, concentrations dropped back below 395 mg L^{-1} (Fig. 3). The seasonal variability in chloride concentrations in the 1918 Marsh was as high as 1000 mg L^{-1} at North Marsh. Large seasonal fluctuations, comparable to those seen in the 1918 Marsh, are more expected in small, shallow water bodies, than deeper lakes, such as Lake Mendota, where extreme salt loading is buffered by the large volume of water. When considering the health of the aquatic ecosystem, these large pulses of salt are a concern (Herbert et al. 2015), even though the summertime concentrations may be below water quality thresholds.

The simple ice formation model is a reasonably good predictor of observed chloride concentrations in the 1918 Marsh. The model fit likely was enhanced by the dry winter conditions of 2014–2015, which increased water residence time in the marsh. Because the model does not account for lateral flow, any overestimates, such as those on 7 Dec 2014, are likely explained by flushing events. In this instance, there was 9 cm of new snow from Nov 25 to Nov 29, followed by air temperatures above 0°C. Dilute snow melt runoff into the Marsh would have diluted and flushed more concentrated water beneath the ice. Following 7th Dec, air temperatures remained below 0°C, and model fit improved. Thus, we speculate that lateral flow over the winter was minimal. We also speculate that the model predictions at Mid and North Marsh were less than observational data, including the 1250 mg L^{-1} concentration at North Marsh, owing to highly saline runoff from an adjacent snow storage site (Fig. 1). The snow pile, which is contaminated with road salt from snow removed from streets and parking lots on the University of Wisconsin campus, drains toward the north end of the marsh (Helmueller et al. pers. comm.).

Our ability to numerically model changes in chloride beneath the ice of 1918 Marsh was limited by the bi-weekly sampling of ice thickness and water depth. Given that the air temperature exceeded 0°C throughout November and December, it is likely our sampling missed sporadic freeze and thaw events. This is evident in our 2-week simulations inability to consistently predict the change in chloride over a single time step (Supporting Information Fig. S2). At the South Marsh site, our model predicted chloride with a $< 1\%$ error at 4 of 10 time points. Conversely, the error was $> 10\%$ for three of the time steps. Model predictions would be improved by better knowledge of water depth, ice thickness, and flow; all of which could be measured at a higher frequency using field deployable sensor technology.

Our model suggests that chloride exclusion rates in 1918 Marsh are likely between 0.6 and 1.0 (complete rejection), but is unable to narrowly constrain these rates given the paucity of observational data. Other field and lab studies of ion rejection in freshwater ice during freezing are rare (Table 1). Furthermore, the rate of ion exclusion in freshwaters likely changes with the rate of freezing and the temperature profile of the ice cover, as it does in seawater (Notz and Worster 2009).

The numeric model demonstrates that chloride concentrations in the marsh are approximately double the theoretical concentration given no ion rejection (Fig. 3). In the 1918 Marsh, this raises chloride levels from ~ 250–400 mg L^{-1} to upward of 800 mg L^{-1}, even with ion exclusion as low as 60%. This doubling of chloride, pushes the 1918 Marsh concentrations to the WI DNR's acute toxicity threshold of 757 mg L^{-1} (Wisconsin DNR 2008). Concentrations of this magnitude would be expected to harm freshwater biota in lakes (Corsi et al. 2010; Hintz et al. 2017; Jones et al. 2017), and there is an extensive literature on the toxicity of chlorides on individual taxa and life stages (EPA 1988; Environment Canada and Health Canada 2001; Evans and Frick 2001). We use "expected to" because published toxicity criteria are based almost exclusively on non-winter water temperatures. In very rare cases, when toxicity was tested at cooler temperatures, there was no consistent pattern to indicate whether toxicity levels at temperatures near 0°C would be higher or lower.

In freshwater systems, the process of ion rejection from ice formation routinely is overlooked as a driver of water quality. In deep lakes, ion accumulation is negligible as a percentage of lake volume; however, most lakes on earth are shallow. A recent evaluation of global lake volumes estimates that the 1.24 million lakes on Earth between 0.1 km^2 to 1 km^2 have a mean depth of 3.5 m (Messager et al. 2016). Combining this with the fact that the highest abundance of lakes are at mid to northern latitudes (Verpoorter et al. 2014), underscores that the process of ion rejection from ice formation and thickening is chemically and biologically relevant for millions of waterbodies. Whether a lake is at risk for under-ice chloride toxicity can be determined from the amount of ice as a percentage of lake depth and the pre-ice chloride concentrations (Fig. 4). This is particularly relevant in Midwest and Northeast North America, where road development and deicing practices have led to chloride concentrations orders of magnitude above background concentrations (Dugan et al. 2017).

The process of chloride rejection discussed here may have broader application in limnology, as nitrogen and

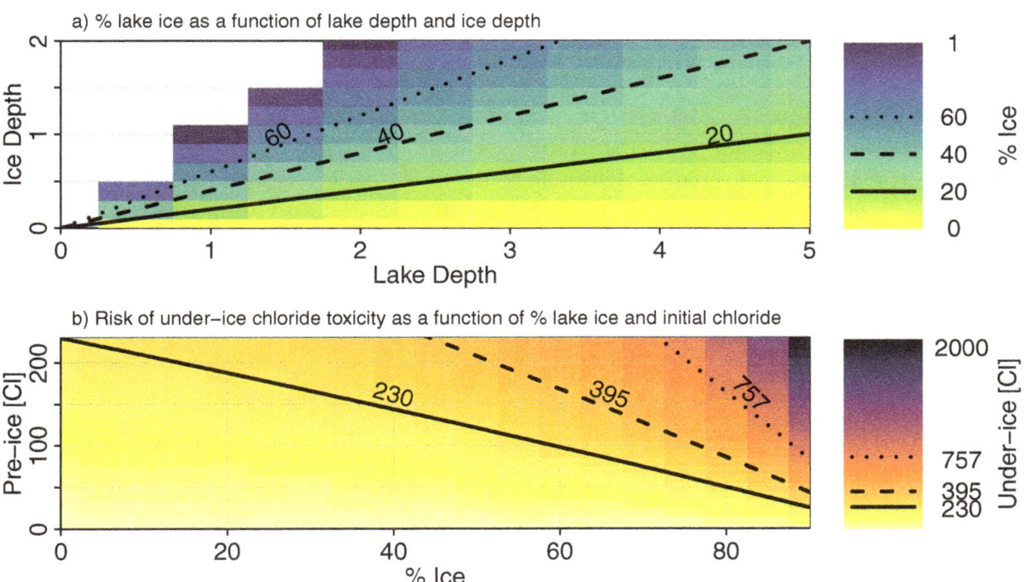

Fig. 4. Risk assessment for chloride toxicity in waterbodies. (**a**) The amount of lake ice as a percentage of total depth (% ice) is determined based on ice depth vs. lake depth. The solid, dashed, and dotted lines represent the graphical area with 20%, 40%, and 60% ice. (**b**) Comparing % ice with pre-ice (or summer time) chloride concentrations can determine the risk of under-ice chloride toxicity. Any depth and concentrations units may be used to interpret this figure. Exclusion rate was set at 90%. If concentrations are measured in mg L^{-1}, lakes that fall to the right of the solid 230 mg L^{-1} line represent increasing risk from no effect with continuous exposure to noticeable effects of exposure. For example, if lake ice is more than 40% of the lake depth, under-ice chloride concentrations would reach toxic levels (> 230 mg L^{-1}) if pre-ice concentrations were greater than 140 mg L^{-1}. As the percent of ice increases, the pre-ice chloride concentration that would develop toxic winter levels, decreases.

phosphorus (Fang et al. 2016), dissolved gases (Wharton et al. 1987; Tyler et al. 1998), and compounds such as dissolved organic matter and dissolved organic carbon (Belzile et al. 2002) are all known to accumulate under ice in lakes. Given that phytoplankton are relatively abundant under lake ice (Hampton et al. 2017), the process of solute rejection from lake ice may be a significant driver of under ice productivity.

Conclusions

In environments where anthropogenic chloride loading is a recognized occurrence, such as in areas of road salt application, shallow water bodies are at a much higher risk of elevated chloride concentrations than are deeper water bodies during ice cover thickening. Future research should investigate the rates of ion and solute rejection in freshwater ice covers, the rate of wintertime flushing in shallow systems, and the thresholds of chloride toxicity at near-freezing temperatures. Most importantly, we encourage winter water samples of chloride to be obtained to compare with summer values where ion exclusion may be an issue.

References

Belzile, C., J. A. E. Gibson, and W. F. Vincent. 2002. Colored dissolved organic matter and dissolved organic carbon exclusion from lake ice: Implications for irradiance transmission and carbon cycling. Limnol. Oceanogr. **47**: 1283–1293. doi:10.4319/lo.2002.47.5.1283

Bluteau, C. E., R. Pieters, and G. A. Lawrence. 2017. The effects of salt exclusion during ice formation on circulation in lakes. Environ. Fluid Mech. **17**: 579–590. doi:10.1007/s10652-016-9508-6

Center for Limnology; NTL LTER. 2012. North Temperate Lakes LTER: Chemical Limnology of Primary Study Lakes: Major Ions 1981 - current. doi: 10.6073/pasta/91fcc6ddf77793be79f88c2b671486fa

Chapra, S. C., A. Dove, and D. C. Rockwell. 2009. Great Lakes chloride trends: Long-term mass balance and loading analysis. J. Great Lakes Res. **35**: 272–284. doi:10.1016/j.jglr.2008.11.013

Corsi, S. R., D. J. Graczyk, S. W. Geis, N. L. Booth, and K. D. Richards. 2010. A fresh look at road salt: Aquatic toxicity and water-quality impacts on local, regional, and national scales. Environ. Sci. Technol. **44**: 7376–7382. doi:10.1021/es101333u

Dugan, H. A., and others. 2017. Salting our freshwater lakes. Proc. Natl. Acad. Sci. USA **114**: 4453–4458. doi:10.1073/pnas.1620211114.

Environment Canada and Health Canada. 2001. Canadian Environmental Protection Act, 1999. Priority substances list assessment report. Road salts. 181 p.

EPA. 1988. Ambient water quality criteria for chloride. US Environmental Protection Agency, Washington DC. 39 p.

Evans, M., and C. Frick. 2001. The effects of road salts on aquatic ecosystems. Environment Canada Series. N.W.R.I. Contribution Series; 02-308.

Fang, Y., L. Changyou, M. Leppäranta, S. Xiaonghong, Z. Shengnan, and Z. Chengfu. 2016 Notable increases in nutrient concentrations in a shallow lake during seasonal ice growth. Water. Sci. Technol. **74**: 2773–2783. doi: 10.2166/wst.2016.433

Findlay, S. E. G., and V. R. Kelly. 2011. Emerging indirect and long-term road salt effects on ecosystems. Ann. N. Y. Acad. Sci. **1223**: 58–68. doi:10.1111/j.1749-6632.2010.05942.x

Hampton S. E., and others. 2017. Ecology under lake ice. Ecology Letters **20**: 98–111. doi:10.1111/ele.12699

Herbert, E., and others. 2015. A global perspective on wetland salinization: Ecological consequences of a growing threat to freshwater wetlands. Ecosphere **6**: 1–43. doi: 10.1890/ES14-00534.1

Hill, A. R., and E. K. Sadowski. 2016. Chloride concentrations in wetlands along a rural to urban land use gradient. Wetlands **36**: 73–83. doi:10.1007/s13157-015-0717-4

Hintz, W. D., B. M. Mattes, M. S. Schuler, D. K. Jones, A. B. Stoler, L. Lind, and R. A. Relyea. 2017. Salinization triggers a trophic cascade in experimental freshwater communities with varying food-chain length. Ecol. Appl. **27**: 833–844. doi:10.1002/eap.1487

Jones, D. K., B. M. Mattes, W. D. Hintz, M. S. Schuler, A. B. Stoler, L. A. Lind, R. O. Cooper, and R. A. Relyea. 2017. Investigation of road salts and biotic stressors on freshwater wetland communities. Environ. Pollut. **221**: 159–167. doi:10.1016/j.envpol.2016.11.060

Kaushal, S. S., P. M. Groffman, G. E. Likens, K. T. Belt, W. P. Stack, V. R. Kelly, L. E. Band, and G. T. Fisher. 2005. Increased salinization of fresh water in the northeastern United States. Proc. Natl. Acad. Sci. USA **102**: 13517–13520. doi:10.1073/pnas.0506414102

Lawrimore, J. H., M. J., Menne, B. E. Gleason, C. N. Williams, D. B. Wuertz, R. S. Vose, and J. Rennie. 2011. An overview of the Global Historical Climatology Network monthly mean temperature data set, version 3. J. Geophys. Res. **116**: D19121. doi:10.1029/2011JD016187

Marsalek, J. 2003. Road salts in urban stormwater: An emerging issue in stormwater management in cold climates. Water Sci. Technol. **48**: 61–70.

Messager, M. L., B. Lehner, G. Grill, I. Nedeva, and O. Schmitt. 2016. Estimating the volume and age of water stored in global lakes using a geo-statistical approach. Nat. Commun. **7**: 13603. doi:10.1038/ncomms13603

Notz, D., and M. G. Worster. 2009. Desalination processes of sea ice revisited. J. Geophys. Res. **114**: C05006. doi: 10.1029/2008JC004885

Novotny, E. V., D. Murphy, and H. G. Stefan. 2008. Increase of urban lake salinity by road deicing salt. Sci. Total Environ. **406**: 131–144. doi:10.1016/j.scitotenv.2008.07.037

Panno, S. V., V. A. Nuzzo, K. Cartwright, B. R. Hensel, and I. G. Krapac. 1999. Impact of urban development on the chemical composition of ground water in a fen-wetland complex. Wetlands **19**: 236–245. doi:10.1007/BF03161753

Panno, S. V., K. C. Hackley, H. H. Hwang, S. E. Greenberg, I. G. Krapac, S. Landsberger, and D. J. O'Kelly. 2006. Characterization and identification of Na-Cl sources in ground water. Ground Water **44**: 176–187. doi:10.1111/j.1745-6584.2005.00127.x

Pieters, R., and G. A. Lawrence. 2009. Effect of salt exclusion from lake ice on seasonal circulation. Limnol. Oceanogr. **54**: 401–412. doi:10.4319/lo.2009.54.2.0401

Resh, V. H., and others. 1988. The role of disturbance in stream ecology. J. North Am. Benthol. Soc. **7**: 433–455. doi:10.2307/1467300

Richburg, J. A., W. A. Patterson, and F. Lowenstein. 2001. Effects of road salt and Phragmites australis invasion on the vegetation of a Western Massachusetts calcareous lake-basin fen. Wetlands **21**: 247–255. doi:10.1672/0277-5212(2001)021[0247:EORSAP]2.0.CO;2]

She, Y., J. Kemp, L. Richards, and M. Loewen. 2016. Investigation into freezing point depression in stormwater ponds caused by road salt. Cold Reg. Sci. Technol. **131**: 53–64. doi:10.1016/j.coldregions.2016.09.003

Thunqvist, E-L. 2004. Regional increase of mean chloride concentration in water due to the application of deicing salt. Sci. Total Environ. **325**: 29–37. doi:10.1016/j.scitotenv.2003.11.020

Tyler, S., P. Cook, A. Butt, J. Thomas, P. Doran, and W. Lyons. 1998. Evidence of deep circulation in two perennially ice-covered Antarctic lakes. Limnol. Oceanogr. **43**: 625–635. doi:10.4319/lo.1998.43.4.0625

Van Meter, R. J., and C. M. Swan. 2014. Road salts as environmental constraints in urban pond food webs. PloS one **9**: e90168. doi:10.1371/journal.pone.0090168

Verpoorter, C., T. Kutser, D. A. Seekell, and L. J. Tranvik. 2014. A global inventory of lakes based on high-resolution satellite imagery. Geophys. Res. Lett. **41**: 6396–6402. doi:10.1002/2014GL060641

Wharton, Jr, R. A., C. P. McKay, R. L. Mancinelli, and G. M. Simmons. 1987. Perennial N2 supersaturation in an Antarctic lake. Nature **325**: 343–345. doi:10.1038/325343a0

Williams, D. D. D., N. E. Williams, and Y. Cao. 2000. Road salt contamination of groundwater in a major metropolitan area and development of a biological index to monitor its impact. Water Res. **34**: 127–138. doi:10.1016/S0043-1354(99)00129-3

Wisconsin, D. N. R. 2008. Chapter NR 105: Surface water quality criteria and secondary values for toxic substances, p. 37–54. *In* Wisconsin Administrative Code: Department of Natural Resources.

Zhang, Y., C. Li, X. Shi, and C. Li. 2012. The migration of total dissolved solids during natural freezing process in Ulansuhai Lake. J Arid Land **4**: 85–94. doi:10.3724/SP.J.1227.2012.00085

Acknowledgments

We thank the Lakeshore Nature Preserve for student funding, Emily Stanley and Elizabeth Runde for chemical analysis of the samples, and Corinna Gries for data management. We thank many other students and staff at the Center for Limnology and the Lakeshore Nature Preserve for assistance in the field. Thank you to the three anonymous reviewers who provided feedback that significantly improved this manuscript. This material is based upon work supported by the National Science Foundation under Cooperative Agreement DEB-1440297.

4

A novel marine bioinvasion vector: Ichthyochory, live passage through fish

Tamar Guy-Haim,[1,2*†] *Orit Hyams-Kaphzan*,[3] *Erez Yeruham*,[1] *Ahuva Almogi-Labin*,[3] *James T. Carlton*,[4]

[1]Israel Oceanographic and Limnological Research, National Institute of Oceanography, Haifa, Israel; [2]Marine Biology Department, The Leon H. Charney School of Marine Sciences, University of Haifa, Mt. Carmel, Haifa, Israel; [3]Geological Survey of Israel, Jerusalem, Israel; [4]Williams College - Mystic Seaport, Maritime Studies Program, Mystic, Connecticut

Scientific Significance Statement

Bioinvasions, the expansions of species into regions outside their native range, threaten biodiversity, and human livelihood worldwide. In the sea, international shipping has been considered the primary vector for the introduction of these non-indigenous species, although in many cases the specific vector is not known nor well understood. We present evidence for a novel bioinvasion vector that can explain the appearance and dispersal of dozens of benthic non-indigenous species in the Mediterranean Sea—that of live passage through fish digestive tracts (ichthyochory). This newly identified vector should be considered when evaluating both the history of prior invasions and the potential vectors of future invasions world-wide.

Abstract

Many species of Indo-Pacific holobenthic foraminifera have been introduced and successfully established sustainable populations in the Mediterranean Sea over the past few decades. However, known natural and anthropogenic vectors do not explain how these species were introduced long distances from their origin. We present evidence for a novel marine bioinvasion vector explaining this long-distance transport and introduction using both contemporary field and historical analyses. In 2015–2016, we found living specimens of 29 foraminiferal species in the fecal pellets of two Red Sea herbivorous rabbitfish—*Siganus rivulatus* and *Siganus luridus* in the Mediterranean. In our historical analysis, we found 34 foraminiferal species in preserved Red Sea rabbitfish specimens, dating between 1967 and 1975. In addition, we found congruent propagation patterns of the non-indigenous rabbitfish and foraminifera, lagging 4–11 yrs between discoveries, respectively. Predation of marine benthos by non-indigenous fish, followed by incomplete digestion and defecation of viable individuals, comprise the main introduction vector of these organisms into novel environments.

---

*Correspondence: tamar.guy-haim@ocean.org.il

†Present address: GEOMAR, Helmholtz Centre of Ocean Research Kiel, Marine Ecology, Germany.

Author Contribution Statement: The study was conceived by TGH and OHK. TGH led and designed the study. TGH and EY performed the experiments. OHK and AAL conducted the taxonomic analysis. TGH and OHK analyzed the data. TGH, OHK, EY, AAL and JTC developed the conceptual framework and drafted the manuscript.

T.G.-H. and O.H.-K contributed equally to this work.

Non-indigenous species (NIS) are increasingly being discovered in terrestrial, freshwater, and marine habitats, altering community structure and ecosystem functions (Simberloff 2001; Sanders et al. 2003; Lowry et al. 2013). Nevertheless, the vectors of NIS introduction often remain obscure. In the sea, human-mediated transport mechanisms, primarily vessel hull fouling and ballast water, fisheries, aquaculture, and canals (Ruiz and Carlton 2003), are known to play important roles in NIS dispersal. However, none of these vectors explain the invasion of at least 68 benthic foraminifera NIS from the Red Sea into the Mediterranean Sea over the past decades (Zenetos et al. 2012; Weinmann et al. 2013).

Since the opening of the Suez Canal in 1869, hundreds of NIS from the Red Sea have been recorded in the Mediterranean Sea (Galil et al. 2015). The unidirectional advancement of Red Sea species via the canal is termed "Lessepsian migration" and the Red Sea NIS are termed "Lessepsian migrants," after Ferdinand de Lesseps, the French diplomat who was in charge of the canal's construction (Por 1978). The Lessepsian migrants moved with currents or under their own volition, as well as by vessel-mediated transport. Many benthic organisms are meroplanktonic, undergoing a larval planktonic phase, and thus are potentially capable of dispersing long distances, including travelling through waterways. Most benthic foraminifera, however, are holobenthic, being permanent bottom-dwellers in sediments, on rocks, or as epiphytes on seaweeds and seagrass, with no planktonic stage (Hyams-Kaphzan et al. 2008). Although capable of active movement across surfaces, benthic foraminifera speed-rates range between 1.8 mm h^{-1} and 8.4 mm h^{-1} (Kitazato 1988). It would thus take a persistent foraminiferal species about 4000 yrs to "walk" the 164 km long Suez Canal.

Several mechanisms have been proposed to explain the range expansions of benthic foraminifera (Alve 1999): ballast water taking up suspended foraminifera stirred up from sediments (McGann et al. 2000), passive dispersal by currents and sediment transport (Alve and Goldstein 2010), attachment to gastropod veliger shells (Nesbitt 2005), ship hull fouling (Gollasch 2002), rafting on marine plants or debris (Winston 2012), and movement with aquaculture products (Cohen 2012). To date, no data have been found to support any of these hypotheses to explain new invasions of rocky substrate or epiphytic foraminifera. First, while foraminifera are known from ballast water and ballast sediments, these are in large part species associated with soft-sediment habitats, as are the few other foraminifera invasions in other regions of the world (McGann et al. 2000). Second, an extensive 10-yr survey of ballast protist communities of cargo vessels arriving in Israeli Mediterranean ports did not reveal any of the known benthic foraminifera NIS (Galil and Hülsmann 1997). Third, dispersal by juvenile drifting or rafting are less probable vectors as the westward propagation of benthic foraminifera NIS in the Mediterranean is both clockwise and anti-clockwise, inconsistent with the Levantine anti-clockwise longshore current (Robinson et al. 1992). Therefore, it is necessary to consider alternate mechanisms of introduction for foraminifera NIS invasion, particularly for the Mediterranean. One phenomenon that has been overlooked is the potential of other non-indigenous species to act as biotic vectors.

Fig. 1. A marbled spinefoot rabbitfish *Siganus rivulatus* grazing on turf algae near Mikhmoret, Israel. The arrow points to a large epiphytic non-indigenous Indo-Pacific foraminifera, *Sorites orbiculus*, magnified at bottom right. Other benthic foraminifera species, smaller than can be seen by the naked eye, are plentiful in the shallow rocky reefs of the Eastern Mediterranean.

In the Mediterranean Sea, the majority of foraminifera NIS inhabit the rocky subtidal, attached to carbonate-rich hard substrates and seaweeds (Hyams-Kaphzan et al. 2014). Over the past decades, these habitats have undergone a profound phase-shift from algal forests to turf barrens, due to over-grazing by invasive Indian Ocean herbivorous rabbitfish [*Siganus rivulatus*, first observed in 1924 (Steinitz 1927), and *Siganus luridus*, first observed in 1955 (Ben-Tuvia 1964)] found in the entire eastern basin (Fig. 1). Rising seawater temperature over the past several decades may have facilitated their expansion westward toward the Atlantic and northward into the Ionian, Tyrrhenian, Aegean, and Adriatic Seas (Vergés et al. 2014) (Fig. 2). These and other herbivorous fish may incidentally ingest living epiphytic foraminifera and then defecate them unharmed, although their calcareous tests may be destroyed in those fish with an acidic digestion phase (Debenay et al. 2011).

In this study, our objective was to compile evidence for the ability of invasive herbivorous fish to transport the large number of benthic foraminifera species from the Red Sea to the Mediterranean Sea and to facilitate their long-distance westward dispersal. We addressed this objective by: (1) testing the viability of benthic foraminifera species in fecal pellets of invasive rabbitfish in the Mediterranean Sea, (2) examining the presence of foraminifera species in the digestive tracts of museum specimens of Red Sea rabbitfish

Fig. 2. Distribution and propagation map of the non-indigenous foraminifera *Amphistegina lobifera* (marked by blue filled area) and the invasive rabbitfish *Siganus luridus* (fish symbols). First record years of *S. luridus* are in black, and of *A. lobifera* in red brackets. Sites of observation are marked by green stars. Modified from Langer et al. (2012). All related references are presented in Supporting Information Table S3.

collected from the Gulf of Suez, and (3) compiling both temporal and spatial records of rabbitfish and benthic foraminifera introductions in the Mediterranean Sea.

Materials and methods

Rabbitfish fecal pellet analysis

Fifty-five specimens of *Siganus luridus* ($n = 20$) and *S. rivulatus* ($n = 35$) were collected by hand at night while free-diving between August 2015 and August 2016 off the coast of Haifa (depth of 2–3 m; 32.8333°N, 034.9738°E), and transferred to the nearby IOLR facilities. The fish were individually placed for 12 h in 5 L circular tanks with a funnel bottom. Seawater was pumped in directly from the Mediterranean at a flow rate of 70 $L{\cdot}h^{-1}$ and sand-filtered to eliminate contamination. The fish were removed from the tanks, and the fecal pellets were funnelled out and taken for analysis.

The number of fecal pellets obtained per fish was 8–45, yielding a defecation rate of 36.2 ± 22.3 fecal $pellets{\cdot}d^{-1}$ (average ± SD). Overall, 1350 fecal pellets were analyzed. Each pellet was isolated, rinsed in filtered seawater, and placed in a 500 mL glass jar. Fecal infauna were manually separated from algal debris and organic material, identified, and counted under a dissecting microscope. Test integrity (the lack of physical damage to the calcareous shell) and symbiont presence in the symbiont-bearing foraminifera species were visually examined using a dissecting microscope and a scanning electron microscope (SEM). The fecal infauna were kept in individual glass jars per pellet for an additional 48 h, to test their movement [using a negative geotaxis assay following Bernhard (2000)]. The foraminifera attached to the glass walls were then manually removed using a fine paintbrush, identified to species, counted under a dissecting microscope, and photographed with a SEM.

Historical data analysis

Forty-six preserved specimens of *Siganus luridus* ($n = 24$) and *S. rivulatus* ($n = 22$) collected from the Gulf of Suez, Red Sea, during 1967–1975 (Supporting Information Fig. S1) were obtained from the Hebrew University Zoological Museum Fish Collection. Upon collection, the fish were fixed in 10% formalin for 7 d and then transferred and stored in 70% ethanol. The preserved fish were dissected and their gastrointestinal tract (gut) was removed, weighed, and volume measured. Gut contents were placed in Rose-Bengal (2 g L^{-1} ethanol 95%) for 14 d to ensure coloring of the entire sample. Foraminifera were examined under a stereomicroscope and SEM and were morphologically identified to species.

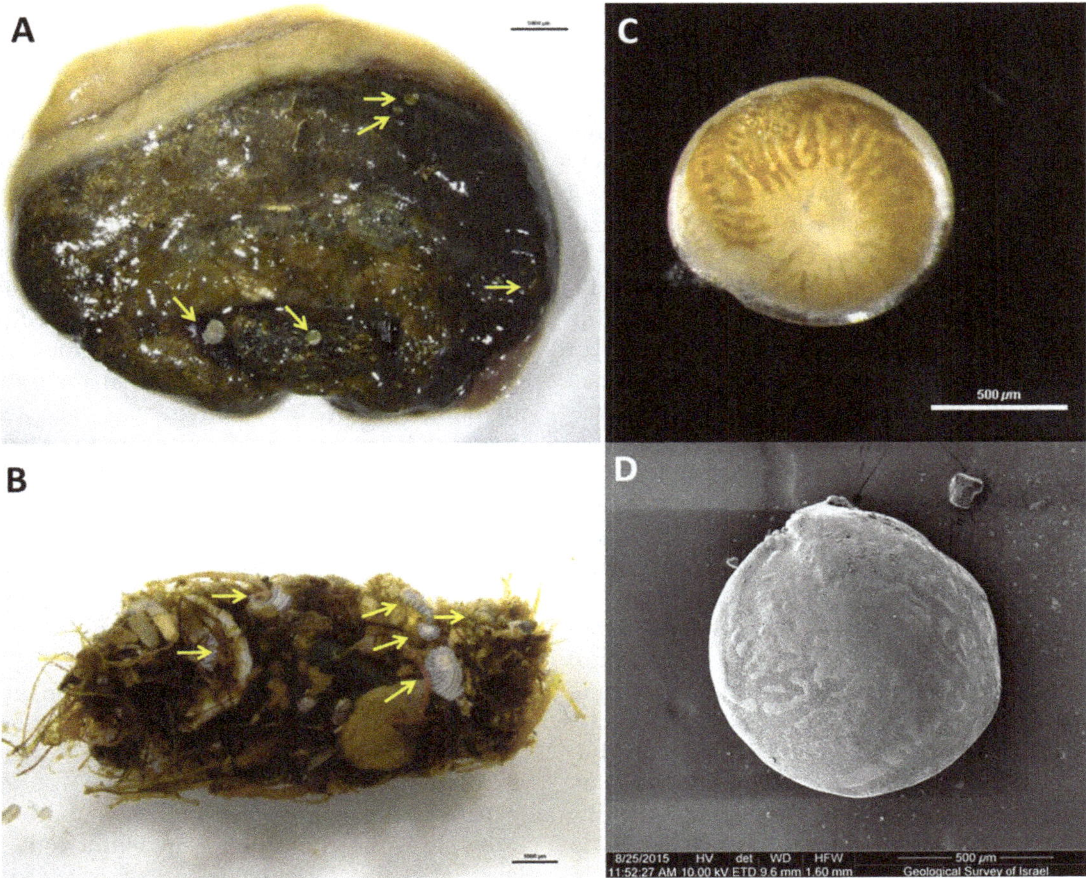

Fig. 3. Living foraminifera in rabbitfish gut and fecal pellets. (**A**) Gastrointestinal tract of *Siganus luridus*. The arrows point to benthic foraminifera, detected through the thin wall. (**B**) Fecal pellet of *S. rivulatus*. The arrows point to benthic foraminifera. (**C**) Live *Amphistegina lobifera* isolated from a fecal pellet by negative geotaxis assay. Endosymbiotic algae appear in dark. (**D**) *A. lobifera* isolated from fecal pellet. Scanning electron microscopy shows intact test. Scale is 5 mm (**A**), 1 mm (**B**), and 0.5 mm (**C**, **D**).

pH analysis of rabbitfish gut

To assess the gastrointestinal pH microenvironment of the rabbitfish, digestive tracts of fresh fish were removed, unfolded, spread horizontally, and then immediately placed in a microprofiling system (Unisense A/S, Denmark), including MM33-2 micromanipulator, microsensor multimeter amplifier, temperature sensor (TP-2000), and a membrane pH electrode (pH-100, tip diameter 90–110 μm, < 10 s response time, precision ± 0.01) with an external reference electrode, and SensorTrace Profiling software. Measurements were taken throughout the length and width of the gastrointestinal tract. All electrodes were calibrated with commercial pH solutions (Sigma-Aldrich, U.S.A.).

Experiments

Animal experiments were performed in compliance with the relevant laws and ethics guidelines.

Results and discussion

Classically recognized natural and anthropogenic dispersal vectors of marine organisms fail to explain how scores of Indo-Pacific holobenthic foraminifera species have rapidly invaded the Mediterranean Sea, successfully establishing sustainable populations. In this study, we examined and tested ichthyochory, live passage of organisms via fish guts, as an alternate mechanism explaining the non-indigenous foraminifera introductions, using analyses of rabbitfish gut and fecal pellet contents, gut analyses of rabbitfish museum specimens collected from the Gulf of Suez, and comparisons of distributions and first records of rabbitfish and foraminifera species in the Mediterranean Sea, since their first appearance there.

Rabbitfish gut content and fecal pellet analyses

Gut content analysis of *S. rivulatus* and *S. luridus*, collected from south-eastern Mediterranean rocky reefs, revealed a high abundance and diversity of live foraminifera, predominantly NIS (Fig. 3A; Supporting Information Table S1). Low gastrointestinal pH ranging from 2.5 to 4.5, is expected in herbivorous fish (Lobel 1981) and may have a detrimental effect on benthic foraminifera, whose tests are characterized by high calcium carbonate concentration. It was therefore important to assess both the rabbitfish gut pH levels and the viability of ingested foraminifera. We measured gut pH levels ranging between 7.05 and 8.41, possibly due to high inorganic carbonate concentration (Wilson et al. 2009). Further analysis of rabbitfish fecal pellets (Fig. 3B) revealed a variety of empty foraminifera tests (2–80 tests·fecal pellet^{-1}) as well as live foraminifera (1–50 individuals·fecal pellet^{-1}). Overall 29 foraminifera species were present in the fecal pellets, 11 of which are known Lessepsian migrants. Most live individuals were found in good physiological condition (Supporting Information Table S1), containing endosymbiotic algae in the symbiont-bearing species (Fig. 3C), and presenting intact tests (Fig. 3D), likely due to the considerably high pH levels in the rabbitfish gut.

We found a variety of additional taxa in the gut and feces of the rabbitfish, with several still being motile, including gastropods, bivalves, copepods, polychaetes, ostracods, and brittle stars (Supporting Information Table S1); as all of these have planktonic larvae (with the possible exception of the gastropods and ophiuroids, not identified to species), determining the possible role of fish predation as a dispersal mechanism of these organisms is challenging. Live gammarid amphipods, which lack larval stages, were also found, but these are typical of tychoplankton and rafting, and thus present the same challenges of distinguishing ichthyochory from drift. Microplastic filaments were also abundant in the gut and feces of the rabbitfish, reflecting a global trend (Cózar et al. 2014).

Viability of ingested benthic foraminifera

We used a locomotion assay to distinguish live from dead foraminifera. Benthic foraminifera move by means of their pseudopodial network, extending from the test along flat surfaces (Travis and Bowser 1991). Many species present a negative geotactic behavior, crawling upwards on substrates (Murray 2006). In this study, over 20% of foraminifers isolated from feces climbed upwards. However, because not all foraminifera species exhibit this behavior, our assessment of the diversity of live benthic foraminifera transferred via rabbitfish ingestion is probably an underestimate.

Historical data

The historical data that we obtained from preserved museum specimens of siganid fish collected in the Gulf-of-Suez and Cyprus, dating back to 1967 (Supporting Information Fig. S1), revealed a large assortment of 34 benthic foraminifera species (Supporting Information Table S2; Figs. 4, 5), seven of which are Lessepsian migrants. The foraminifera, included large symbiont-bearing species, originated from both Mediterranean and Red Sea fishes. Prominent among them was *Amphistegina lobifera* Larsen (Fig. 3C,D), the most abundant Lessepsian migrant in the Mediterranean Sea (Weinmann et al. 2013). It is among the most prolific and ubiquitous foraminifera on coral reefs, seagrass beds, and carbonate shelves, and is known to be a major carbonate producer, with a contribution of at least 260 g $CaCO_3$ m^{-2} yr^{-1} (Hyams-Kaphzan et al. 2014). *A. lobifera* was first observed on the Israeli shelf during the late 1950s in low numbers (Reiss 1961). Today, up to 90% of the rocky reef foraminifera assemblage off northern Israel is comprised of *A. lobifera* (Hyams-Kaphzan et al. 2014).

While recent geological studies have shown that *A. lobifera*, at least as a morphospecies, was present in the Mediterranean Sea up until the Middle Pleistocene (Meriç et al. 2016), its modern day presence in the Mediterranean is due to its dispersal in the 20th century through the Suez Canal. There is no evidence to suggest that prehistoric populations of *A. lobifera* in the Mediterranean persisted into modern times: detailed studies documenting the Mediterranean foraminiferal fauna have been conducted since the 1800s, and no living *A. lobifera* were detected until the 1950s, even as many very rare foraminifera were reported. No environmental triggers are known that would have lead a previously unknown species of foraminifera to suddenly become abundant at the end of the 20th century, nor do we have evidence that any known native, but previously extraordinarily rare, species, has now become a predominant member of the Mediterranean biota. Indeed, geological or prehistoric occurrences of species that went extinct and then reappeared in modern times (due to anthropogenic dispersal or natural expansion) is not uncommon. A well-documented marine example is the end-of- Pliocene extinction of the clam *Mya arenaria* in the North Pacific Ocean, followed by its introduction in the 1870s by human activity from the North Atlantic to the Pacific coast of North America (MacNeil 1965).

Non-indigenous rabbitfish and foraminifera propagation pattern

Based on parasitological data, the invasion of the Mediterranean rabbitfish involved actively swimming adults, rather than passively dispersed larvae (Diamant 2010). Adult rabbitfish can swim long distances (Kaunda-Arara and Rose 2004), and in doing so we suggest that they spread benthic foraminifera from the Red Sea, via the Suez Canal, to the Levantine basin and westward. Assuming that passive dispersal by rabbitfish is a major vector in determining the distribution of benthic foraminifera NIS into the Mediterranean Sea, we can expect that the propagation of both will be reflected in one another. Indeed, when we compare the propagation of *S. luridus* with that of *A. lobifera* (Fig. 2; Supporting Information Table S3) we find an overall correspondence in the extent and timing of first reported sightings in new biogeographic regions.

Ichthyochory – dispersal by live passage via fish guts

Dispersal of organisms by "hitchhiking" in the digestive tract of another species is well-known in terrestrial and freshwater habitats (Darwin 1859). Seed dispersal via ingestion by vertebrates (mostly birds and mammals), termed

Plate 1

Fig. 4. Assortment of benthic foraminifera isolated from gastrointestinal tracts of preserved *S. luridus* and *S. rivulatus* collected at the Gulf of Suez between 1967 and 1975 (the Hebrew University Zoological Museum, Jerusalem). Living specimens are indicated by Rose Bengal stain. Possible Lessepsian species (Red Sea species that are found in the Mediterranean Sea) are marked by asterisk. **1–2.** *Ammonia convexa*, **1.** spiral side, **2.** umbilical side; **3–4.** *Amphistegina lobifera*, **3.** spiral side, **4.** umbilical side; **5–6.** *Challengerella bradyi*, **5.** spiral side, **6.** umbilical side; **7.** *Elphidium advenum* subsp. *limbatum*, lateral view; **8.** *Elphidium fichtellianum*, lateral view; **9.** *Elphidium striatopunctatum*, lateral view; **10–12.** *Neorotalia calcar*, **10–11.** spiral side, **12.** umbilical side; **13–14.** *Pararotalia* sp., **13.** spiral side, **14.** umbilical side; **15.** *Affinetrina planciana*, lateral view; **16.** *Hauerina diversa*, lateral view; **17.** *Peneroplis pertusus*, lateral view; **18.** *Peneroplis planatus*, lateral view; **19.** *Pseudoschlumbergerina ovata*, oblique lateral view; **20.** *Sigmoihauerina bradyi*, lateral view; **21.** *Sorites orbiculus*, lateral view; **22.** *Spiroloculina communis*, lateral view; **23.** *Triloculina trigonula*, lateral view. Scale bar: 3–4, 9, 18, 21 = 500 μm; 19 = 300 μm; 1–2, 5–7, 10–17, 16, 20, 22–23 = 200 μm; 8 = 100 μm.

Plate 2

Fig. 5. Scanning electron micrographs of benthic foraminifera isolated from gastrointestinal tracts of preserved *S. luridus* and *S. rivulatus* collected at the Gulf of Suez between 1967 and 1975 (the Hebrew University Zoological Museum, Jerusalem). Possible Lessepsian species (Red Sea species that are found in the Mediterranean Sea) are marked by asterisk. **1–2.** *Ammonia convexa*, **1.** spiral side, **2.** umbilical side; **3.** *Amphistegina lobifera*, spiral side; **4.** *Challengerella bradyi*, oblique spiral side; **5.** *Elphidium advenum* subsp. *limbatum*, lateral view; **6.** *Elphidium fichtellianum*, lateral view; **7.** *Elphidium striatopunctatum*, lateral view; **8–9.** *Neorotalia calcar*, **8.** spiral side, **9.** umbilical side; **10–11.** *Hauerina diversa*, **10.** lateral view, **11.** enlargement of the apertural view, showing diatoms; **12.** *Peneroplis planatus*, lateral view; **13.** *Peneroplis pertusus*, lateral view; **14.** *Pseudomassilina* sp., lateral view; **15.** *Pseudoschlumbergerina ovata*, oblique lateral view, **16.** *Pseudotriloculina* sp. lateral view, **17–18.** *Sorites orbiculus*, **17.** lateral view, **18.** profile view; **19.** *Spiroloculina communis*, lateral view; **20.** *Spiroloculina* sp., lateral view; **21.** *Triloculina terquemiana*, oblique lateral view; **22.** *Triloculina trigonula*, oblique lateral view; **23.** *Textularia agglutinans*, lateral view. Scale bar: 7, 12, 17, 23 = 500 μm; 10, 14–15, 21–22, 18 = 400 μm; 3–4, 8, 13, 16, 19–20 = 300 μm; 1–2, 5–6, 9 = 200 μm; 11 = 50 μm.

"endozoochory," is the most common seed dispersal mechanism in trees. Ichthyochory, dispersal by fish, plays a significant role in the dispersal of seeds of freshwater and riparian plants (Pollux 2011). Survival of fish gut passage has been shown in numerous freshwater zoobenthic groups, and has been suggested to mediate NIS introductions (Gatlin et al. 2013). In marine environments, little is known about the role of ichthyochory in species dispersal and introduction. The evidence presented here suggests that this vector has a major role in structuring marine benthic communities and may accelerate "invasional meltdown," whereby a group of NIS facilitates one another's establishment, spread and impacts (Simberloff and Von Holle 1999). Interestingly, the time lag between the discovery of the fish and the later discovery of the foraminifera was about 4–11 yr (Fig. 2; Supporting Information Table S2). While this could be a consequence of the invasion of foraminifera in multiple time-steps (Arim et al. 2006), it is more likely attributable to the relative ease of discovering and identifying exotic fish vs. the much more intensive effort required to detect and identify non-indigenous foraminifera.

Implications of ichthyochory as a bioinvasion and dispersal vector

The invasion and subsequent dominance of *A. lobifera* in the rocky shores of the Eastern Mediterranean Sea, suppressing the native inhabitants of these environments (Hyams-Kaphzan et al. 2008, 2014), combined a prime example of how species migration from the Red Sea is drastically and rapidly changing the biota of the native invaded communities. High turnover rates make foraminifera tests a dominant component of sands worldwide. *Amphistegina* can account for up to 90% of sand biomass in certain locations (McKee et al. 1959). In the Aegean Sea, *A. lobifera* predominates on the rocky reefs at formidable densities of 230–310 K individuals$\cdot$m^{-2} (Yokeş and Meriç 2009). This overgrowth not only outcompetes the native foraminifera species (Hyams-Kaphzan et al. 2008; Zenetos et al. 2012), but it also dramatically changes the rocky reef habitats. The thickness of the deposited tests at multiple locations along the Turkish coast has reached 60–80 cm (Yokeş and Meriç 2009), with a consequent "desertification" of the rocky reefs and a shift in the biodiversity from hard- to soft-bottom-dwelling species. Biodiversity shifts and potential losses are inevitable results of such a process.

Benthic foraminifera are commonly used as paleogeographical proxies for models reconstructing ancient environmental conditions, including temperature, oxygen, pH, nutrients, productivity, organic matter flux, and ocean circulation (Sen Gupta 1999; Lea 2006). These models often correlate distinctive species assemblages with specific paleoconditions, assuming limited connectivity of foraminiferal communities. The fish-mediated long distance dispersal of benthic foraminifera suggested by our study may alter the underlying assumptions behind paleogeographical modelling, and motivate integrating this vector in future revised models.

In August 2015, the newly expanded Suez Canal was opened. This project is designed to double the shipping capacity of the canal, thereby intensifying the inflow of Lessepsian migrants into the Mediterranean (Galil et al. 2015). Furthermore, a now-warming Mediterranean Sea increases habitat suitability for both invasive fish (Vergés et al. 2014) and Indo-Pacific benthic foraminifera (Langer et al. 2012), further facilitating their westward propagation.

The light shed here upon the mystery of the mechanisms of benthic foraminifera invasion and their rapid spread throughout the Mediterranean illustrates the importance of understanding the intricacies of species interactions and their role in species invasion. The discovery of ichthyochory as a new trans-provincial ocean vector signals a critical shift in thinking on the dispersal of marine species. The ingestion and egestion of live prey by highly mobile predators in the sea—noting that both prey and predators may represent, in modern times, combinations of native and introduced species—portends to be as global a phenomenon as the dispersal of viable plant seeds by birds and mammals on land. We argue that retrospective studies will reveal that the spatially and temporally linked biogeography of predator–prey patterns that we suggest here may in fact have already been manifested but gone unnoticed in other regions of the world. Ichthyochory has profound implications for the interpretation and re-interpretation of modern and historical biogeography, ecology, and evolutionary biology of the world's seas.

References

Alve, E. 1999. Colonization of new habitats by benthic foraminifera: A review. Earth Sci. Rev. **46**: 167–185. doi:10.1016/S0012-8252(99)00016-1

Alve, E., and S. T. Goldstein. 2010. Dispersal, survival and delayed growth of benthic foraminiferal propagules. J. Sea Res. **63**: 36–51. doi:10.1016/j.seares.2009.09.003

Arim, M., S. R. Abades, P. E. Neill, M. Lima, and P. A. Marquet. 2006. Spread dynamics of invasive species. Proc. Natl. Acad. Sci. USA **103**: 374–378. doi:10.1073/pnas.0504272102

Ben-Tuvia, A. 1964. Two siganid fishes of Red Sea origin in the eastern Mediterranean. Ministry of Agriculture, Department of Fisheries, Sea Fisheries Research Station.

Bernhard, J. M. 2000. Distinguishing live from dead foraminifera: Methods review and proper applications. Micropaleontology **46**: 38–46. http://www.jstor.org/stable/1486179

Cohen, A. 2012. Aquatic invasive species vector risk assessments: Live saltwater bait and the introduction of non-native species into California. California Ocean Science Trust.

Cózar, A., and others. 2014. Plastic debris in the open ocean. Proc. Natl. Acad. Sci. USA **111**: 10239–10244. doi:10.1073/pnas.1314705111

Darwin, C. 1859. On the origin of species by means of natural selection, or the preservation of favoured races in the struggle for life. John Murray.

Debenay, J.-P., A. Sigura, and J.-L. Justine. 2011. Foraminifera in the diet of coral reef fish from the lagoon of New Caledonia: Predation, digestion, dispersion. Rev. Micropaléontol. **54**: 87–103. doi:10.1016/j.revmic.2010.04.001

Diamant, A. 2010. Red-Med immigration: A fish parasitology perspective, with special reference to the Myxosporea, p. 85–97. *In* D. Golani and B. Appelbaum-Golani [eds.], Fish imvasions of the Mediterranean Sea: Change and renewal. Pensoft Publishers, Sofia-Moscow.

Galil, B. S., and N. Hülsmann. 1997. Protist transport via ballast water—biological classification of ballast tanks by food web interactions. Eur. J. Protistol. **33**: 244–253. doi:10.1016/S0932-4739(97)80002-8

Galil, B. S., and others. 2015. 'Double trouble': The expansion of the Suez Canal and marine bioinvasions in the Mediterranean Sea. Biol. Invasions **17**: 973–976. doi:10.1007/s10530-014-0778-y

Gatlin, M. R., D. E. Shoup, and J. M. Long. 2013. Invasive zebra mussels (*Dreissena polymorpha*) and Asian clams (*Corbicula fluminea*) survive gut passage of migratory fish species: Implications for dispersal. Biol. Invasions **15**: 1195–1200. doi:10.1007/s10530-012-0372-0

Gollasch, S. 2002. The importance of ship hull fouling as a vector of species introductions into the North Sea. Biofouling **18**: 1051–1121. doi:10.1080/08927010290011361

Hyams-Kaphzan, O., A. Almogi-Labin, D. Sivan, and C. Benjamini. 2008. Benthic foraminifera assemblage change along the southeastern Mediterranean inner shelf due to fall-off of Nile-derived siliciclastics. Neues Jahrb. Geol. Paläontol. Abh. **248**: 315–344. doi:10.1127/0077-7749/2008/0248-0315

Hyams-Kaphzan, O., L. Perelis Grossowicz, and A. Almogi-Labin. 2014. Characteristics of benthic foraminifera inhabiting rocky reefs in northern Israeli Mediterranean shelf. Isr. Geol. Surv., Rep. GSI/36/2014.

Kaunda-Arara, B., and G. A. Rose. 2004. Long-distance movements of coral reef fishes. Coral Reefs **23**: 410–412. doi:10.1007/s00338-004-0409-7

Kitazato, H. 1988. Locomotion of some benthic foraminifera in and on sediments. J. Foraminiferal Res. **18**: 344–349. doi:10.2113/gsjfr.18.4.344

Langer, M. R., A. E. Weinmann, S. Lötters, and D. Rödder. 2012. "Strangers" in paradise: modeling the biogeographic range expansion of the foraminifera Amphistegina in the Mediterranean Sea. The Journal of Foraminiferal Research **42**: 234–244. doi:10.2113/gsjfr.42.3.234

Lea, D. 2006. Elemental and isotopic proxies of past ocean temperatures, p. 365–390. *In* J. I Drever Ed., Treatise on geochemistry, v. **6**, Elsevier.

Lobel, P. S. 1981. Trophic biology of herbivorous reef fishes: Alimentary pH and digestive capabilities. J. Fish Biol. **19**: 365–397. doi:10.1111/j.1095-8649.1981.tb05842.x

Lowry, E., E. J. Rollinson, A. J. Laybourn, T. E Scott, M. E. Aiello-Lammens, S. M. Gray, J. Mickley, and J. Gurevitch. 2013. Biological invasions: A field synopsis, systematic review, and database of the literature. Ecol. Evol. **3**: 182–196. doi:10.1002/ece3.431

MacNeil, F. S. 1965. Evolution and distribution of the genus *Mya*, and Tertiary migrations of Mollusca, 51 pp. United States Geological Survey Professional Paper 483-G.

McGann, M., D. Sloan, and A. N. Cohen. 2000. Invasion by a Japanese marine microorganism in western North America. Hydrobiologia **421**: 25–30. doi:10.1023/A:1003808517945

Mckee, E. D., J. Chronic, and E. B. Leopold. 1959. Sedimentary belts in lagoon of Kapingamarangi Atoll. AAPG Bull. **43**: 501–562.

Meriç, E., and others. 2016. Did *Amphistegina lobifera* Larsen reach the Mediterranean via the Suez Canal? Quat. Int. **401**: 91–98. doi:10.1016/j.quaint.2015.08.088

Murray, J. W. 2006. Ecology and applications of benthic foraminifera. Cambridge Univ. Press.

Nesbitt, E. A. 2005. A novel trophic relationship between cassid gastropods and mysticete whale carcasses. Lethaia **38**: 17–25. doi:10.1080/00241160510013132

Pollux, B. 2011. The experimental study of seed dispersal by fish (ichthyochory). Freshw. Biol. **56**: 197–212. doi:10.1111/j.1365-2427.2010.02493.x

Por, F. D. 1978. Lessepsian migration: The influx of Red Sea biota into the Mediterranean by way of the Suez Canal. Springer.

Reiss, Z. 1961. Recent foraminifera from the Mediterranean and the Red Sea coasts of Israel. Bull. Geol. Survey Israel **32**: 27–28.

Robinson, A. R., and others. 1992. General circulation of the Eastern Mediterranean. Earth Sci. Rev. **32**: 285–309. doi:10.1016/0012-8252(92)90002-B

Ruiz, G., and J. Carlton. 2003. Invasive species: Vectors and management strategies. Island Press.

Sanders, N. J., N. J. Gotelli, N. E. Heller, and D. M. Gordon. 2003. Community disassembly by an invasive species. Proc. Natl. Acad. Sci. USA **100**: 2474–2477. doi:10.1073/pnas.0437913100

Sen Gupta, B. K. 1999. Modern foraminifera. Kluwer Academic Publishers.

Simberloff, D. 2001. Biological invasions—how are they affecting us, and what can we do about them? West. N. Am. Nat. 308–315. http://scholarsarchive.byu.edu/wnan/vol61/iss3/7?utm_source=scholarsarchive.byu.edu%2Fwnan%2Fvol61%2Fiss3%2F7&utm_medium=PDF&utm_campaign=PDFCoverPages

Simberloff, D., and B. Von Holle. 1999. Positive interactions of nonindigenous species: Invasional meltdown? Biol. Invasions **1**: 21–32. doi:10.1023/A:1010086329619

Steinitz, W. 1927. Beiträge zur Kenntnis der Küstenfauna Palästinas, p. 311–353. *In* U. Hoepli Ed., Pubblicazioni della Stazione Zoologica di Napoli, v. **8**.

Travis, J. L., and S. S. Bowser. 1991. The motility of foraminifera, p. 91–155. *In* J.J. Lee, and O. R. Anderson Ed., Biology of Foraminifera. Academic Press.

Vergés, A., and others. 2014. The tropicalization of temperate marine ecosystems: Climate-mediated changes in herbivory and community phase shifts. Proc. R. Soc. Lond. B Biol. Sci. **281**: 20140846. doi:10.1098/rspb.2014.0846

Weinmann, A. E., D. Rödder, S. Lötters, and M. R. Langer. 2013. Traveling through time: The past, present and future biogeographic range of the invasive foraminifera *Amphistegina* spp. in the Mediterranean Sea. Mar. Micropaleontol. **105**: 30–39. doi:10.1016/j.marmicro.2013.10.002

Wilson, R., F. J. Millero, J. R. Taylor, P. J. Walsh, V. Christensen, S. Jennings, and M. Grosell. 2009. Contribution of fish to the marine inorganic carbon cycle. Science **323**: 359–362. doi:10.1126/science.1157972

Winston, J. E. 2012. Dispersal in marine organisms without a pelagic larval phase. Integr. Comp. Biol. **52**: 447–457. doi:10.1093/icb/ics040

Yokeş, M. B., and E. Meriç. 2009. Drowning in sand: Invasion by foraminifera. *In* C. P. Wilcox and R. B. Turpin [eds.], Invasive species: Detection, impact and control. Nova Science Publishers.

Zenetos, A., and others. 2012. Alien species in the Mediterranean Sea by 2012. A contribution to the application of European Union's Marine Strategy Framework Directive (MSFD). Part 2. Introduction trends and pathways. Mediterr. Mar. Sci. **13**: 328–352. doi:10.12681/mms.327

Acknowledgments

We thank Gil Rilov for the use of the research facilities in IOLR, Haifa, Dani Golani for access to fish specimens in the National History Collections of the Hebrew University of Jerusalem, Dan Tchernov for the use of microsensors, Jacob Silverman, Jonathan Erez, Arik Diamant, Gil Rilov, Martin Wahl, Dor Edelist, Eyal Rahav and Yair Achituv for providing insights on the study and valuable comments on the manuscript, Chana Netzer-Cohen for aid in processing the adapted map, Michael Kitin and Raanan Bodzin for technical aid and for acquiring the SEM images. We would also like to thank Isabelle Côté and the anonymous reviewer for their highly constructive review of the manuscript. We dedicate this paper to the memory of Dr. Nechama Ben-Eliyahu (1935-2014), marine biologist, invertebrate zoologist, and a pioneer in Lessepsian migration research.

5

Key differences between lakes and reservoirs modify climate signals: A case for a new conceptual model

Nicole M. Hayes,[1]* *Bridget R. Deemer*,[2,a] *Jessica R. Corman*,[3] *N. Roxanna Razavi*,[4] *Kristin E. Strock*[5]

[1]Department of Biology, University of Regina, Regina, Saskatchewan, Canada; [2]School of the Environment, Washington State University-Vancouver, Vancouver, Washington; [3]Center for Limnology, University of Wisconsin-Madison, Madison, Wisconsin; [4]Finger Lakes Institute, Hobart and William Smith Colleges, Geneva, New York; [5]Environmental Science Department, Dickinson College, Carlisle, Pennsylvania

Scientific Significance Statement

Climate change poses a significant threat to freshwater ecosystems, though the exact nature of these threats can vary by waterbody type. An existing conceptual model describes how altered fluxes of mass and energy will affect standing waterbodies, but it does not differentiate reservoirs from lakes. Here, we synthesize evidence suggesting that lakes and reservoirs differ in fundamental ways that are likely to influence their response to climate change. We then present a revised conceptual model that contrasts climate change effects on reservoirs versus lakes.

Abstract

Lakes and reservoirs are recognized as important sentinels of climate change, integrating catchment and atmospheric climate change drivers. Climate change conceptual models generally consider lakes and reservoirs together despite the possibility that these systems respond differently to climate-related drivers. Here, we synthesize differences between lake and reservoir characteristics that are likely important for predicting waterbody response to climate change. To better articulate these differences, we revised the energy mass flux framework, a conceptual model for the effects of climate change on lentic ecosystems, to explicitly consider the differential responses of lake versus reservoir ecosystems. The model predicts that catchment and management characteristics will be more important mediators of climate effects in reservoirs than in natural lakes. Given the increased reliance on reservoirs globally, we highlight current gaps in our understanding of these systems and suggest research directions to further characterize regional and continental differences among lakes and reservoirs.

*Correspondence: Nicole.Hayes@uregina.ca

[a]Present address: U.S. Geological Survey, Southwest Biological Science Center, Flagstaff, Arizona

Author Contribution Statement: NMH and BRD co-led the manuscript effort and contributed equally. JRC and BRD conducted the statistical analyses. KES and JRC designed the lake pairing analysis. NMH, NRR, and KES developed the climate change conceptual model. This paper was a highly collaborative effort and all authors contributed equally to the development of the research question and study design as well as the writing of the paper.

Nicole M. Hayes and Bridget R. Deemer are joint first authors.

Climate change and freshwaters

Climate change is one of the greatest threats to aquatic ecosystems (Blenckner 2005; Hayhoe et al. 2008). The effects of climate change range from direct changes in water level (Smol and Douglas 2007) and surface-water temperature (O'Reilly et al. 2015) to indirect, complex ecological shifts that alter trophic interactions (Winder and Schindler 2004), and have been observed in many different regions (e.g., Quayle et al. 2002; Schindler and Smol 2006; Schneider and Hook 2010). At the same time, increased human demand for water-related ecosystem services has resulted in the construction and operation of over 1 million dams globally (Lehner et al. 2011). As a result, human-made lakes (i.e., reservoirs) have come to comprise anywhere between 6% and 11% of global lentic surface area (Downing et al. 2006; Lehner et al. 2011; Verpoorter et al. 2014). While the global expansion of reservoirs has increased access to drinking water, irrigation, navigation, flood control, and hydropower, it has also fundamentally changed the movement of water, sediment, nutrients, and biota through aquatic networks. Globally, reservoirs are estimated to increase the standing stock of natural river water by over 700% (Vörösmarty et al. 1997), reduce sediment flux to the ocean by over 1 billion metric tons of sediment per year (Syvitski et al. 2005), reduce phosphorus transport to the coast by approximately 12% (Maavara et al. 2015), contribute more than 30% of all lentic nitrogen and silica retention (Harrison et al. 2009, 2012), and emit methane at higher per area rates than any natural aquatic ecosystem (Deemer et al. 2016). These findings are consistent with the notion that inland waters are not "passive pipes" (Cole et al. 2007) and that the ecological role of reservoirs is unique from lakes.

As low points on the landscape, lentic ecosystems also serve a unique role as integrators of atmospheric and catchment scale climate signals (Williamson et al. 2009). These signals, or sentinel responses, are shaped by a number of factors including large-scale geographic patterns and internal waterbody processes. Climate change conceptual frameworks have previously lumped reservoirs with natural lakes (Williamson et al. 2009) or excluded them from efforts to develop broadly applied sentinel response metrics (Adrian et al. 2009). Reservoirs and lakes are generally thought to share a number of similarities. Reservoirs are often divided into three zones for the purposes of ecological study: river, transitional, and lacustrine—with the lacustrine zone having slower water velocities and pronounced thermal stratification much like a lake ecosystem (Thornton et al. 1990). As a result, the lacustrine or lentic zone of a reservoir is thought to be similar to a lake in terms of planktic production, nutrient limitation of phytoplankton growth, and biogeochemical cycling (Wetzel 2001). Despite these similarities, reservoirs and lakes also differ in a number of ways that lead to differences in ecosystem functioning. Given the growing influence of reservoir ecosystems on the global hydrologic system (Zarfl et al. 2015), and recent evidence that reservoirs may serve ecological roles distinct from lakes even in the lacustrine zone (Beaulieu et al. 2013), we argue that these human-made systems should not be lumped with natural lakes in climate change conceptual models. An improved understanding of the interaction between reservoirs and climate may have broad scale implications for water quality from headwaters to coasts.

Key differences between lakes and reservoirs

Teasing apart the ecologically relevant differences between reservoirs and natural lakes can be a daunting task given the background variability in lentic ecosystem types. For example, one of the most common lake typological divisions is based on water source (i.e., relative contribution of groundwater versus surface water), of which reservoirs are a single lake type (Hutchinson 1957; Wetzel 2001; Figs. 1, 2). While natural lakes are generally subdivided based on hydrology (i.e., seepage, glacial, oxbow, intermittent, etc.), reservoirs are often categorized based on their size or their designed purpose (i.e., their primary reason for being constructed; Thornton et al. 1990; Poff and Hart 2002). Currently, the most common classification scheme for reservoirs divides these ecosystems into two groups: storage and run-of-river (Poff and Hart 2002). Storage reservoirs typically store large volumes of water and have large hydraulic heads, long hydraulic residence times, and allow for relatively fine-tuned control over the rate at which water is released from the dam. Run-of-river reservoirs, on the other hand, typically store less water and have relatively small hydraulic heads, short hydraulic residence times, and little or no control over the rate that water is released from the dam (Poff and Hart 2002). In many ways, these categories represent two extremes on a spectrum of reservoirs with larger "lacustrine" zones that are more like lakes (storage) to reservoirs with larger "river" zones that are more like rivers (run-of-river). To foster a more detailed discussion of reservoir type, we have depicted reservoir types based on position within river network (Fig. 1). While reservoirs created by damming a pre-existing lake are likely to share characteristics with the storage reservoir category, reservoirs created by damming a pre-existing river can resemble either a storage or a run-of-river system. In addition, reservoirs created to store water outside of a river network (e.g., farm ponds, pump storage systems, etc.) may function quite differently than those receiving surface water inputs via stream or river inlets.

In this article, we define reservoirs broadly as any human-made lake, whether it be embedded within a river network or not (Fig. 1). However, for the purpose of our analysis, we restricted our comparison to *reservoirs within river networks* (Fig. 1). This selection was made to harmonize our reservoir comparison with the types of reservoirs considered in the

What are reservoirs and lakes?
Reservoirs are defined as an intermediary between lakes and rivers, but several other characteristics, including outlet control, origin, and placement in a river network, more specifically define these systems. Similarly, lakes vary with regards to water source. Black bars indicate a dam and stars indicate reservoir types considered in this study.

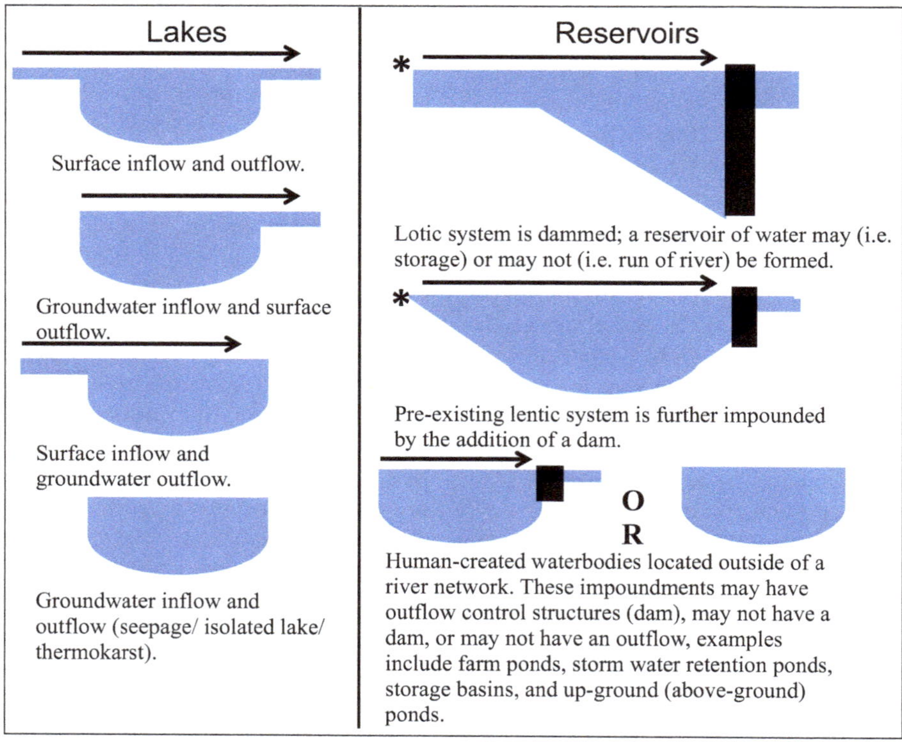

Reservoir Definition: Broadly, we define reservoirs as human made lakes-- whether they be embedded in a river network or not. Still, data availability limited the types of reservoirs considered in this study. The synthesis and conceptual model developed here applies to reservoirs >0.04 km^2 that are located within a river system (indicated by the stars). Excluded reservoir systems include artificially constructed lakes and ponds placed outside of river networks (which may or may not include a dammed outflow).

Lake Definition: We broadly define lakes as naturally occurring low points in the landscape that contain standing water, predominantly in the form of open water habitat, year round. Still, data availability limited the types of lakes considered in this study. The synthesis and conceptual model developed here applies to lakes >0.04 km^2.

Fig. 1. Schematic and definition of lake and reservoir types included in this study.

Environmental Protection Agency's National Lake Assessment (NLA) and other surveys in our literature analysis (Thornton et al. 1980; Harrison et al. 2009; Powers et al. 2015). Site selection in the NLA was limited to lakes and reservoirs greater than 0.04 km^2 in size. Smaller systems, which are also often located outside of river networks (i.e., farm ponds and stormwater retention ponds), are ecologically important (Downing et al. 2009; Holgerson and Raymond 2016), however, we do not have the data necessary to consider them in our conceptual model. Despite the diverse array of lake and reservoir types, we argue that broad differences between reservoirs and natural lakes can be distinguished at the landscape scale given the unique human-made and human-operated aspects of reservoir ecosystems.

In order to assess the differences between lakes and reservoirs, we synthesize evidence from the literature and quantify evidence using a dataset from the 2007 NLA (Supporting Information Material; U.S. Environmental Protection Agency 2007; U.S. Environmental Protection Agency 2009; Corman et al. 2016). We focus our comparison of lakes and reservoirs on characteristics that have the potential to affect ecosystem response to climate change. These physical and chemical attributes are separated into four basic categories: (1)

Alternative reservoir definitions:

Environmental Protection Agency National Lakes Assessment (EPA NLA): Reservoirs include any open waters resulting from impoundment created after European settlement (e.g. post-1900).

Global Reservoirs and Dam Database (GRanD): Lakes that are explicitly classified as manmade, however there are some caveats. Some reservoirs do not have dams, for example when water is stored in natural or artificial depressions. Not all dams create reservoirs, for example run-of-river hydropower stations may not impound water and thus may not form reservoirs.

Hutchinson (1957): Lake "type 73, dams built by man, e.g., Lake Mead." This falls under the broader category, "Lakes Produced by the Complex Behavior of Higher Organisms" and is not to be confused with Type 72, "beaver dams" or Type 74, "excavations by man, as the abandoned diamond mines at Kimberley, South Africa.".

International Commission on Large Dams (ICOLD): An artificial, human-made lake, basin or tank in which a large quantity of water can be stored.

Thornton: No explicit definition provided, however reservoirs are described as having a lotic, transitional, and lentic zone, supporting the notion that a reservoir is a limnological intermediate between a lake and a river.

United States Bureau of Reclamation: (1) An artificially impounded body of water to store, regulate, or control water. (2) Body of water, such as a natural or constructed lake, in which water is collected and stored for use.

United States Geological Survey (USGS): A pond, lake, or basin, either natural or artificial, for the storage, regulation, and control of water.

Wetzel (2001): Impounded waters resulting from construction of a dam across a river.

Fig. 2. Selected alternative definitions of reservoirs.

catchment characteristics, (2) waterbody characteristics, (3) management, and (4) geographic distribution (Table 1). We use categories 1–3 to link our synthesis and NLA analysis to a modified E*m* flux conceptual model (*see* description below). Management (category 3) includes human decision-making at both the planning (hereafter referred to as "design") and post-construction (hereafter referred to as "management") stages. At the design stage, water managers strategically place dams to meet human need, usually by damming a pre-existing river or lake. Dam placement ultimately determines the catchment and waterbody characteristics discussed above as well as the regional positioning and watershed land use context, which we do not specifically address in our analysis. In the management post construction stage, reservoirs can be managed via specific decisions regarding dam withdrawal and water level management regimes, but are also subject to within-waterbody management techniques that overlap with strategies employed in natural lakes (Table 2). For the purposes of this study, we focus on within-waterbody management of lakes and reservoirs and we do not assess watershed land use management. We also do not specifically address category 4 in our analysis, as the broad-scale differences in the geographic distribution of lakes and reservoirs are addressed elsewhere in the literature (e.g., Lehner and Döll 2004 and Fig. 3) and can hinder our ability to characterize and quantify the mechanistic differences between lakes and reservoirs in a single region. We expect that many of the differences in lake and reservoir catchment management are confounded with differences in geographic distribution and we note that differences in the management of lake and reservoir catchments within the same region are not well quantified.

We identified studies that quantify key differences between lake and reservoir catchment, waterbody, and/or management characteristics (Table 1). Based on these studies and the proposed differences described in a seminal work on Reservoir Limnology (Thornton et al. 1990), we designed a targeted analysis of a subset of NLA lakes and reservoirs. We paired lake and reservoir systems based on their geographic proximity and tested for significant differences in catchment and waterbody characteristics by system type (*see* Supporting Information Material for more detail). We discuss our findings in the subsections that follow and then place the key differences we identified in the context of a new conceptual framework for climate change in lake versus reservoir ecosystems.

Differences in catchment characteristics

The literature synthesis and NLA analysis both suggest that catchment properties differ for lakes and reservoirs (Fig. 4; Table 1). The ratio of catchment area to surface area (CA : SA) was 3–4 times greater in reservoirs than lakes in a U.S. dataset (Thornton et al. 1980) and in a global database (Harrison et al. 2009). Within the subsetted NLA database, median CA and CA : SA were substantially larger in reservoirs compared to lakes (6.5 and 3 times larger respectively, Fig. 4, Supporting Information Material Table S1). Thornton et al. (1990) also proposed that reservoirs would be located lower in the landscape (Table 1). While explicit testing for differences in landscape position based on waterbody

Table 1. Summary of probable differences between lakes and reservoirs, adapted from Thornton et al. (1990), with new citations in italics.

Property	Natural lakes	Reservoirs	Evidence of differences	Citation
(1) Catchment characteristics				
Catchment area (CA): surface area (SA)	CA>SA	CA>>>SA	Reservoirs have larger CA and CA:SA	Thornton et al. (1980), *Harrison et al. (2009), This study*
Landscape position	Natural placement; fed by lower order streams; diffuse inputs	Lower in the catchment; fed by higher order streams; surface water dominated	Higher mass throughput and transfer to reservoir beds versus lake-beds from: (1) higher nitrogen and silica retention in reservoirs than lakes, and (2) high carbon burial and sedimentation rates in reservoirs	Canfield and Bachman (1981), *Syvitski et al. (2005), Harrison et al. (2009), Tranvik et al. (2009), Harrison et al. (2012), Clow et al. (2015)*
Reservoir morphometry	Circular basin; less complex perimeter	Longer, narrower basin; more complex perimeter	Reservoirs have greater perimeter areas than lakes	*This study*
(2) Waterbody characteristics				
Secchi depth	Generally deeper	Generally more shallow	Secchi depth shallower in reservoirs compared to lakes	Thornton et al. (1980), *This study*
Temperature	Somewhat lower	Somewhat higher	Similar surface waters; bottom waters are warmer in reservoirs than lakes	*This study*
(3) Management				
Waterbody management	Less managed	More managed	*See* Table 2, however management is likely geographically specific and is thus an area for future study. Smaller winter drawdowns and lower estimates of overall management stress in lakes versus reservoirs and greater water level instability in reservoirs.	*Keto et al. (2008), Kolding and van Zwieten (2012), This study (in italics)*
Catchment management	Watersheds both developed and undeveloped	Watersheds more developed	Higher P and N load in reservoirs from greater human activity in the catchment and higher shoreline development	Thornton et al. (1980)
(4) Geographic distribution				
Broad-scale distribution	Predominantly glaciated regions	Predominantly non glaciated regions	Greater fraction of lakes found in the boreal region and greater fraction of reservoirs found in the north temperate region	*Lehner and Döll (2004)*
Regional distribution (and function)	Dictated by geomorphology	Influenced by regional demand for dam-related ecosystem services	Primary designed purpose of dams varies by geographic region, with possible implications for morphometry and watershed placement (e.g., run of river versus storage dams)	*Poff and Hart (2002)*

Table 2. Common types of aquatic ecosystem management and implications for energy and mass processing.

Type of management	Purpose of management	Energy (E) and/or mass (*m*) implications	Citation
Selective withdrawal (predominantly reservoirs)	In reservoirs with outlet gates at multiple heights, selective hypolimnetic withdrawal can control downstream temperatures and/or improve water quality within the reservoir.	E: Enhances "net energy" per water volume by preferentially spilling colder water, may also lead to de-stratification.	Nürnberg (2007)
	Density flow releases shunt runoff plumes downstream to avoid reservoir sedimentation and reduce particulate associated pollutants. Increasing flow velocity through bottom outlets can also flush deposited sediments out of the reservoir by scouring the bed.	*m*: Can reduce the effect of the catchment on mass inputs via shunting of turbidity flows out the dam.	Huang et al. (2014), Tiğrek and Aras (2012)
Pool drawdowns (predominantly reservoirs)	Water level drawdowns can create flood storage, control for nuisance macrophytes, generate hydropower (e.g., hydropower peaking), and/or meet demands for consumptive water uses.	E: May either enhance or destabilize stratification regime. *m*: May lead to shifts in water column chemistry, species composition and community metabolism.	Blenckner (2005), Moreno-Ostos et al. (2008), Baldwin et al. (2008), Bell et al. (2008), Zohary and Ostrovsky (2011), Valdespino-Castillo et al. (2014)
Temperature curtains	Engineered curtains can control inflow or outflow ecosystem hydrology.	E: Can enhance "net energy" per water volume by preferentially spilling colder water. Can change stratification dynamics by preventing plunging inflows.	Vermeyen (2000)
Aeration and oxygenation systems	Air bubblers and oxygen diffusers can oxygenate bottom waters.	E: Can weaken stratification. *m*: Can stir up sediments and change mass processing by changing redox status of bottom waters.	Beutel and Horne (2014), Beutel et al. (1999)
Chemical treatments	Aluminum sulfate additions can strip phosphorus from the water column and increase water clarity.	E: Possible change to stratification regime due to increased waterbody clarity. *m*: May alter the effects of mass inputs via artificial flocculation and precipitation of nutrients and via unintentional effects on benthic communities and biogeochemical cycles.	Kennedy and Cooke (1982), Nogaro et al. (2013)
	Metals, photosensitizers, and herbicides can eliminate cyanobacterial blooms.	*m*: May change the biotic pool (including toxic effects on nontarget species).	Jančula and Maršálek (2011)
Sediment dredging (predominantly reservoirs)	Dredging may clear navigation channels, improve water quality, reduce eutrophication, and/or enhance bottom water oxygen saturation.	E: Greater volumes of water preserve the energy (and mass) effects for a longer duration. *m*: Takes away the "mass legacy" and changes nature of waterbody filter	Sahoo and Schladow (2008); Blenckner (2005)
Eco-technology (biomanipulation)	Removal of benthivorous/cyprinid fish can increase water clarity and decrease Chl *a*.	E: Possible change to stratification regime due to increased waterbody clarity. *m*: Takes away a portion of the biotic pool.	Meijer et al. (1999), Søndergaard et al. (2007)
Coverings	Light limiting coverings can reduce evaporation and algal growth.	E: Block solar heating and evaporative cooling, limit water column turbulence. *m*: Reduce autochthonous production.	Álvarez et al. (2006), Maestre-Valero et al. (2011)

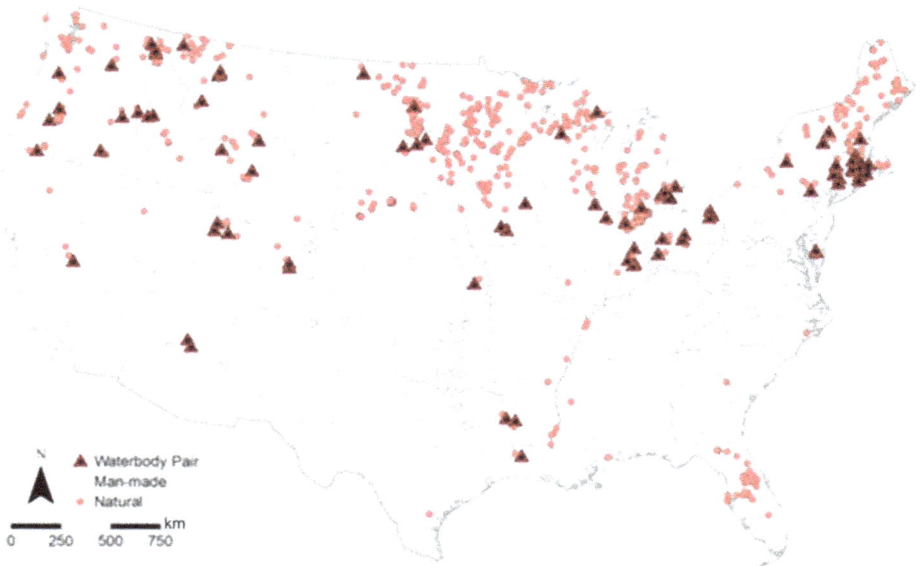

Fig. 3. Regional distribution of lakes (light red) and reservoirs (white) in the NLA in 2007. Lake and reservoir pairs selected for analysis in this study are shown as red triangles.

elevation did not reveal ecologically significant differences between lakes and reservoir pairs in the NLA (Supporting Information Material Table S1), greater CA : SA for reservoirs suggests they are indeed located lower in the landscape (Soranno et al. 1999). Given that waterbodies with lower landscape position have higher material inputs from the watershed (Soranno et al. 1999; Kratz et al. 1997), we hypothesize that reservoirs generally receive higher mass inputs than natural lakes. Although we did not find any studies that directly compared material inputs in lakes versus reservoirs, several global and national-scale studies report that reservoirs tend to have greater retention of watershed material than natural lakes (Harrison et al. 2009; Harrison et al. 2012; Clow et al. 2015). Finally, Thornton et al. (1990) predicted that reservoirs would differ in their morphometry; lakes would be more circular while reservoirs would be long and narrow with complex perimeters. Our analysis of the NLA paired lake-reservoir dataset supported this prediction with a median reservoir perimeter two times greater than the median lake perimeter (Fig. 4, Supporting Information Material Table S1).

Differences in waterbody characteristics

Reservoirs also differ from lakes with respect to within waterbody characteristics (Fig. 4; Table 1). For example, Secchi depths were found to be shallower in reservoirs than lakes in a U.S. dataset (Thornton et al. 1980) as well as in our analysis of the subsetted NLA dataset (Fig. 4; Table 1 and Supporting Information Material Table S1). Greater CA : SA and associated nutrient and sediment inputs likely lead to higher production and suspended sediments in reservoirs, decreasing water clarity (Supporting Information Material

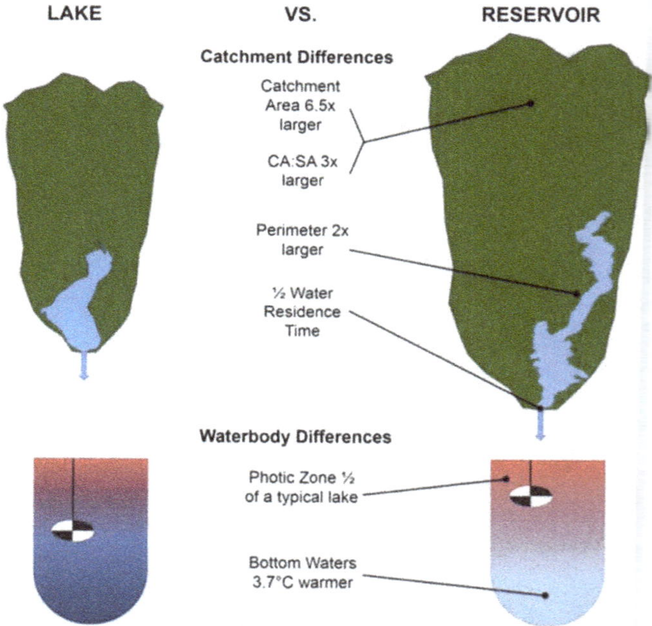

Fig. 4. Generalized watershed and waterbody schematics for a lake versus a reservoir. Differences are summarized from the literature synthesis and NLA analysis.

Table S1). Additionally, residence times were shorter in reservoirs as compared to lakes (approximately half as long, Supporting Information Material Table S1, Fig. 4). We also found that surface-water temperatures were similar in NLA lakes and reservoirs; however, bottom water temperatures were warmer in reservoirs (Supporting Information Material Table S1, Fig. 4). While differences in bottom water temperatures could result from differences in flushing rates and associated stratification regimes, these differences have not been fully explored in the literature. Future research efforts should focus on determining differences in lake and reservoir waterbody characteristics but also identifying the mechanisms driving these differences and the implications for ecosystem processes.

Differences in within-system management

Within-system management can alter the hydrology, chemistry, biology, and/or light regime of the waterbody to improve water quality, fish production, or other characteristics necessary for human uses (e.g., recreation, hydropower generation, flood control, erosion reduction, water supply, navigation, etc.). We identified eight common lake and reservoir management strategies of broadscale ecological significance (Table 2). Three of the strategies (pool drawdown, selective withdrawal, and sediment dredging) are common in reservoirs whereas the other five (temperature curtains, aeration/oxygenation systems, biomanipulation, chemical treatments, and covering) can be employed in either system type. While not an exhaustive list, these management techniques highlight the extent to which human activities may mask or amplify climate signals. For example, within-system management strategies can interfere with biological life cycles (e.g., fish stranding mortality associated with hydropower peaking, Bell et al. 2008; and elimination of cyanobacteria blooms via chemical treatments, Jančula and Maršálek 2011), seasonal hydrologic dynamics (e.g., via pool drawdowns, Zohary and Ostrovsky 2011; and via the reduced evaporation associated with coverings, Álvarez et al. 2006), and water column chemistry (reduced water column phosphorus concentrations associated with alum treatments, Kennedy and Cooke 1982; Nogaro et al. 2013; and higher concentrations of reduced solutes associated with pool drawdown, Baldwin et al. 2008; Zohary and Ostrovsky 2011; Harrison et al. 2017).

Based on our synthesis of available literature, we expect reservoirs to experience more within-system management than natural lakes for two main reasons. First, dams are constructed with some human use in mind and are thus likely to experience altered hydrology, biology, chemistry, and/or light regimes in support of these intended uses. Second, an outlet structure is a characteristic component of many reservoir ecosystems that can exert significant control on waterbody conditions. While lakes can have managed outflow structures (i.e., temperature curtains, Vermeyen 2000; and water withdrawal pumps), many reservoirs necessitate a water outlet that prevents over-filling. In fact, reservoirs were excluded from one recent study of lake sentinel responses due to their "anthropogenically controlled" hydrology alone (Adrian et al. 2009). Still, large-scale information about lake and reservoir management regimes is quite limited. For example, the NLA category for management contains qualitative information about perceived management stress where available, but is often left blank (U.S. Environmental Protection Agency 2009). Of the management pressure that was noted in the NLA paired dataset, reservoirs experienced approximately double the management stress of natural lakes (mean lake management stressor score of 2.2 for natural lakes and 4.2 for reservoirs out of a total possible score of 5). NLA-based visual estimates also suggest that reservoirs experience water level fluctuations of significantly higher amplitude than those in natural lakes (mean of 4 m and 0.7 m of fluctuation in reservoirs and lakes, respectively). Quantifying the relative importance of different management strategies in lakes as compared to reservoirs is beyond the scope of this synthesis but is an important area for future work that may also complement efforts to better classify reservoir systems.

Existing conceptual models for climate change effects in lentic ecosystems

Several models have been developed to predict the effects of climate change on lake ecosystems (e.g., Blenckner 2005; Leavitt et al. 2009) and these models are often uniformly applied to both lakes and reservoirs. There is a general notion that lakes and reservoirs act similarly as regulators (Tranvik et al. 2009), integrators, and sentinels of climate change (Williamson et al. 2009); however, studies that explicitly compare the sensitivity of lakes versus reservoirs to climate change are rare (e.g., Nowlin et al. 2004; Beaulieu et al. 2013). This lack of comparative information limits the inclusion of reservoirs into conceptual models. Yet, incorporating basic differences between reservoirs and lakes in terms of catchment, waterbody, and management characteristics help make existing frameworks more useful for predicting climate change effects.

The Energy (E) mass (*m*) flux framework proposed by Leavitt et al. (2009) may be particularly useful in distinguishing the effects of climate change on reservoirs as compared to lakes. In this framework, the effects of climate change on lentic ecosystems are modeled via a consideration of the transfer of both E (irradiance, heat, kinetic E of wind) and *m* (water, solutes, particles) through landscape and "lake" filters. We refer to the lake filter as the "waterbody" filter so as to facilitate discussion of both lakes and reservoirs. These filters function to transform E and *m* inputs and can thus influence the way that climate drivers influence ecosystems. The capacity for human-mediated processing of E and *m* in these systems is an important topic for research and is particularly relevant for reservoir ecosystems (which by definition are human-designed and human-managed). While Leavitt and colleagues emphasize the important role of human disturbance in mediating the landscape filter (e.g., with respect to land use), they do not explicitly consider aquatic ecosystem management. In the sections that follow, we discuss

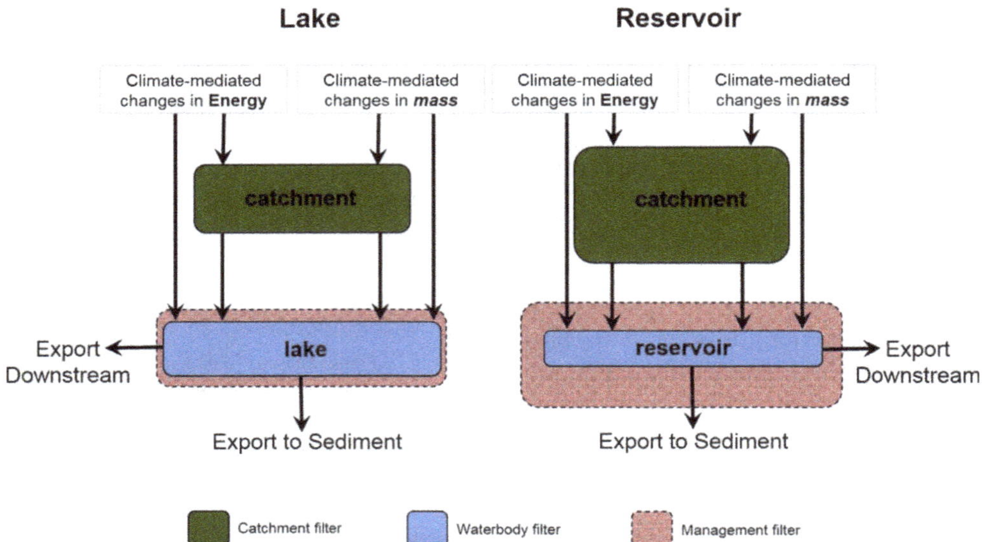

Fig. 5. A conceptual model of climate related effects on lake ecosystems (left) and reservoirs (right). This model builds on the Energy-mass (E*m*) flux framework proposed by Leavitt et al. (2009) and includes the four pathways by which climate drivers (factors that cause changes in an ecosystem, shown in black) enter environmental filters and move between the filters. Filters are environmental features that transduce and transform Energy (E) and mass (*m*). The catchment and waterbody act as filters (Blenckner 2005; green and blue boxes respectively) in this model as does the new management filter proposed as a result of this study (red boxes). The relative size of the catchment and waterbody filters represents the magnitude of the transduction and transformation based on the differences identified between lake and reservoir ecosystems in the NLA data analysis. Because differences in lake and reservoir management were not tested directly in this study, we show the management filter with a dashed border and suggest that it should be the focus of future research. Ultimately, the catchment, waterbody, and management filters interact to determine the relative proportion of the climate signal exported to the sediments or to downstream river networks and estuaries.

how reservoir design and aquatic management strategies alter the way in which E and *m* fluxes are transformed within the catchment and the resulting effects on abiotic and biotic variables within the waterbody. We predict that these differences will alter the magnitude and fraction of E and *m* exported to lentic sediments versus downstream river networks and coastal environments.

A new model for climate change in reservoirs and lakes

Conceptual ecosystem model

We constructed a conceptual model that incorporates key differences between lakes and reservoirs (Figs. 4, 5) to visualize variability in the export of climate-mediated changes in E and *m* to the sediment or downstream river networks. Our model builds on previous models proposed by Blenckner (2005) and Leavitt et al. (2009) and considers the pathways by which E and *m* are transformed by environmental filters. Our model incorporates three environmental filters: catchment, waterbody, and management. Filter size represents the magnitude of E or *m* transformed within each filter. Although previous versions of the model consider an atmospheric filter (Leavitt et al. 2009), our sample design does not allow for comparisons beyond the regional level. The catchment filter (or landscape filter; Blenckner 2005) alters the properties and magnitudes of E and *m* flux into waterbodies. Our waterbody filter affects the magnitude of E and *m* fluxes exported to sediments or downstream river networks, which varies considerably from previous models.

The arrows in our conceptual diagram represent direct and indirect transfer of E and *m* to waterbodies (Fig. 5). Direct inputs of E to the waterbody filter include solar irradiance, heat, and wind while indirect E pathways, those that move through the catchment filter first, influence catchment characteristics such as soil and vegetation development and subsequent terrestrial subsidies to lakes (Leavitt et al. 2009). Mass directly enters the waterbody via precipitation, particles, and solutes (e.g., wet and dry deposition) and indirectly via run-off of water and associated dissolved and particulate matter from the catchment. These indirect pathways are subject to environmental filtering and thus the properties of E and *m* that pass through the environmental filters are altered (Leavitt et al. 2009). Quantifying the extent to which filters alter the magnitude of each pathway is beyond the scope of this study, thus each line carries the same thickness (Fig. 5).

E*m* flux in lakes and reservoirs

Our conceptual model proposes a larger catchment filter for reservoirs than lakes (Table 1; Fig. 5). The larger catchment filter for reservoirs represents their propensity to drain larger catchments, their more complex shorelines, and their higher influx: content ratio (Fig. 4). With larger CA : SAs and larger perimeters, reservoirs are expected to experience a greater interaction with the catchment and thus increased catchment-mediated influx of E and *m*. These higher catchment-mediated influxes do not appear to be received by larger waterbodies as we found no evidence that reservoirs are larger than natural lakes (Supporting Information Material Table S1). Thus, larger inputs of *m* relative to content (e.g., greater inputs of water from inflow relative to reservoir water content) lead to shorter residence times and higher influx to content ratios in reservoirs. Mean residence time exerts an important control on processing time within the water column (Soranno et al. 1999), thus while it is a function of catchment characteristics, we discuss it in terms of the waterbody filter.

Key differences between lake and reservoir waterbody characteristics highlight the importance of the waterbody filter in processing E and *m* from the catchment in ways that will affect export to sediment or downstream ecosystems (Table 1; Fig. 4). Differences in water clarity can alter water chemistry and lead to changes in the productivity and diversity of lake ecosystems (Berger et al. 2006). Thermal regime can also affect the extent to which a waterbody reflects climate forcing. A study of three medium-sized German lakes with different mixing regimes and thermal structures found very different thermal responses to the North Atlantic Oscillation (NAO) among lakes, wherein the NAO signal was most persistent in the deep, dimictic system with stable summer stratification (Gerten and Adrian 2001). Additionally, shorter residence times may reduce E and *m* processing time. For example, Brooks et al. (2014) concluded that waterbody residence time was negatively correlated with nitrogen, phosphorus, and chlorophyll *a* concentrations, suggesting increased *m* processing in systems with longer residence times. We propose a smaller waterbody filter for reservoirs given their shorter residence times; however, more work is needed to tease apart the role of the differences we report between lakes and reservoirs and the resulting effects on E and *m* processing and export from the waterbody filter.

In addition to the catchment and waterbody filter, we propose a management filter that mediates functioning of the waterbody filter as well as E and *m* flux leaving the lake (Fig. 5). This was motivated by the literature review (described above, Table 2) wherein we identified eight common lake and reservoir management strategies that are likely to have direct effects on E and/or *m* fluxes. As discussed above, we expect management differences to be particularly pronounced in reservoirs given the presence of outlet control features on many dams and the fact that reservoirs were designed for human use (Table 2). Thus, we propose a larger management filter for reservoirs than for lakes (Fig. 5). These management strategies can have important consequences for both E and *m* fluxes (Table 2) although quantifying the magnitude of these effects is beyond the scope of this synthesis. Still, several recent studies have highlighted the capacity for

Summary of Research Recommendations:

1. Determine the extent to and mechanisms by which catchment properties and/or morphological factors alter the magnitude of E and *m* processing in lakes as compared to reservoirs.
2. Quantify the relative importance of different management strategies in buffering or intensifying climate change signals.
3. Describe connections between expected shifts in E and *m* processing and implications for ecosystem processes (e.g. carbon burial, biomass production, nutrient transformations, etc.).
4. Directly compare effects of climate change on lake as compared to reservoir limnological properties through field observations, experiments, or using compiled datasets.
5. Test conceptual model proposed here in tropical regions.
6. Improve the global mapping of reservoir systems. Quantify and map the global coverage of various reservoir (and lake) types and examine the potential role of typology in defining differences between systems (e.g. for small ponds).

Fig. 6. Recommended directions for future research.

within-system management to alter whole-ecosystem ecology. For example, hypolimnetic oxygenation was found to alter the fraction of *m* that was exported to sediments versus potentially transported downstream in a temperate reservoir (Gerling et al. 2016) and shade coverings were found to affect ecosystem E distribution (i.e., stratification) and reduce water column turbidity and dissolved oxygen (Maestre-Valero et al. 2011). Given the ubiquity of outlet control features on dams and the variety of effects that outlet control can have on E and *m* fluxes (Table 2), future research should focus on the propensity for outlet management techniques to either buffer or intensify climate signals (Fig. 6).

Model limitations

The model we present identifies mechanistic differences in how lakes and reservoirs process climate change by focusing on geographically paired lakes, which allows for comparison of lakes and reservoirs experiencing similar climate forcing. However, this approach cannot be used to identify differences in climate forcing between lakes and reservoirs at a broader scale. In the United States, reservoirs have a more southern distribution than lakes (Fig. 3) and this pattern of regional distribution will determine the type of climate forcing these ecosystems experience. For example, the southwestern United States has experienced recent regional tendencies toward more severe droughts (Kunkel et al. 2008), likely leading to decreased indirect *m* inputs and increased direct E inputs compared to the more northerly distributed lakes. These differences in the distribution of lakes and reservoirs and the geographic difference in climate change will lead to differences in E and *m* exposure between lakes and reservoirs. Despite the geographic limitations, our model provides evidence for key differences in how geographically co-located reservoirs and lakes respond to climate change.

In addition, we simplify what constitutes the catchment itself. The filtering effect of the catchment varies based on land use (Blenckner 2005), for example, climate drove differential effects in nutrient loading between forested and agricultural watersheds (Hayes et al. 2015). Although there were no significant differences in land use in the subset of the NLA analyzed in this study (Supporting Information Material Table S2), observations from a global dataset of 115 lakes and reservoirs suggest that reservoir catchments are more human-dominated (higher nitrogen loading rates in reservoirs than lakes, Harrison et al. 2009). Future research should focus on how catchment land use affects the transport and transformation of E and *m* in lakes versus reservoirs.

Broadscale implications of the model

The E*m* flux framework predicts that the effect of a climate driver is determined by the ratio of input to content, suggesting that reservoirs, with higher input to content ratios, will be more affected by catchment mediated E and *m* inputs. Previous work has established the importance of *m* inputs to ecosystem climate sensitivity (Vogt et al. 2011) and has documented greater inputs of *m* (including sediment, nutrients, and water) in waterbodies with larger catchments (Soranno, et al. 1999). Recent work finds that reservoirs retain larger quantities of carbon, nitrogen, silica, and phosphorus than lakes (e.g., Harrison et al. 2009; Harrison et al. 2012; Clow et al. 2015; Maavara et al. 2015) and many models of lake and reservoir *m* retention find that *m* export is positively related to *m* loading (Saunders and Kalff 2001; Harrison et al. 2009). Thus, we suggest that reservoirs may be especially sensitive to climate drivers that increase *m* flux, such as storm events.

While our conceptual diagram suggests higher indirect E and *m* loading to reservoirs than lakes, it is less clear how different ecosystem parameters such as residence time and waterbody morphology interact with *m* loading to determine the *fraction* of inflow *m* that is buried (either as biomass or particulate matter) versus exported from the waterbody. For example, we suggest that low residence times in reservoirs may result in less processing time for E and *m* within the waterbody (e.g., lower fraction of E and *m* converted to biomass and exported to sediment), despite relatively large *m* burial on a total mass basis (due to the high magnitude of *m* loading and the potentially greater interface between sediments and water as a result of increased perimeter). In addition to residence time, other ecosystem characteristics (e.g., depth and temperature at sediment water interface) can also affect *m* processing rates. In a study of global nitrogen retention in lakes and reservoirs, reservoirs had higher settling velocities (function of residence time, depth, and fraction of *m* retained) than lakes (Harrison et al. 2009) indicating that a higher fraction of *m* is exported to sediments in reservoirs than in lakes. The fraction of inflowing *m* that is exported to sediments can also vary based on the type of *m* in question. For example, a regional analysis of river networks in agricultural basins found consistent N retention, but variable P retention behind dams (Powers et al. 2015). More work is needed to tease apart the role of E and *m* type, inflow rate, residence time, and other morphological factors in determining the efficiency of E and *m* processing in the waterbody filter (e.g., how E and *m* are either exported downstream or to sediments).

The waterbody filter has important implications not just for lake and reservoir ecology, but also for downstream and coastal ecosystems. Processing within waterbodies affects both the relative fraction and absolute magnitude of E and *m* exported downstream (as compared to sediments). While reservoirs are known to retain high fractions of *m* (e.g., bioavailable elements and sediment), the effect of climate drivers on the transport of *m* through the waterbody can fundamentally alter downstream and coastal ecosystems.

This study also highlights the important role of management, especially in reservoir systems. Under a changing climate, waterbody management may be re-assessed to address any of the following broad categories: (1) management to support designed purpose, (2) management for climate adaptation, or (3) management for climate change mitigation. Management for the designed purpose may require modification to comply with the Endangered Species Act or to accommodate other socio-ecological considerations that were not apparent when the dam was constructed. In the case of management for climate adaptation, reservoir management may be modified compensate for a changing climate (Eum and Simnovic 2010). Reservoirs, especially those with highly disturbed catchments, are extremely likely to require management to adapt to the effects of climate change including higher peak flows and improved water conservation under drought (Palmer et al. 2008). Finally, the capacity for reservoir management to either mitigate or enhance greenhouse gas emissions is a topic of current research given the important role of reservoirs in contributing to anthropogenic CH_4 emissions (Deemer et al. 2016). In the Pacific Northwest U.S.A., a study of six reservoirs found that water level drawdowns were associated with significantly higher methane emissions (Harrison et al. 2017), suggesting that water level management may affect the contribution of these systems toward radiative forcing in the atmosphere. Catchment management may also affect CH_4 emissions from lakes and reservoirs by altering lentic nutrient loading and associated primary production. High rates of primary production have been linked to high lake and reservoir CH_4 emissions in mesocosm (Davidson et al. 2015), regional (West et al. 2015), and global studies (Deemer et al. 2016). Given the significant global push to construct new dams (Zarfl et al. 2015), the propensity for reservoir design (e.g., landscape placement) to determine adaptive and mitigative capacity is an important area for future work.

Latitude matters: addressing the temperate lake bias

The conceptual model presented here does not consider differences in the geographic distribution of lakes and reservoirs, but instead describes the mechanistic differences expected between a lake and reservoir found in the same region. In addition, the evidence we used to formulate our conceptual model is based largely on north temperate and boreal systems. This is a common bias as global analyses of lakes and reservoirs generally have disproportionately less data from tropical systems than temperate and boreal systems (three tropical systems of 27 in Harrison et al. 2012, and 28 tropical systems of 115 in Harrison et al. 2009). Similarly, the reservoir pairs from the U.S. NLA dataset used to supplement this synthesis are temperate biased; only three of 66 lake-reservoir pairs are located below 35° north latitude (Fig. 3). While the majority of global lakes are situated at northern latitudes (Verpoorter et al. 2014), the same is not true for reservoirs, which are disproportionately located at lower latitudes (Lehner and Döll 2004). Future studies that compare lakes and reservoirs in tropical regions are needed to better verify the validity of the conceptual model proposed here for tropical systems. With increasing dam construction in tropical areas (Zarfl et al. 2015), this dearth of data for tropical and subtropical reservoirs becomes increasingly urgent to correct.

Reservoirs at lower latitudes may differ in fundamental ways from those at higher latitudes. Tropical lakes and reservoirs generally experience larger water level fluctuations than in temperate zones (Kolding and van Zwieten 2012), such that hydrology-driven changes in E and *m* processing may be amplified in these systems. Tropical lakes and reservoirs are known to experience higher sediment loading than their temperate and boreal counterparts, a pattern that has been largely attributed to tropical watershed deforestation (Syvitski et al. 2005 and citations therein). While analysis of the paired NLA dataset did not find significant land use differences between lakes and reservoirs (Supporting Information Material Table. S2), other studies suggest that reservoirs may be located in more developed watersheds (e.g., higher shoreline development indices, Thornton et al. 1980). In the tropics, deforestation constitutes a large portion of watershed "development." Reservoir management for fish production may also be disproportionately important in tropical systems. Since most freshwater fish production comes from the tropics, management of these systems often includes introduction of new species adapted to the reservoir environment, stocking, and management of fishing effort (van Zwieten et al. 2011). Our proposed conceptual model is thus likely to differ for tropical ecosystems; however, more research is needed to formulate specific hypotheses about differential fates of E and *m* in these systems (Fig. 6).

Important typological considerations when comparing lakes and reservoirs

This synthesis focuses on reservoirs that are formed by damming pre-existing lakes and rivers, and compares these human-created ecosystems to natural lakes > 0.04 km^2 in size. Still, there are other types of artificial and natural lakes that deserve further attention (Fig. 1). For example, small ponds (< 0.01 km^2), both natural and human made are estimated to represent upward of 20% of global lake and reservoir surface area (Verpoorter et al. 2014; Holgerson and Raymond 2016) and can be disproportionately active with respect to ecosystem functioning such as having higher per-area greenhouse gas emissions (Holgerson and Raymond 2016) and rates of sediment deposition (Downing et al. 2008) than larger lentic systems do. Still, these smaller systems are often ignored in ecological studies (Downing 2010), making it difficult to include them in a synthesis such as this one. One might not expect the same dichotomies between natural and human-made ponds as the ones we report here for larger systems. For example, human-made farm ponds and stormwater retention ponds may not necessarily have higher catchment areas and CA : SAs than natural ponds. Future effort should be made to quantify and map the global coverage of various reservoir (and lake) types and to examine the potential role of typology in defining differences between systems (e.g., for small ponds, Fig. 6).

Similarly, the comparisons between lakes and reservoirs made here are subject to other biases in how lakes and reservoirs are surveyed and studied. For example, the NLA does not include some less common lake types (e.g., saline lakes, etc.) nor do they consider mine ponds, or cooling ponds. The NLA sites are also selected based on the National Hydrography Dataset which may not accurately represent the full suite of reservoir types. While the papers synthesized here do not generally focus on or discuss detailed system typology, it is likely that they only represent a subset of system types. Some less studied natural lake types may also have properties analogous to human made reservoirs (i.e., lakes formed behind travertine dams, beaver ponds, and floodplain lakes) and these comparisons could be instructive to study in the future.

Conclusions

Here, we identify fundamental differences between lake and reservoir systems likely to yield different responses to climate forcing. The analysis and synthesis presented above support the notions that reservoirs receive more catchment-mediated E and *m* inputs per unit volume than lakes and that E and *m* is likely to be processed differently within the waterbody of a reservoir than of a lake. We stress the important role that the catchment and management filters may play in determining the ultimate fate of E and *m* in reservoirs and, consequently, the extent to which reservoirs are functioning as sentinels or buffers of climate change drivers. While this synthesis focuses on temperate lakes and reservoirs > 0.04 km^2, we stress the need to consider both latitude and typology in future efforts to compare lake and reservoir climate change responses. Given the rapid response of lentic ecosystems to current climate forcing and the continued global construction of reservoirs, an improved understanding of how reservoirs are mediating climate effects has important implications for lake and reservoir ecology as well as downstream ecosystems.

References

Adrian, R., and others. 2009. Lakes as sentinels of climate change. Limnol. Oceanogr. **54**: 2283–2297. doi:10.4319/lo.2009.54.6_part_2.2283

Álvarez, V. M., A. Baille, J. M. M. Martínez, and M. M. González-Real. 2006. Efficiency of shading materials in reducing evaporation from free water surfaces. Agric. Water Manag. **84**: 229–239. doi:10.1016/j.agwat.2006.02.006

Baldwin, D. S., H. Gigney, J. S. Wilson, G. Watson, and A. N. Boulding. 2008. Drivers of water quality in a large water storage reservoir during a period of extreme drawdown. Water Res. **42**: 4711–4724. doi:10.1016/j.watres.2008.08.020

Beaulieu, M., F. Pick, and I. Gregory-Eaves. 2013. Nutrients and water temperature are significant predictors of cyanobacterial biomass in a 1147 lakes data set. Limnol. Oceanogr. **58**: 1736–1746. doi:10.4319/lo.2013.58.5.1736

Bell, E., S. Kramer, D. Zajanc, and J. Aspittle. 2008. Salmonid fry stranding mortality associated with daily water level fluctuations in Trail Bridge Reservoir, Oregon. N. Am. J. Fish. Manag. **28**: 1515–1528. doi:10.1577/M07-026.1

Berger, S. A., and others. 2006. Water temperature and mixing depth affect timing and magnitude of events during spring succession of the plankton. Oecologia **150**: 643–654. doi:10.1007/s00442-006-0550-9

Beutel, M., and others. 2014. Effects of hypolimnetic oxygen addition on mercury bioaccumulation in Twin Lakes, Washington, USA. Sci. Total Environ. **496**: 688–700. doi:10.1016/j.scitotenv.2014.06.117

Beutel, M. W., and A. J. Horne. 1999. A review of the effects of hypolimnetic oxygenation on lake and reservoir water quality. Lake Reserv. Manag. **15**: 285–297. doi:10.1080/07438149909354124

Blenckner, T. 2005. A conceptual model of climate-related effects on lake ecosystems. Hydrobiologia **533**: 1–14. doi:10.1007/s10750-004-1463-4

Brooks, J. R., J. J. Gibson, S. J. Birks, M. H. Weber, K. D. Rodecap, and J. L. Stoddard. 2014. Stable isotope estimates of evaporation: Inflow and water residence time for lakes across the United States as a tool for national lake water quality assessments. Limnol. Oceanogr. **59**: 2150–2165. doi:10.4319/lo.2014.59.6.2150

Canfield, Jr., D. E., and R. W. Bachman. 1981. Prediction of total phosphorus concentrations, chlorophyll a, and secchi depths in natural and artificial lakes. Can. J. Fish. Aquat. Sci. **38**: 414–423. doi:10.1139/f81-058

Clow, D. W., S. M. Stackpoole, K. L. Verdin, D. E. Butman, Z. Zhu, D. P. Krabbenhoft, and R. G. Striegl. 2015. Organic carbon burial in lakes and reservoirs of the conterminous United States. Environ. Sci. Technol. **49**: 7614–7622. doi:10.1021/acs.est.5b00373

Cole, J. J., and others. 2007. Plumbing the global carbon cycle: Integrating inland waters into the terrestrial carbon budget. Ecosystems **10**: 172–185. doi:10.1007/s10021-006-9013-8

Corman, J., B. Deemer, K. Strock, N. Hayes, and R. Razavi. 2016. Geographically paired lake-reservoir dataset derived from the 2007 USA EPA national lakes assessment. Long Term Ecological Research Network Information System, [accessed 2016 July 27]. Available from https://doi.org/10.6073/pasta/17cb7958c74f8bfc135f3e7f04ee944e.

Davidson, T. A., J. Audet, J. C. Svenning, T. L. Lauridsen, M. Søndergaard, F. Landkildehus, S. E. Larsen, and E. Jeppesen. 2015. Eutrophication effects on greenhouse gas fluxes from shallow-lake mesocosms override those of climate warming. Glob. Chang. Biol. **21**: 4449–4463. doi:10.1111/gcb.13062

Deemer, B. R., and others. 2016. Greenhouse gas emissions from reservoir water surfaces: A new global synthesis. BioScience **66**: 949–964. doi:10.1093/biosci/biw117

Downing, J. A. 2010. Emerging global role of small lakes and ponds: Little things mean a lot. Limnetica **29**: 9–24.

Downing, J. A., J. J. Cole, J. J. Middelburg, R. G. Striegl, C. M. Duarte, P. Kortelainen, Y. T. Prairie, and K. A. Laube. 2008. Sediment organic carbon burial in agriculturally eutrophic impoundments over the last century. Global Biogeochem. Cycles **22**: 1–10. doi:10.1029/2006GB002854

Downing, J. A., Y. T. Prairie, J. J. Cole, C. M. Duarte, L. J. Tranvik, R. G. Striegl, W. H. McDowell, P. Kortelainen, N. F. Caraco, J. M. Melack, and J. J. Middelburg. 2006. The global abundance and size distribution of lakes, ponds, and impoundments. Limnol. Oceanogr. **51**: 2388–2397. doi:10.4319/lo.2006.51.5.2388

Eum, H. I., and S. P. Simnovic. 2010. Integrated reservoir management system for adaptation to climate change: The Nakdong Reservoir basin in Korea. Water Resour. Manag. **24**: 3397–3417. doi:10.1007/s11269-010-9612-1

Gerling, A. B., Z. W. Munger, J. P. Doubek, K. D. Hamre, P. A. Gantzer, J. C. Little, and C. C. Carey. 2016. Whole-catchment manipulations of internal and external loading reveal the sensitivity of a century-old reservoir to hypoxia. Ecosystems. **19**: 555–571. doi:10.1007/s10021-015-9951-0

Gerten, D., and R. Adrian. 2001. Differences in the persistency of the North Atlantic Oscillation signal among lakes. Limnol. Oceanogr. **46**: 448–455. doi:10.4319/lo.2001.46.2.0448

Harrison, J. A., and others. 2009. The regional and global significance of nitrogen removal in lakes and reservoirs. Biogeochemistry **93**: 143–157. doi:10.1007/s10533-008-9272-x

Harrison, J. A., P. J. Frings, A. H. Beusen, D. J. Conley, and M. L. McCrackin. 2012. Global importance, patterns, and controls of dissolved silica retention in lakes and reservoirs. Global Biogeochem. Cycles **26**: GB2037. doi:10.1029/2011GB004228

Harrison, J. A., B. R. Deemer, M. K. Birchfield, and M. T. O'Malley. 2017. Reservoir water-level drawdowns accelerate and amplify methane emission. Environ. Sci. Technol. **51**: 1267–1277. doi:10.1021/acs.est.6b03185

Hayes, N. M., M. J. Vanni, M. J. Horgan, and W. H. Renwick. 2015. Climate and land use interactively affect lake phytoplankton nutrient limitation status. Ecology **96**: 392–402. doi:10.1890/13-1840.1

Hayhoe, K., and others. 2008. Regional climate change projections for the Northeast USA. Mitig. Adapt. Strategies Glob. Chang. **13**: 425–436. doi:10.1007/s11027-007-9133-2

Holgerson, M. A., and P. A. Raymond. 2016. Large contribution to inland water CO2 and CH4 emissions from very small ponds. Nat. Geosci. **9**: 222–226. doi:10.1038/ngeo2654

Huang, T., X. Li, H. Rijnaarts, T. Grotenhuis, W. Ma, X. Sun, and J. Xu. 2014. Effects of storm runoff on the thermal regime and water quality of a deep, stratified reservoir in a temperate monsoon zone, in Northwest China. Sci. Total Environ. **485–486**: 820–827. doi:10.1016/j.scitotenv.2014.01.008

Hutchinson, G. E. 1957. A treatise on limnology. V. 1. Geography, physics, and chemistry, p. 1015. John Wiley and Sons.

Jančula, D., and B. Maršálek. 2011. Critical review of actually available chemical compounds for prevention and management of cyanobacterial blooms. Chemosphere **85**: 1415–1422. doi:10.1016/j.chemosphere.2011.08.036

Kennedy, R. H., and G. D. Cooke. 1982. Control of lake phosphorus with aluminum sulfate- dose determination and application techniques. Water Resour. Bull. **18**: 389–395. doi:10.1111/j.1752-1688.1982.tb00005.x

Keto, A., A. Tarvainen, M. Marttunen, and S. Hellsten. 2008. Use of the water-level fluctuation analysis tool (Regcel) in hydrological status assessment of Finnish lakes. Hydrobiologia **613**: 133–142. doi:10.1007/s10750-008-9478-x

Kolding, J., and P. A. M. van Zwieten. 2012. Relative lake level fluctuations and their influence on productivity and resilience in tropical lakes and reservoirs. Fish. Res. **115–116**: 99–109. doi:10.1016/j.fishres.2011.11.008

Kratz, T., and others. 1997. The influence of landscape position on lakes in northern Wisconsin. Freshwater Biol **37**: 209–213. doi:10.1046/j.1365-2427.1997.00149.x

Kunkel, K. E., and others. 2008. Observed changes in weather and climate extremes in weather and climate extremes in a changing climate. Regions of focus: North America, Hawaii, Caribbean, and U.S. Pacific Islands. *In* T. R. Karl, G. A. Meehl, C. D. Miller, S. J. Hassol, A. M. Waple, and W. L. Murray [eds.], A report by the U.S. Climate Change Science Program and the Subcommittee on Global Change Research. pp. 35–80.

Leavitt, P. R., and others. 2009. Paleolimnological evidence of the effects on lakes of energy and mass transfer from climate and humans. Limnol. Oceanogr. **54**: 2330–2348. doi:10.4319/lo.2009.54.6_part_2.2330

Lehner, B., and P. Döll. 2004. Development and validation of a global database of lakes, reservoirs and wetlands. J. Hydrol. **296**: 1–22. doi:10.1016/j.jhydrol.2004.03.028

Lehner, B., and others. 2011. High-resolution mapping of the world's reservoirs and dams for sustainable river-flow management. Front. Ecol. Environ. **9**: 494–502. doi:10.1890/100125

Maavara, T., C. T. Parsons, C. Ridenour, S. Stojanovic, H. H. Dürr, H. R. Powley, and P. Van Cappellen. 2015. Global phosphorus retention by river damming. Proc. Natl. Acad. Sci. USA. **112**: 15603–15608. doi:10.1073/pnas.1511797112

Maestre-Valero, J. F., V. Martínez-Alvarez, B. Gallego-Elvira, and P. Pittaway. 2011. Effects of a suspended shade cloth cover on water quality of an agricultural reservoir for irrigation. Agric. Water Manag. **100**: 70–75. doi:10.1016/j.agwat.2011.08.020

Meijer, M.-L., I. de Boois, M. Scheffer, R. Portielje, and H. Hosper. 1999. Biomanipulation in shallow lakes in The Netherlands: An evaluation of 18 case studies. Hydrobiologia. **408**: 13–30. doi:10.1023/A:1017045518813

Moreno-Ostos, E., R. Marcé, J. Ordóñez, J. Dolz, and J. Armengol. 2008. Hydraulic management drives heat budgets and temperature trends in a Mediterranean reservoir. Int. Rev. Hydrobiol. **93**: 131–147. doi:10.1002/iroh.200710965

Nogaro, G., A. J. Burgin, V. A. Schoepfer, M. J. Konkler, K. L. Bowman, and C. R. Hammerschmidt. 2013. Aluminum sulfate (alum) application interactions with coupled metal and nutrient cycling in a hypereutrophic lake ecosystem. Environ. Pollut. **176**: 267–274. doi:10.1016/j.envpol.2013.01.048

Nowlin, W. H., J.-M. Davies, R. N. Nordin, and A. Mazumder. 2004. Effects of water level fluctuation and short-term climate variation on thermal and stratification regimes of a British Columbia reservoir and lake. Lake Reserv. Manag. **20**: 91–109. doi:10.1080/07438140409354354

Nürnberg, G. 2007. Lake responses to long-term hypolimnetic withdrawal treatments. Lake Reserv. Manag. **23**: 388–409.

O'Reilly, C. M., and others. 2015. Rapid and highly variable warming of lake surface waters around the globe. Geophys. Res. Lett. **42**: 10773–10781. doi:10.1002/2015GL066235

Palmer, M. A., C. A. R. Liermann, C. Nilsson, M. Flörke, J. Alcamo, P. S. Lake, and N. Bond. 2008. Climate change and the world's river basins: Anticipating management options. Front. Ecol. Environ. **6**: 81–89. doi:10.1890/060148

Poff, N. L., and D. D. Hart. 2002. How dams vary and why it matters for the emerging science of dam removal an ecological classification of dams is needed to characterize how the tremendous variation in the size, operational mode, age, and number of dams in a river basin influences the potential for restoring regulated rivers via dam removal. BioScience **52**: 659–668. doi:10.1641/0006-3568(2002)052[0659:HDVAWI]2.0.CO;2

Powers, S. M., J. L. Tank, and D. M. Robertson. 2015. Control of nitrogen and phosphorus transport by reservoirs in agricultural landscapes. Biogeochemistry **124**: 417–439. doi:10.1007/s10533-015-0106-3

Quayle, W. C., L. S. Peck, H. Peat, J. C. Ellis-Evans, and P. R. Harrigan. 2002. Extreme responses to climate change in Antarctic lakes. Science **295**: 645–645. doi:10.1126/science.1064074

Sahoo, G. B., and S. G. Schladow. 2008. Impacts of climate change on lakes and reservoirs dynamics and restoration policies. Sustain Sci **3**: 189–199. doi:10.1007/s11625-008-0056-y

Saunders, D. L., and J. Kalff. 2001. Nitrogen retention in wetlands, lakes and rivers. Hydrobiologia **443**: 205–212. doi:10.1023/A:1017506914063

Schindler, D. W., and J. P. Smol. 2006. Cumulative effects of climate warming and other human activities on freshwaters of Arctic and subarctic North America. Ambio. **35**: 160–168. doi:10.1579/0044-7447(2006)35[160:CEOCWA]2.0.CO;2

Schneider, P., and S. J. Hook. 2010. Space observations of inland water bodies show rapid surface warming since 1985. Geophys. Res. Lett. **37**: 1–5. doi:10.1029/2010GL045059

Smol, J. P., and M. S. V. Douglas. 2007. Crossing the final ecological threshold in high Arctic ponds. Proc. Natl. Acad. Sci. USA. **104**: 12395–12397. doi:10.1073/pnas.0702777104

Søndergaard, M., E. Jeppesen, T. L. Lauridsen, C. Skov, E. H. Van Nes, R. Roijackers, E. Lammens, and R. Portielje. 2007. Lake restoration: Successes, failures and long-term effects. J. Appl. Ecol. **44**: 1095–1105. doi:10.1111/j.1365-2664.2007.01363.x

Soranno, P. A., and others. 1999. Spatial variation among lakes within landscapes: Ecological organization along lake chains. Ecosystems **2**: 395–410. doi:10.1007/s100219900089

Syvitski, J. P., C. J. Vörösmarty, A. J. Kettner, and P. Green. 2005. Impact of humans on the flux of terrestrial sediment to the global coastal ocean. Science **308**: 376–380. doi:10.1126/science.1109454

Thornton, K. W., R. H. Kennedy, J. H. Carroll, W. W. Walker R. C. Gunkel, and S. Ashby. 1980. Reservoir sedimentation and water quality- an heuristic model, p. 654–661. *In* H. G Stefan [ed.], Proceedings of the symposium on surface water impoundments. Amer. Soc. Civil Engr.

Thornton, K. W., B. L. Kimmel, and F. E. Payne [eds.]. 1990. Reservoir limnology: Ecological perspectives. John Wiley & Sons.

Tiğrek and Aras. 2012. Reservoir sediment management. CRC Press, Taylor & Francis Group.

Tranvik, L. J., and others. 2009. Lakes and reservoirs as regulators of carbon cycling and climate. Limnol. Oceanogr. **54**: 2298–2314. doi:10.4319/lo.2009.54.6_part_2.2298

U.S. Environmental Protection Agency. 2007. National aquatic resource surveys. National Lakes Assessment, [accessed 2016 July 27]. Available from http://www.epa.gov/national-aquatic-resource-surveys/data-national-aquatic-resource-surveys.

U.S. Environmental Protection Agency. 2009. National lakes assessment: A collaborative survey of the nation's lakes. United States Environmental Protection Agency.

Valdespino-Castillo, P. M., M. Merino-Ibarra, J. Jiménez-Contreras, F. S. Castillo-Sandoval, and J. A. Ramírez-Zierold. 2014. Community metabolism in a deep (stratified) tropical reservoir during a period of high water-level fluctuations. Environ. Monit. Assess. **186**: 6505–6520. doi:10.1007/s10661-014-3870-y

van Zwieten, P. A. M., C. Béné, J. Kolding, R. Brummett, and J. Valbo-Jorgensen. 2011. Review of tropical reservoirs and their fisheries - the cases of Lake Nasser, Lake Volta, and Indo-Gangetic Basin reservoirs, p. 148. FAO Fisheries and Aquaculture Technical Paper. No. 557. FAO.

Vermeyen, T. 2000. Application of flexible curtains to control mixing and enable selective withdrawal in reservoirs. Presented at the 5th International Symposium on Stratified Flows, IAHR; July 10-13, 2000, Vancouver, Canada, PAP-847.

Verpoorter, C., T. Kutser, D. A. Seekell, and L. J. Tranvik. 2014. A global inventory of lakes based on high-resolution satellite imagery. Geophys. Res. Lett. **41**: 6396–6402. doi:10.1002/2014GL060641

Vogt, R. J., J. A. Rusak, A. Patoine, and P. R. Leavitt. 2011. Differential effects of energy and mass influx on the landscape synchrony of lake ecosystems. Ecology **92**: 1104–1114 doi:10.1890/10-1846.1

Vörösmarty, C. J., K. P. Sharma, B. M. Fekete, A. H. Copeland, J. Holden, J. Marble, and J. A. Lough. 1997. The storage and aging of continental runoff in large reservoir systems of the world. Ambio **26**: 210–219.

West, W. E., K. P. Creamer, and S. E. Jones. 2015. Productivity and depth regulate lake contributions to atmospheric methane. Limnol. Oceanogr. **61**: S51–S61. doi:10.1002/lno.10247

Wetzel, R. G. 2001. Limnology lake and reservoir ecosystems. Academic Press.

Williamson, C. E., J. E. Saros, W. F. Vincent, and J. P. Smol. 2009. Lakes and reservoirs as sentinels, integrators, and regulators of climate change. Limnol. Oceanogr. **54**: 2273–2282. doi:10.4319/lo.2009.54.6_part_2.2273

Winder, M., and D. E. Schindler. 2004. Climate change uncouples trophic interactions in an aquatic ecosystem. Ecology **85**: 2100–2106. doi:10.1890/04-0151

Zarfl, C., A. E. Lumsdon, J. Berlekamp, L. Tydecks, and K. Tockner. 2015. A global boom in hydropower dam construction. Aquat. Sci. **77**: 161–170. doi:10.1007/s00027-014-0377-0

Zohary, T., and I. Ostrovsky. 2011. Ecological impacts of excessive water level fluctuations in stratified freshwater lakes. Inland Waters **1**: 47–59. doi:10.5268/IW-1.1.406

Acknowledgments

We thank the EcoDAS organizers: L. Baker, P. Kemp, and E. Wood-Charlson as well as the sponsors of the event: the Center for Microbial Oceanography: Research and Education (C-MORE), the University of Hawai'i (UH) School of Ocean and Earth Science Technology (SOEST), and the UH Department of Oceanography. This manuscript benefitted from additional data graciously provided by J. R. Brooks and from GIS consultation provided by J. Ciarrocca. We also thank L. Baker, J. Harrison, P. Kemp, P. Leavitt, and P. Soranno for valuable feedback on manuscript drafts. Finally, we thank the many people who designed, collected, and analyzed data from the National Lakes Assessment. This work is the product of a collaboration formed during the Ecological Dissertations in the Aquatic Sciences (Eco-DAS) XI symposium and was funded by the National Science Foundation (award OCE-1356192) and the Association for the Sciences of Limnology and Oceanography.

6

Copiotrophic marine bacteria are associated with strong iron-binding ligand production during phytoplankton blooms

Shane L. Hogle,[1a] Randelle M. Bundy,[1b] Jessica M. Blanton,[2] Eric E. Allen,[2] Katherine A. Barbeau[1]*

[1]Geosciences Research Division, Scripps Institution of Oceanography, La Jolla, California; [2]Marine Biology Research Division, Scripps Institution of Oceanography, La Jolla, California

Scientific Significance Statement

Organic ligands shape the marine biogeochemical cycle of iron by controlling its solubility in the ocean. The strongest of the detectable marine organic ligands are presumed to be produced by iron-limited marine microbes, but some field studies have shown that these same strong ligands paradoxically increase in concentration after iron fertilization when microbes are thought to be iron replete. Currently, the specific microbes and mechanisms responsible for increases in strong iron-binding ligands remain unknown. We present evidence that bacterial taxa typically associated with nutrient-rich environments are robustly associated with strong iron-binding ligands detected during the early stages of phytoplankton bloom collapse in experimental incubations. These results may potentially explain observations of spikes in strong ligand concentrations after iron additions in the field.

Abstract

Although marine bacteria were identified nearly two decades ago as potential sources for strong iron-binding organic ligands detected in seawater, specific linkages between ligands detected in natural water and the microbial community remain unclear. We compared the production of different classes of iron-binding ligands, dissolved iron and macronutrient concentrations, and phytoplankton and bacterioplankton assemblages in a series of iron amended 6-d incubations. Incubations with high iron additions had near complete macronutrient consumption and higher phytoplankton biomass compared with incubations with low iron additions, but both iron treatments were dominated by diatoms. However, we only detected the strongest ligands in high-iron treatments, and strong iron-binding ligands were generally correlated with an increased abundance of copiotrophic bacteria, particularly *Alteromonas* strains. Ultimately, these robust correlations suggest a potential linkage between copiotrophic bacteria and strong iron-binding ligand production after iron fertilization events in the marine environment.

*Correspondence: shogle@mit.edu

Present address:

[a]Department of Civil and Environmental Engineering, Massachusetts Institute of Technology, Cambridge, Massachusetts, U.S.A
[b]Department of Marine Chemistry and Geochemistry, Woods Hole Oceanographic Institution, Woods Hole, Massachusetts, U.S.A

Author Contribution Statement: SLH, RMB, EEA, and KAB designed research; SLH, RMB, JMB, and KAB performed research; SLH, JMB, and RMB analyzed data; SLH, RMB, and KAB wrote the article.

The concentrations, chemical forms, and spatial/temporal distributions of oceanic iron are important factors shaping the ecology of marine phytoplankton and the overall productivity of marine ecosystems (Boyd et al. 2007). Dissolved iron (dFe) that has been regenerated from less kinetically labile forms, such as particles, is important in sustaining surface ocean productivity in iron-limited regions (Bowie et al. 2001). The mechanisms by which particulate iron is recycled back to dissolved iron and into the wider microbial food web, a process herein referred to as iron remineralization, are largely uncharacterized and represent major unknowns in global iron budgets (Tagliabue et al. 2016). The activity of microzooplankton (Barbeau et al. 2001) and viruses (Strzepek et al. 2005) are known to facilitate iron remineralization, and heterotrophic bacteria can also play a fundamental role (Boyd et al. 2010) as has been observed for other nutrients (Bidle and Azam 1999).

DFe is predominantly bound by organic ligands of mostly uncharacterized composition (Gledhill and van den Berg 1994; Rue and Bruland 1995). The concentrations and conditional stability constants, a measure of iron-binding strength, of natural marine ligands are typically measured using an electrochemical technique known as competitive ligand exchange—adsorptive cathodic stripping voltammetry (CLE-ACSV). Historically, organic ligands have been operationally partitioned into two classes based on their binding affinity for Fe (Gledhill and van den Berg 1994; Rue and Bruland 1995), but additional classes can be specified by varying the concentrations of a well-characterized competing ligand during electrochemical titrations, in a methodology termed Multiple Analytical Windows (MAW) (Bundy et al. 2014, 2015). Here, we partitioned organic ligands into three distinct classes in decreasing order of binding strength ($L_1 > L_2 > L_3$) using MAW and novel data processing techniques.

Some portion of the strong organic iron-binding ligands from seawater (Mawji et al. 2008) or marine cultures (Boiteau and Repeta 2015) are siderophores, bacterial secondary metabolites produced as an iron acquisition strategy. Intriguingly, the binding strengths of L_1 ligands measured in the field can be similar to or higher than those of known siderophores, leading to hypotheses that at least some portion of L_1 ligands in seawater may in fact be siderophores (Rue and Bruland 1995, 1997; Witter et al. 2000). Furthermore, electrochemical studies have demonstrated a positive correlation between L_1 ligand concentrations and biological activity as reviewed in Gledhill and Buck (2012), but the precise biological sources of those ligands are difficult to ascertain through electrochemical methods alone. Ultimately, it is clear that strong L_1 organic ligands are present in marine systems and are probably connected to the metabolic activities of marine bacteria and phytoplankton, but direct functional and mechanistic linkages between L_1 and biological community structure at this point remain elusive. Here, we couple high-throughput 16S rRNA marker gene surveys with MAW CLE-ACSV chemical analysis in incubation experiments in order to explore how phytoplankton and bacterioplankton assemblage composition is connected to the production of different classes of iron-binding ligands.

Methods

Oceanographic setting and incubation setup

We collected trace-metal clean seawater for incubation experiments from 35 m depth in August 2012 approximately 250 km from the Southern California coast (33.879° N, 123.306° W). We sampled using 5 L Teflon-coated external-spring Niskin bottles on a rosette deployed on nonmetallic hydroline. After sampling, we moved Niskin bottles to a positive pressure clean van and dispensed unfiltered water into an acid-cleaned 50 L carboy. We added 10.0 μmol L^{-1}, NO_3^- 1.0 μmol L^{-1} PO_4^{3-}, and 9.3 μmol L^{-1} $Si(OH)_4$ to the dispensed water and divided it between six 2.7 L polycarbonate bottles. Prior to use, we pretreated all macronutrients with Chelex resin to remove contaminating trace metals. We added 1 nmol L^{-1} $FeCl_3$ to Low Fe bottles and 5 nmol L^{-1} $FeCl_3$ to High Fe bottles then moved all bottles to an on-deck flow-through incubator screened to 30% incident light levels. We sampled bottles at 0 and 144 h (6 d) for dFe, dFe-binding ligands, chlorophyll *a* (chl *a*), macronutrients (NO_3^-, PO_4^{3-}, and $Si(OH)_4$), pigment concentrations, phytoplankton cell counts, and DNA and for only nutrients and chl *a* at 72, 120, and 144 h (days 3, 5, and 6). We sampled by rapidly moving incubation bottles to a trace metal clean van, subsampling bottles in a laminar flow hood, then returning bottles to the incubator.

Nutrients, pigments, and phytoplankton

We filtered subsamples (0.2 μm Acropak) of incubations, stored the flow through at −80°C, and measured $Si(OH)_4$, PO_4^{3-}, and NO_3^- (nitrate + nitrite) onshore using a Lachat QuickChem 8000 flow injection analysis system. We collected samples for chl *a* and phytoplankton pigments onto GF/F filters, extracted the chl *a* filters in acetone for 24 h, and analyzed chl *a* concentrations fluorometrically onboard. We stored filters for accessory pigments in liquid N_2 until analysis by high performance liquid chromatography onshore. We preserved phytoplankton for cell counts in formalin and counted cells onshore using light microscopy. We classified cells by morphology into categories of *Chaetoceros* spp., *Pseudo-nitzschia* spp., other diatoms (>10 μm), dinoflagellates, flagellates (<10 μm), and ciliates.

Dissolved Fe and dFe-binding organic ligands

We analyzed total dFe using flow-injection analysis (FIA) as described by King and Barbeau (2007), and measured the concentration and conditional binding strengths of organic dFe-binding ligands using MAW CLE-CSV methods (Bundy et al. 2014, 2015, 2016). We resolved ligand classes using the ProMCC software and first attempted to fit our titration data to three different ligand classes in a mass balance optimized speciation model (Omanović et al. 2015). If the model did not converge after 500 iterations, we fit the model again but for only two ligand classes. We characterized ligand classes as L_1 when $\log K^{cond}_{FeL1,Fe'} \geq 12.0$, L_2 when $\log K^{cond}_{FeL2,Fe'} = 11.0$–$12.0$, and L_3 when $\log K^{cond}_{FeL3,Fe'} = 10.0$–$11.0$. See Supporting Information for full sampling and methodological details.

DNA extraction, library preparation, and sequencing

We filtered 100 mL of seawater onto 0.22 μm Sterivex-GV filters (Millipore) without prefiltration and froze filters at −80°C. We extracted DNA onshore using the methods described by DeLong et al. (2006). We PCR amplified the V3-V4 region of the 16S ribosomal RNA gene from our samples using Q5 polymerase (New England Biolabs) and primers, S-D- Bact-0341-b-S-17/S-D-Bact-0785-a-A-21, (Klindworth et al. 2013) modified with Illumina adapters. We sequenced samples using 300 bp paired-end sequencing on an Illumina MiSeq machine running v3 chemistry, which generated an average of 1.2×10^5 reads per sample ($n = 6$) (Supporting Information Table S1).

OTU processing and analysis

Briefly an OTU or "Operational Taxonomic Unit" is a grouping of sequenced 16S ribosomal RNA read fragments with 97% nucleotide similarity. This classification cutoff has been traditionally assumed to correspond to microbial "species." We merged paired reads from each sample using USEARCH v8.1.1861_i86linux32 (Edgar 2013), allowing no more than 1.0 total expected errors for all bases in the merged read (default parameters). We pooled, de-replicated, and grouped merged reads into OTUs at 97% sequence identity and removed chimeras by comparison to the UCHIME gold standard reference database. We then used the RDP Naïve Bayesian classifier (Wang et al. 2007) to taxonomically classify OTUs and remove unclassifiable OTUs at the kingdom level and OTUs matching eukaryotic, cyanobacterial, or archaeal sequences (~20% of reads). We used the PhyloSeq package in R to process the resulting OTU abundance matrix (McMurdie and Holmes 2013) and tested for differential abundance of OTUs between treatments using DESeq2 (Love et al. 2014; McMurdie and Holmes 2014). See Supporting Information for data analysis procedures. All unix shell and R scripts for reproducing analyses can be downloaded from https://dx.doi.org/10.6084/m9.figshare.3184534.v1.

Results

Nutrients

The initial water for incubation experiments contained both low nitrate concentrations (0.94 μmol L^{-1}) and dFe concentrations (0.31 nmol $L^{-1} \pm 0.04$ nmol L^{-1}) (Supporting Information Tables S1, S2), indicating that the initial phytoplankton community was likely limited by NO_3^- (King and Barbeau 2007). For the nutrient amended incubations, we

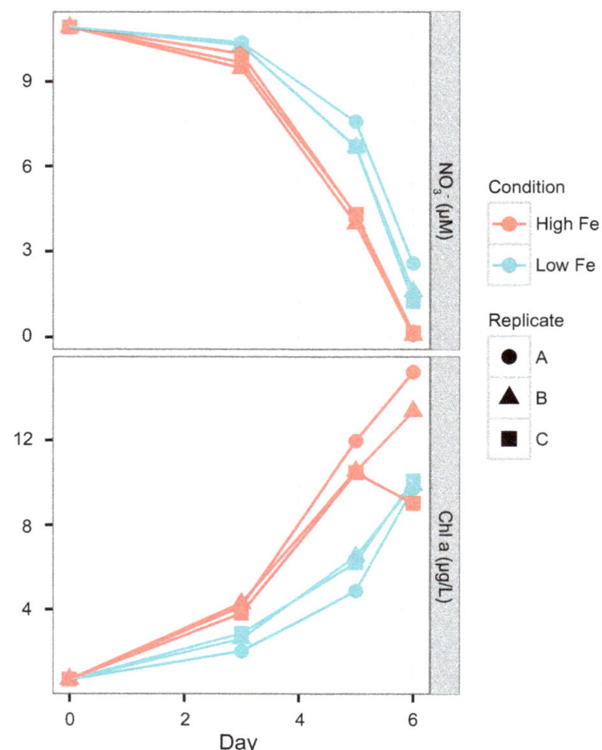

Fig. 1. Progression of Chlorophyll a (chl *a*) and nitrate (NO_3^-) concentrations during the course of the incubations. Nitrate concentrations at Day 0 are after the initial nutrient spike. High Fe incubations are in red, while Low Fe are in blue. Biological replicates are represented by different shapes. Note different scales in each subplot.

selected a low Low Fe treatment (1 nmol L^{-1} dFe, NO_3^-: dFe = 9) to simulate initial conditions approaching phytoplankton Fe-limitation and a High Fe treatment (5 nmol L^{-1} dFe, NO_3^-: dFe = 2) to simulate Fe-replete conditions. King and Barbeau (2007) determined that a nitrate to dFe ratio approaching 10 (μM NO_3^-: nM Fe) strongly indicated insufficient Fe to support complete phytoplankton nitrate consumption and consequently could be a reliable proxy for community Fe limitation in the field. After 6 d, average ratios of NO_3^-: dFe were elevated in Low Fe (NO_3^-: dFe ≅ 6) relative to High Fe treatments (NO_3^-: dFe ≅ 0.2) suggesting that Low Fe incubations were likely Fe stressed (Supporting Information Table S2).

Phytoplankton assemblage

The chl *a* concentration was initially 0.75 μg L^{-1}, and increased to 9.91 ± 0.18 μg L^{-1} in Low Fe treatments and 12.50 ± 3.17 μg L^{-1} in High Fe treatments (Fig. 1). In both treatments, most of the biomass gain was due to an increase in the abundance of diatoms as indicated by the increase in fucoxanthin in Low Fe and High Fe incubations (Supporting Information Table S3) and direct observation in the cell counts (Supporting Information Table S4). Before Fe addition, the eukaryotic phytoplankton assemblage was composed

Table 1. Dissolved Fe-binding ligand concentrations.

Treatment	Day	DFe (nM)	DFe SD (nM)	L_1 (nM)	logK1	L_2 (nM)	logK2	L_3 (nM)	logK3
Initial	0	0.31	0.04	nd	nd	2.19	11.81	2.51	10.76
Low Fe A	6	0.31	0.02	nd	nd	2.55	11.82	3.79	10.20
Low Fe B	6	0.42	0.03	nd	nd	3.67	11.20	2.58	10.10
Low Fe C	6	0.21	0.02	nd	nd	4.10	11.74	3.65	10.20
High Fe A	6	0.61	0.04	4.39	12.02	2.97	11.38	6.58	10.10
High Fe B	6	0.62	0.17	3.65	12.36	3.07	11.00	7.54	9.39
High Fe C	6	0.60	0.06	2.54	12.08	3.57	11.16	4.60	10.16

Concentrations of DFe and DFe-binding ligands (L_x) in the six incubations after 6 d and at the start of the experiment (Day 0). Initial values are from in situ seawater used to start the incubations and do not include biological replicates. DFe is the total dissolved Fe concentration in nM, DFe SD is the standard deviation of three technical FIA replicates for the dissolved Fe measurements, $L_{(x)}$ is the concentration in nM of the three ligand classes defined in this study, LogK(x) displays $\log_{10}$ of the conditional stability constant measured for each ligand class determined at the 95% confidence interval in chemical speciation mode in ProMCC with less than 10% root mean square error.

primarily of *Pseudo-nitzschia* species (67%) and other large diatoms. *Pseudo-nitzschia* spp. responded to Fe addition in both treatments, increasing to 78% of the assemblage on average in Low Fe bottles and 82% in High Fe bottles. Cell counts of specific taxonomic groups were not statistically different between High and Low Fe treatments when normalized to total cell counts, indicating that phytoplankton assemblage composition was largely unchanged between Fe treatments although total biomass increased in High Fe samples.

Dissolved Fe-binding ligands

We determined several classes of dFe-binding ligands in each experimental treatment, and Low Fe and High Fe treatments displayed different patterns after 6 d. L_2 and L_3 ligands were detected in situ in the initial water mass, while the strongest L_1 ligands were conspicuously absent (Table 1). After 6 d, Low and High Fe treatments displayed similar L_2 concentrations to initial CCE waters. L_3 concentrations increased at day 6 in High Fe compared with Low Fe treatments (Student's *t*-test, $p < 0.05$). However, the most striking result was the presence of strong L_1 ligands exclusively in High Fe treatments at day 6, while no Fe-binding ligands of this strength were detected in Low Fe treatments. In some treatments, the calculated logK of the L_1 class was similar to that of the L_2 class (within logK of 0.2). Although our rigorous data processing methodology resolved three High Fe ligand classes, it is possible that the L_1 and L_2 classes as defined in our study may have partially overlapped in High Fe samples. Regardless of overlapping binding strengths, there would still be significant ligand concentration difference between Low and High Fe treatments. These results suggest that L_1 production was dependent on Fe treatment and the resulting biological dynamics in these incubations.

Heterotrophic bacterial assemblage

We examined the composition of the heterotrophic bacterial assemblage in High and Low Fe treatments using 16S rRNA marker gene sequencing. We first explored broad patterns of taxonomic diversity using a nonmetric multidimensional scaling (NMDS) analysis based on diversity and abundance of bacterial OTUs. This analysis indicated that Low Fe treatments were of a similar taxonomic composition, while High Fe treatments were largely variable between

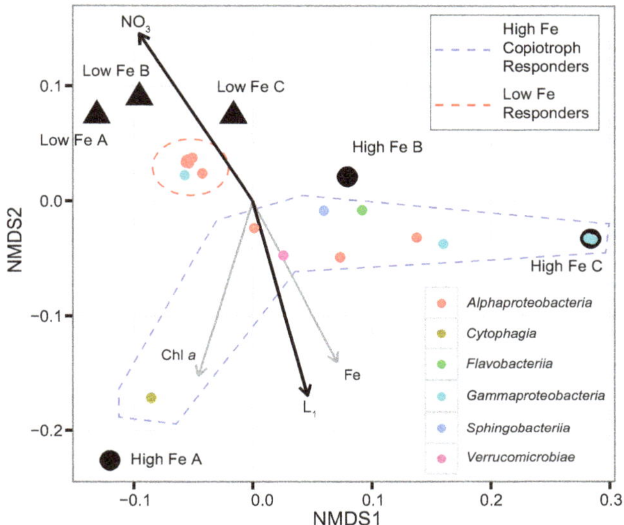

Fig. 2. Ordination plot of a NMDS analysis based on the diversity and abundance of bacterial OTUs detected at Day 6 of the experiment. The three High Fe samples (circles) and Low Fe samples (triangles) are plotted in addition to the ordination of individual OTUs (small colored circles). OTUs are colored according to the five most abundant taxonomic classes and only OTUs from Table 1 and Fig. 3 are displayed. Copiotroph OTUs responding to High Fe conditions are enclosed with a blue dashed line, while the red dashed line encompasses OTUs enriched in Low Fe samples. Arrows represent fitted vectors of continuous associated environmental variables and show the direction of the increasing gradient. Variables with a correlation *p* value < 0.1 are shown in black while those with $p > 0.1$ are shown in gray. Arrow length is proportional to the correlation between the variable and ordination. L_1, $R^2 = 0.77$; Fe, $R^2 = 0.62$; chl *a*, $R^2 = 0.64$; NO_3^- $R^2 = 0.75$.

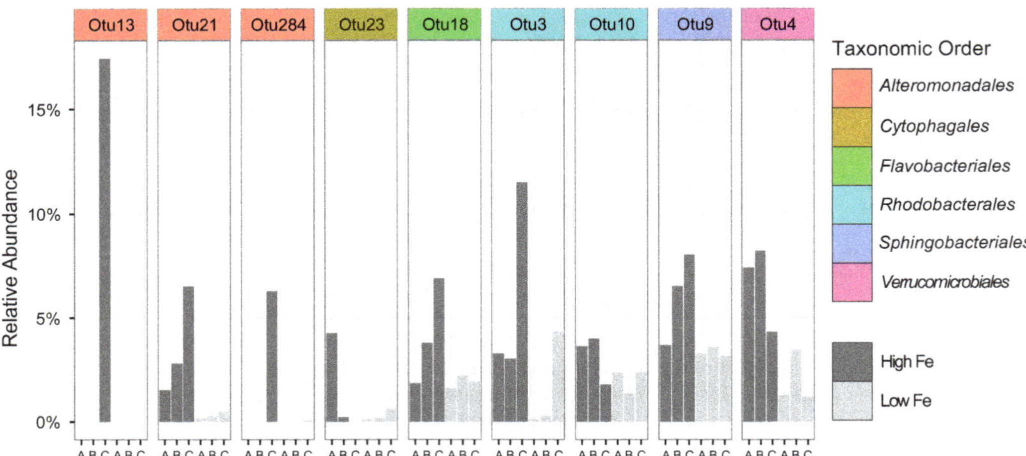

Fig. 3. Bar plot of copiotroph OTUs responding to High Fe conditions. Copiotroph OTUs have a mean relative abundance greater than 1% of all reads in each replicate and the High Fe mean is at least 1.5 times greater than the Low Fe mean. High Fe samples are dark gray while Low Fe samples are light gray.

Table 2. Differentially abundant OTUs between Fe treatments.

OTU	Base Mean	Log_2 Fold Change	FDR adjusted *p* value	Class	Family	Genus
*Otu21	3340.45	−3.66	0.0006	*Gammaproteobacteria*	*Alteromonadaceae*	*Aestuariibacter*
*Otu78	133.98	−2.74	0.0001	*Alphaproteobacteria*	*Hyphomonadaceae*	*Hyphomonas*
*Otu4	5061.43	−2.17	0.0006	*Verrucomicrobiae*	*Rubritaleaceae*	*Rubritalea*
Otu2	7038.59	0.73	0.02	*Alphaproteobacteria*	*SAR11*	*Pelagibacter*
Otu245	109.48	1.06	0.03	*Alphaproteobacteria*	*SAR11*	*Pelagibacter*
Otu45	248.55	1.08	0.04	*Alphaproteobacteria*	*Rhodobacteraceae*	*Roseibacterium*
Otu32	1146.14	1.18	0.01	*Alphaproteobacteria*	*SAR11*	*Pelagibacter*
Otu57	119.82	1.21	0.002	*Alphaproteobacteria*	*Kordiimonadaceae*	*Kordiimonas*
Otu56	212.94	1.23	0.02	*Gammaproteobacteria*	*Oleiphilaceae*	*Oleiphilus*
Otu15	1776.33	1.52	0.01	*Flavobacteriia*	*Flavobacteriaceae*	*Corallibacter*

Significant differential abundances of OTUs between high and low Fe conditions after 6 d. Asterisks denote taxa considered to be copiotrophic responders to High Fe treatments. Only OTUs with false discovery rate corrected *p* values less than 0.05 and with base means greater than 100 are displayed. Base mean indicates the mean OTU abundance across all samples. Log_2 Fold Change is the fold change from high Fe to low Fe samples (i.e., a negative value indicates enrichment in high Fe samples) and OTUs are ordered by increasing fold change, FDR adjusted *p* value is the False Discovery Rate adjusted *p* value from negative binomial Wald Tests for significant differential abundance between Fe treatments, and the remaining columns designate taxonomic class, family, and genus of each OTU.

replicates but also taxonomically distinct from Low Fe incubations (Fig. 2). We then fit continuous environmental variables to the OTU ordination and tested for correlations at a significance level of $\alpha = 0.1$. Increasing NO_3^- concentrations were positively correlated with the bacterial OTU composition in Low Fe incubations ($R^2 = 0.75$, $p < 0.05$), while L_1 concentrations were strongly positively correlated with OTUs more abundant in High Fe treatments (L_1, $R^2 = 0.77$, $p < 0.05$). The alpha diversity, or the total species richness in each incubation, was significantly lower in High Fe treatments (Supporting Information Table S5).

Because we were struck by the differences in L_1 concentrations between Fe treatments and the general alignment of the OTU composition with increasing L_1, we searched for potential indicator OTUs that were abundant and overrepresented in High Fe incubations. We narrowed our search to OTUs that each comprised greater than 1% of the total reads in each sample and also had average abundances that were at minimum 1.5 times greater in High Fe samples than in Low Fe samples. Nine OTUs matched these criteria, and all were similar to known copiotrophic strains particularly from the orders *Alteromonadales* and *Rhodobacterales* (Fig. 3, Supporting Information Table S6). We also tested for differentially abundant OTUs using a mixture model framework (Love et al. 2014) which identified ten OTUs (Table 2) with a statistically significant difference in abundance between

High and Low Fe treatments. Three of these OTUs were enriched in High Fe incubations and had strong similarity to copiotrophic strains. Some highly abundant copiotrophic OTUs, for example *Alteromonas* OTU13, co-occurred almost exclusively with a single replicate from High Fe treatments (Fig. 3) and were missed by the mixed model approach due to their large variability.

Discussion

After 6 d of incubation, both High and Low Fe treatments stimulated phytoplankton biomass and produced diatom blooms. However, the phytoplankton assemblages in High and Low Fe treatments were largely of the same taxonomic composition, while the Fe-binding ligand pool, macronutrient concentrations, and the heterotrophic microbial assemblage differed substantially. The strongest L_1 ligands were particularly distinct between Fe treatments, and the L_1 ligand classification, defined in this study as having a $\log K \geq 12$, is comparable with the binding affinities of siderophores found in cultures of marine bacteria (Vraspir and Butler 2009). Siderophores are a probable constituent of the operationally defined L_1 class defined here, but we stress that the L_1 class should not be considered to be entirely composed of siderophores.

One possible explanation for the emergence of strong L_1 ligands in High Fe treatments is that they were produced actively by phytoplankton. However, sequenced eukaryotic phytoplankton genomes lack siderophore biosynthesis pathways and do not appear directly take up siderophores (Morrissey and Bowler 2012). An alternative explanation for differences in L_1 is from variation in the composition and activities of bacteria. The bacterial assemblage in High Fe incubations was broadly aligned with L_1 concentrations, and these incubations were dominated by a handful of OTUs related to copiotrophic strains (Supporting Information Table S6). This copiotroph enrichment was also reflected by the reduced species richness in High Fe treatments (Supporting Information Table S5). Although we only detected L_1 in High Fe incubations, replicates had variable L_1 concentrations as well as variable abundances of copiotroph OTUs (Figs. 2, 3), which may reflect unsampled ecological dimensions from our study. This could include differences in the concentrations and types of bacterial growth substrates in each replicate (Goutx et al. 2007; Mayali et al. 2015), potentially antagonistic microbial interactions (Long and Azam 2001), and/or that multiple copiotrophic groups may produce strong organic ligands. Despite the intravariability of High Fe replicates, broad patterns in the composition of the heterotrophic microbial assemblage appeared to align with the High/Low Fe treatment experimental structure (Fig. 2).

Our observations are consistent with other studies reporting an enrichment of generalist, copiotrophic bacterial taxa after dissolved organic matter additions, and biogenic particle additions to seawater (Nelson and Carlson 2012; Mayali et al. 2015). Other studies have demonstrated that the activities of mostly rare but transiently abundant and transcriptionally active marine copiotrophic bacteria have disproportionately large impacts on biogeochemical cycling (Pedler et al. 2014). Copiotrophic microbial taxa tend to have large genomes with greater regulatory capacity and a greater diversity of Fe acquisition pathways including siderophore biosynthesis and/or siderophore uptake (Hogle et al. 2016). Copiotrophs can also rapidly adapt to patchy nutrient conditions and exploit diverse growth substrates. It may be that copiotrophs also have a large Fe demand in order to fuel their rapid growth and carbon consumption after episodic nutrient pulses. For example, hydroxamate siderophore production by *Pseudoalteromonas haloplanktis* is sensitive to carbon availability (Sijerčić and Price 2015), and the *Alteromonas* sp. ALT199 genome contains a petrobactin biosynthesis pathway (Pedler et al. 2014).

We suggest that the most likely scenario accounting for the emergence of L_1 in High Fe treatments is that copiotrophic bacteria were directly producing strong Fe-binding organic ligands in order to scavenge iron from lysing algal cells. The almost complete drawdown of nitrate in all High Fe replicates and the chl *a* collapse in High Fe C suggests that after 6 d the phytoplankton assemblage in High Fe treatments had entered or was beginning to enter the initial phases of remineralization. We postulate that phytoplankton senescence was coupled with a release of dissolved organic matter (DOM) which stimulated endemic bacterial copiotrophs (Buchan et al. 2014). We suspect the differences between Low and High Fe treatments in the bacterial assemblages were primarily driven by DOM, while the differences in L_1 concentrations reflected siderophore and other strong organic ligand production by copiotrophs responding to DOM enrichment. Although this study provides only correlative evidence and limited temporal resolution, our results do suggest a potential link between specific copiotrophic bacteria and strong dFe-binding ligand production and further study is warranted.

Large scale Fe fertilization studies have demonstrated increases in strong dFe-binding ligands when microbial communities were presumably Fe replete (Rue and Bruland 1997) as well as during phytoplankton bloom decline after Fe fertilization (Kondo et al. 2008). Our experiments suggest that spikes of L_1 production in these mesoscale studies may have been related to shifts in copiotrophic bacterial abundance during phytoplankton blooms. We hypothesize that high Fe conditions initially stimulate Fe-limited photoautotrophs, particularly diatoms, but not endemic heterotrophic bacteria due to organic carbon limitation (Kirchman 1990; Church et al. 2000). As the diatom bloom exhausts inorganic nutrients and progresses to senescence, newly released diatom-derived DOM associated with algal senescence stimulates ambient carbon-limited copiotroph bacteria (Seyedsayamdost et al. 2011; Sule and Belas 2013). Because aerobic respiration requires many Fe-containing enzymes (Hogle et al. 2014), rapidly growing copiotrophs may quickly shift to Fe limitation relative to excess algal-derived organic carbon. The most abundant local Fe source in this context would likely be algal-derived metalloproteins, and copiotroph L_1 production may function to extract Fe from these algal sources and make it bioavailable to the wider bacterial community. If this phenomenon is widespread it may serve as a significant source of strong organic ligands in marine waters. Strong ligand production by heterotrophic bacteria during early bloom senescence may thus be important for overall iron recycling efficiency in microbial ecosystems and may serve to reduce Fe loss due to particle export in the upper ocean.

References

Barbeau, K., E. Kujawinski, and J. Moffett. 2001. Remineralization and recycling of iron, thorium and organic carbon by heterotrophic marine protists in culture. Aquat. Microb. Ecol. **24**: 69–81.

Bidle, K. D., and F. Azam. 1999. Accelerated dissolution of diatom silica by marine bacterial assemblages. Nature **397**: 508–512.

Boiteau, R. M., and D. J. Repeta. 2015. An extended siderophore suite from Synechococcus sp. PCC 7002 revealed by LC-ICPMS-ESIMS. Metallomics **7**: 877–884. doi:10.1039/C5MT00005J

Bowie, A. R., M. T. Maldonado, R. D. Frew, and others. 2001. The fate of added iron during a mesoscale fertilisation experiment in the Southern Ocean. Deep Sea Res. Part 2 Top. Stud. Oceanogr. **48**: 2703–2743.

Boyd, P. W., T. Jickells, C. S. Law, and others. 2007. Mesoscale iron enrichment experiments 1993-2005: Synthesis and future directions. Science **315**: 612–617.

Boyd, P. W., E. Ibisanmi, S. G. Sander, K. A. Hunter, and G. A. Jackson. 2010. Remineralization of upper ocean particles: Implications for iron biogeochemistry. Limnol. Oceanogr. **55**: 1271–1288.

Buchan, A., G. R. LeCleir, C. A. Gulvik, and J. M. González. 2014. Master recyclers: Features and functions of bacteria associated with phytoplankton blooms. Nat. Rev. Microbiol. **12**: 686–698.

Bundy, R. M., D. V. Biller, K. N. Buck, K. W. Bruland, and K. A. Barbeau. 2014. Distinct pools of dissolved iron-binding ligands in the surface and benthic boundary layer of the California current. Limnol. Oceanogr. **59**: 769–787.

Bundy, R. M., H. A. N. Abdulla, P. G. Hatcher, D. V. Biller, K. N. Buck, and K. A. Barbeau. 2015. Iron-binding ligands and humic substances in the San Francisco Bay estuary and estuarine-influenced shelf regions of coastal California. Mar. Chem. **173**: 183–194.

Bundy, R. M., M. Jiang, M. Carter, and K. A. Barbeau. 2016. Iron-binding ligands in the Southern California Current System: Mechanistic studies. Front. Mar. Sci. **3**: article 27, 1–17. doi:10.3389/fmars.2016.00027

Church, M. J., D. A. Hutchins, and H. W. Ducklow. 2000. Limitation of bacterial growth by dissolved organic matter and iron in the Southern ocean. Appl. Environ. Microbiol. **66**: 455–466. doi:10.1128/AEM.66.2.455-466.2000

DeLong, E. F., C. M. Preston, T. Mincer, and others. 2006. Community genomics among stratified microbial assemblages in the ocean's interior. Science **311**: 496–503.

Edgar, R. C. 2013. UPARSE: Highly accurate OTU sequences from microbial amplicon reads. Nat. Methods **10**: 996–998.

Gledhill, M., and C. M. G. van den Berg. 1994. Determination of complexation of iron(III) with natural organic complexing ligands in seawater using cathodic stripping voltammetry. Mar. Chem. **47**: 41–54.

Gledhill, M., and K. N. Buck. 2012. The organic complexation of iron in the marine environment: A review. Front. Microbiol. **3**: 69.

Goutx, M., S. G. Wakeham, C. Lee, and others. 2007. Composition and degradation of marine particles with different settling velocities in the northwestern Mediterranean Sea. Limnol. Oceanogr. **52**: 1645–1664.

Hogle, S. L., K. A. Barbeau, and M. Gledhill. 2014. Heme in the marine environment: From cells to the iron cycle. Metallomics **6**: 1107–1120.

Hogle, S. L., J. Cameron Thrash, C. L. Dupont, and K. A. Barbeau. 2016. Trace metal acquisition by marine heterotrophic bacterioplankton with contrasting trophic strategies. Appl. Environ. Microbiol. **82**: 1613–1624.

King, A. L., and K. Barbeau. 2007. Evidence for phytoplankton iron limitation in the southern California Current System. Mar. Ecol. Prog. Ser. **342**: 91–103.

Kirchman, D. L. 1990. Limitation of bacterial growth by dissolved organic matter in the subarctic Pacific. Mar. Ecol. Prog. Ser. **62**: 47–54.

Klindworth, A., E. Pruesse, T. Schweer, J. Peplies, C. Quast, M. Horn, and F. O. Glöckner. 2013. Evaluation of general 16S ribosomal RNA gene PCR primers for classical and next-generation sequencing-based diversity studies. Nucleic Acids Res. **41**: e1.

Kondo, Y., S. Takeda, J. Nishioka, H. Obata, K. Furuya, W. K. Johnson, and C. S. Wong. 2008. Organic iron (III) complexing ligands during an iron enrichment experiment in the western subarctic North Pacific. Geophys. Res. Lett. **35**: L12601.

Long, R. A., and F. Azam. 2001. Antagonistic interactions among marine pelagic bacteria. Appl. Environ. Microbiol. **67**: 4975–4983.

Love, M. I., W. Huber, and S. Anders. 2014. Fully formatted moderated estimation of fold change and dispersion for RNA-seq data with DESeq2. Genome Biol. **15**: 1–21.

Mawji, E., M. Gledhill, J. A. Milton, and others. 2008. Hydroxamate siderophores: Occurrence and importance in the Atlantic Ocean. Environ. Sci. Technol. **42**: 8675–8680. doi:10.1021/es801884r

Mayali, X., B. Stewart, S. Mabery, and P. K. Weber. 2015. Temporal succession in carbon incorporation from macromolecules by particle-attached bacteria in marine microcosms. Environ. Microbiol. Rep. **8**: 68–75. doi:10.1111/1758-2229.12352

McMurdie, P. J., and S. Holmes. 2013. Phyloseq: An R package for reproducible interactive analysis and graphics of microbiome census data. PLoS One **8**: e61217. doi:10.1371/journal.pone.0061217

McMurdie, P. J., and S. Holmes. 2014. Waste not, want not: Why rarefying microbiome data is inadmissible. PLoS Comput. Biol. **10**: e1003531. doi:10.1371/journal.pcbi.1003531

Morrissey, J., and C. Bowler. 2012. Iron utilization in marine cyanobacteria and eukaryotic algae. Front. Microbiol. **3**: 43.

Nelson, C. E., and C. A. Carlson. 2012. Tracking differential incorporation of dissolved organic carbon types among diverse lineages of Sargasso Sea bacterioplankton. Environ. Microbiol. **14**: 1500–1516.

Omanović, D., C. Garnier, and I. Pižeta. 2015. ProMCC: An all-in-one tool for trace metal complexation studies. Mar. Chem. **173**: 25–39.

Pedler, B. E., L. I. Aluwihare, and F. Azam. 2014. Single bacterial strain capable of significant contribution to carbon cycling in the surface ocean. Proc. Natl. Acad. Sci. USA **111**: 7202–7207.

Rue, E. L., and K. W. Bruland. 1995. Complexation of iron(-III) by natural organic ligands in the Central North Pacific as determined by a new competitive ligand equilibration/adsorptive cathodic stripping voltammetric method. Mar. Chem. **50**: 117–138.

Rue, E. L., and K. W. Bruland. 1997. The role of organic complexation on ambient iron chemistry in the equatorial Pacific Ocean and the response of a mesoscale iron addition experiment. Limnol. Oceanogr. **42**: 901–910.

Seyedsayamdost, M. R., G. Carr, R. Kolter, and J. Clardy. 2011. Roseobacticides: Small molecule modulators of an algal-bacterial symbiosis. J. Am. Chem. Soc. **133**: 18343–18349.

Sijerčić, A., and N. M. Price. 2015. Hydroxamate siderophore secretion by Pseudoalteromonas haloplanktis during steady-state and transient growth under iron limitation. Mar. Ecol. Prog. Ser. **531**: 105–120.

Strzepek, R. F., M. T. Maldonado, J. L. Higgins, J. Hall, K. Safi, S. W. Wilhelm, and P. W. Boyd. 2005. Spinning the "Ferrous Wheel": The importance of the microbial community in an iron budget during the FeCycle experiment. Glob. Biogeochem. Cycles **19:1**–14.

Sule, P., and R. Belas. 2013. A novel inducer of roseobacter motility is also a disruptor of algal symbiosis. J. Bacteriol. **195**: 637–646.

Tagliabue, A., O. Aumont, R. DeAth, and others. 2016. How well do global ocean biogeochemistry models simulate dissolved iron distributions? Glob. Biogeochem. Cycles **30**: 149–174. doi:10.1002/2015GB005289

Vraspir, J. M., and A. Butler. 2009. Chemistry of marine ligands and siderophores. Ann. Rev. Mar. Sci. **1**: 43–63.

Wang, Q., G. M. Garrity, J. M. Tiedje, and J. R. Cole. 2007. Naïve Bayesian classifier for rapid assignment of rRNA sequences into the new bacterial taxonomy. Appl. Environ. Microbiol. **73**: 5261–5267.

Witter, A. E., D. A. Hutchins, A. Butler, and G. W. Luther. 2000. Determination of conditional stability constants and kinetic constants for strong model Fe-binding ligands in seawater. Mar. Chem. **69**: 1–17.

Acknowledgments

We thank Christopher Rivera for help with Illumina library preparation, Jeff Hasty at UCSD for use of his MiSeq system, and Melissa Carter at SIO for phytoplankton cell counts. This work was funded by NSF GRFP grant DGE-144086 to SLH and NSF grants OCE-1061068 and OCE-1558841 to KAB. We thank the participants of the California Current Ecosystem Long Term Ecological Research program (NSF OCE-1026607) and the captain and the crew of the R/V Melville. The funders had no role in study design, data collection and interpretation, or the decision to submit the work for publication.

7

Centennial-long trends of lake browning show major effect of afforestation

*Emma S. Kritzberg**
Department of Biology/Aquatic Ecology, Lund University, Lund, Sweden

Scientific Significance Statement

Monitoring of northern lakes in recent decades has revealed strong increases in water color and organic carbon concentrations, with important implications for the structure and function of freshwater systems. It has been argued that this browning of lake waters is a result of declining acid deposition, and that lakes were unnaturally clear during the period of high atmospheric sulfur deposition, since enhanced acidity in soils reduces organic matter mobility. However, this study presents historical water color data in 50 lakes for the last 80 yr and shows that recovery from acidification is not sufficient to explain trends in water color for a majority of the lakes, and that browning may be more related to changes in land-use, from agriculture to modern forestry.

Abstract

Observations of increasing water color and organic carbon concentrations in lakes are widespread across the Northern Hemisphere. The drivers of these trends are debated. Declining atmospheric sulfur deposition has been put forward as an important underlying factor, since recovery from acidification enhances mobility of organic matter from surrounding soils. This would suggest that the current browning represents a return to a more natural state. This study explores historical lake data from Sweden—1935 to 2015—providing a unique opportunity to see how and why water color has varied during almost a century. The data shows that sulfur deposition has not been the primary driver of water color trends over this period. I propose that the observed browning is to a large extent driven by a major transition from agriculture to forestry.

During the last decades, dissolved organic carbon (DOC) concentrations have increased dramatically in temperate surface waters throughout the Northern hemisphere (Monteith et al. 2007), paralleled by increases in water color (Roulet and Moore 2006; Kritzberg and Ekström 2012). Understanding the driver/s of these trends is imperative, for three major reasons. (1) Water color is an important factor determining the structure and function of aquatic food webs as the light climate governs processes ranging from photosynthesis and system productivity (Karlsson et al. 2009) to predator-prey interactions (Ranåker et al. 2012). (2) DOC concentrations in freshwaters both *reflect* and *affect* important processes in the global carbon cycle. For instance, export of DOC from the catchment corresponds to roughly half of the global terrestrial net ecosystem production (Battin et al. 2009). Moreover, sediment burial of organic carbon in lakes is estimated to be fourfold higher than that in the entire global ocean (Downing et al. 2008) and, importantly, it is increasing with the increasing DOC concentrations that we observe today (Ferland et al. 2012). (3) Finally, increasing DOC levels have serious negative effects on important ecosystem services, such as

*Correspondence: emma.kritzberg@biol.lu.se

Author Contribution Statement: The study was conceived by ESK, who also analyzed the data and wrote the paper.

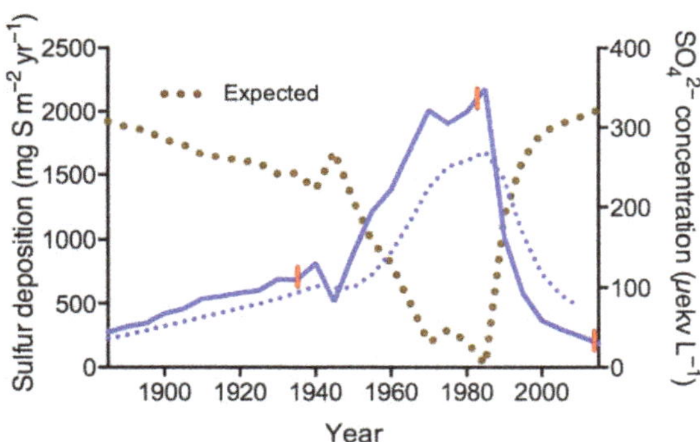

Fig. 1. Historical changes in sulfur deposition in the area (solid blue line) and sulfate concentration in lake water (dotted blue line, L. Skärshultssjön), based on deposition data from EMEP processed as described in (Moldan et al. 2013). The dotted brown line (inverse of sulfur deposition) shows the expected variation in water color if sulfur deposition was the only and instant driver of water color (relative change, no unit). The red transverse lines denote the years 1935, 1982, and 2015, for which comparisons in Fig. 2 are made.

the provisioning of drinking water, with increasing costs for drinking water treatment and concerns for the risk of carcinogenic by-products (Chow et al. 2007).

While there is strong evidence for a dramatic increase in water color and DOC, there are diverging views on the underlying mechanisms. Some studies argue that the observed trends are driven by processes associated to climate change, such as rising CO_2 concentration, temperature and precipitation—factors that enhance terrestrial productivity, alter vegetation communities, and affect the hydrological control on production and transport of DOC (Larsen et al. 2011; Laudon et al. 2012; Weyhenmeyer et al. 2016). Others point to the role of land-use and land management (Meyer-Jacob et al. 2015). A third proposed driver, which has gained ground in recent years, is the long-term change in atmospheric deposition, where in particular the strong reductions in anthropogenic sulfur (S) emissions allow for recovery from acidification and increasing solubility and transport of organic matter from soils (Monteith et al. 2007; Ekström et al. 2011).

Climate change forcing suggests that current DOC concentrations and water color are without precedent and will continue to rise. In contrast, the argument that historical changes in S deposition is the principal control of long-term dynamics of DOC and water color, implies that the browning observed in the last decades represents a return to levels of pre-industrial times (Monteith et al. 2007). The recovery-hypothesis is supported by experimental studies that verify the proposed mechanism, i.e., pH and ionic strength exert a control on soil organic matter solubility (Ekström et al. 2011; Evans et al. 2012), and further, spatial and temporal patterns in S deposition are proportional to rising DOC concentrations (Monteith et al. 2007). However, a notable problem is that the temporal coherence between S and DOC has only been tested for a limited period of strongly declining acid deposition, since monitoring of surface waters generally started in the 1980s.

To the extent that decadal changes in aquatic DOC concentrations are explained by atmospheric deposition, the sharp increase in S deposition during the accelerating industrialization from the mid-1940s should have been paralleled by a sharp decline in lake DOC and water color (Fig. 1). Further, since we are now returning to pre-industrial levels of S deposition, the lake DOC concentrations of today should be similar to those prior to the 1940s. In order to test these predictions from the recovery-hypothesis, I analyzed historical data from 50 lakes in the area around the Aneboda Field Station in the south of Sweden. Limnologists have studied lakes in this area far back in time, starting with Einar Naumann in the beginning of the 20th century, providing a unique opportunity to assess what factors control water color and DOC during the past 80 yr. Moreover, since this region has received high S deposition (> 2 g S m^{-2} yr^{-1} in the 1980s) and is dominated by acid sensitive soils, water quality in these lakes should be highly responsive to atmospheric deposition.

Methods

The study area is part of the boreo-nemoral zone characterized by a mixture of coniferous and deciduous woods, with *Picea* and *Pinus* as the dominant trees (Fredh et al. 2012). Current land-use is dominated by forestry, but grass and crop cultivation occur on the most suitable soils. I found data on water color, sampled in July/August and reported as mg Pt L^{-1}, for 50 lakes from 1935 (Thunmark 1937), 1971, 1977, 1982, 1987, 1993 (Lessmark 1997), and I had the lakes sampled in August 2015. Synchronous dynamics were tested by correlating water color from those seven time points for all the 50 lakes.

For 18 out of the 50 lakes, water color estimates were found for ≥ 10 occasions including at least one measurement between 1940 and 1970 and at least one measurement between 1995 and 2015. The data is extracted from (Thunmark 1937; Åberg and Rodhe 1942; Thunmark 1945; Malmer 1961; Björk 1967; Andersson 1971; Lessmark 1997), some unpublished data, and from the open database (http://web-star.vatten.slu.se/db.html) of the national Swedish monitoring program run by the Swedish University of Agricultural Sciences (SLU). To capture the general pattern of water color in the individual lakes over time, LOESS (locally weighted scatterplot smoothing) curve fitting was applied, using the freely available LOESS Utility for Excel (peltiertech.com/loess-smoothing-in-excel, developed by Jon Peltier) with the smoothing parameter set to 0.5. To derive a general pattern of water color over time among all the lakes, water color for the individual lakes was normalized to water color in 2015 (according to the LOESS fit), and the relative changes were then averaged across lakes and fit by LOESS smoothing.

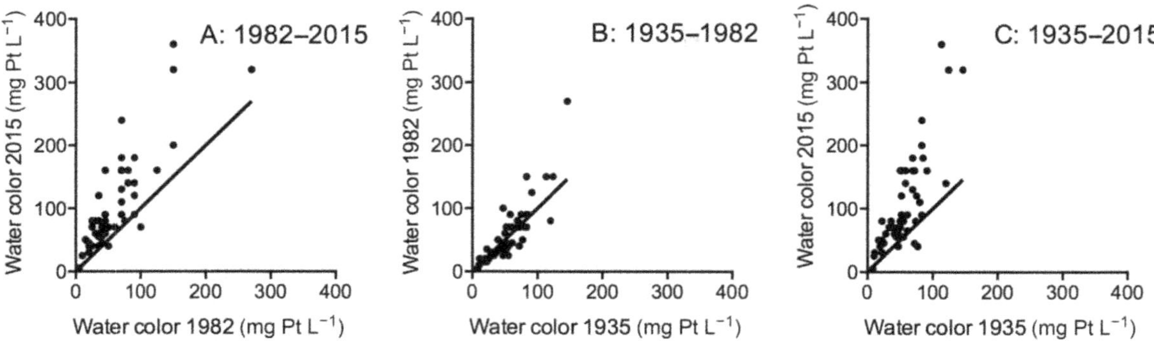

Fig. 2. Comparisons of water color in 50 lakes between 1982–2015 (**A**), 1935–1982 (**B**), and 1935–2015 (**C**). The lines denote the 1 : 1 relationship.

When water color was reported as absorbance at 420 nm (Abs_{420}), a conversion was applied to allow for comparison with water color measured in reference to a Pt-Co-Cl solution or disc and reported as mg Pt L^{-1}. This conversion factor was estimated by correlating parallel measurements of water color (mg Pt L^{-1} and Abs_{420}) across the 50 lakes sampled in 2015, as well as parallel measurements from the three river catchments that the majority of the lakes belong to (Rivers Mörrum, Helge, and Lagan, sampled as part of the national monitoring program). Conversion from Abs_{420} (5 cm^{-1}) to mg Pt L^{-1} was based on the regression water color (mg Pt L^{-1}) = 353.6 × Abs_{420} + 13.86 ($R^2 = 0.90$, $p < 0.001$). Comparison of parallel measures of water color and DOC was based on data across the 50 lakes in 2015 and over time in Lake Bolmen 1987–2015 ($n = 89$).

Data on S deposition was based on deposition data from EMEP, processed as described in (Moldan et al. 2013) and provided by the Swedish Environmental Research Institute IVL. Climate data was provided by the Swedish Meteorological and Hydrological Institute. Temperature was measured in Jönköping ~80 km from Aneboda. Growing degree days were calculated as the number of days in each year that had an average temperature >5°C. Discharge data comes from Lake Allgunnen. Information on land-use was extracted from Fredh et al. (2012) and based on sediment sampling of Lake Fiolen and reconstruction of vegetation cover in a 50-km radius (encompassing 37 of the 50 lakes of this study) based on pollen analyses.

Differences in water color across lakes for individual years were tested by paired *t*-test. Significant trends in the time series were detected by Mann–Kendall trend test using an Excel macro, MULTMK/PARTMK developed by Anders Grimvall and Claudia Libiseller, Linköping University, Sweden, in collaboration with SLU. The Theil-Sen slope estimator was used to determine size and direction of changes in variables over time. The Theil-Sen slope estimations were performed using a version of the Mann–Kendall macro, extended and modified by Jens Fölster and Jan Seibert at SLU. Relative rates of change (% yr^{-1}) were determined by dividing the Theil-Sen slope by the median water color and multiplying by 100.

Results and discussion

The lakes included in this study encompassed a wide gradient of water color, varying from 8 mg Pt L^{-1} to 146 mg Pt L^{-1} in 1935 and 5 mg Pt L^{-1} to 360 mg Pt L^{-1} in 2015 (Fig. 2). Water color was well correlated with DOC where parallel measures existed ($r = 0.84$, $p < 0.001$, $n = 89$), corroborating that water color is a reliable proxy for DOC. It should be acknowledged, however, that water color is also affected by Fe concentration and the structural characteristics of the organic matter, which in turn may be modified by land-use, climate and acid deposition (Wilson and Xenopoulos 2009; Ekström et al. 2011; Jane et al. 2017). Water color across the 50 lakes is markedly higher today than during the period of peak acidification, e.g., water color was significantly higher in 2015 than in 1982 by 90% ± 74% (mean, 1 SD; $t_{49} = 6.884$, $p < 0.001$; Fig. 2A). This is in line with the predictions of the recovery-hypothesis, and similar to results from other studies on recent browning of lakes in the boreal regions (Monteith et al. 2007; Kritzberg and Ekström 2012). However, assuming that S deposition is the main driver of long-term water color dynamics, we expect a decline in water color in response to the sharply increasing atmospheric loading from the 1940s. At odds with that expectation, no difference in water color between 1935 and 1982 was detected, when comparing across lakes ($t_{49} = -1.014$, $p = 0.32$; Fig. 2B). Even though S deposition was around threefold higher in 1982 than in 1935, the water color was almost similar (water color in 1982 was 3% ± 36% [mean, 1 SD] higher than that in 1935). Further, S deposition has now returned to levels from before industrialization, and while there is a lag in the decline of sulfate in the catchment, sulfate concentrations in lake waters of this region are down to levels of the 1920s (Fig. 1). If the last decades of browning of surface waters was primarily a sign of recovery, water color before the accelerated S emissions should have been similar as today. However, water color

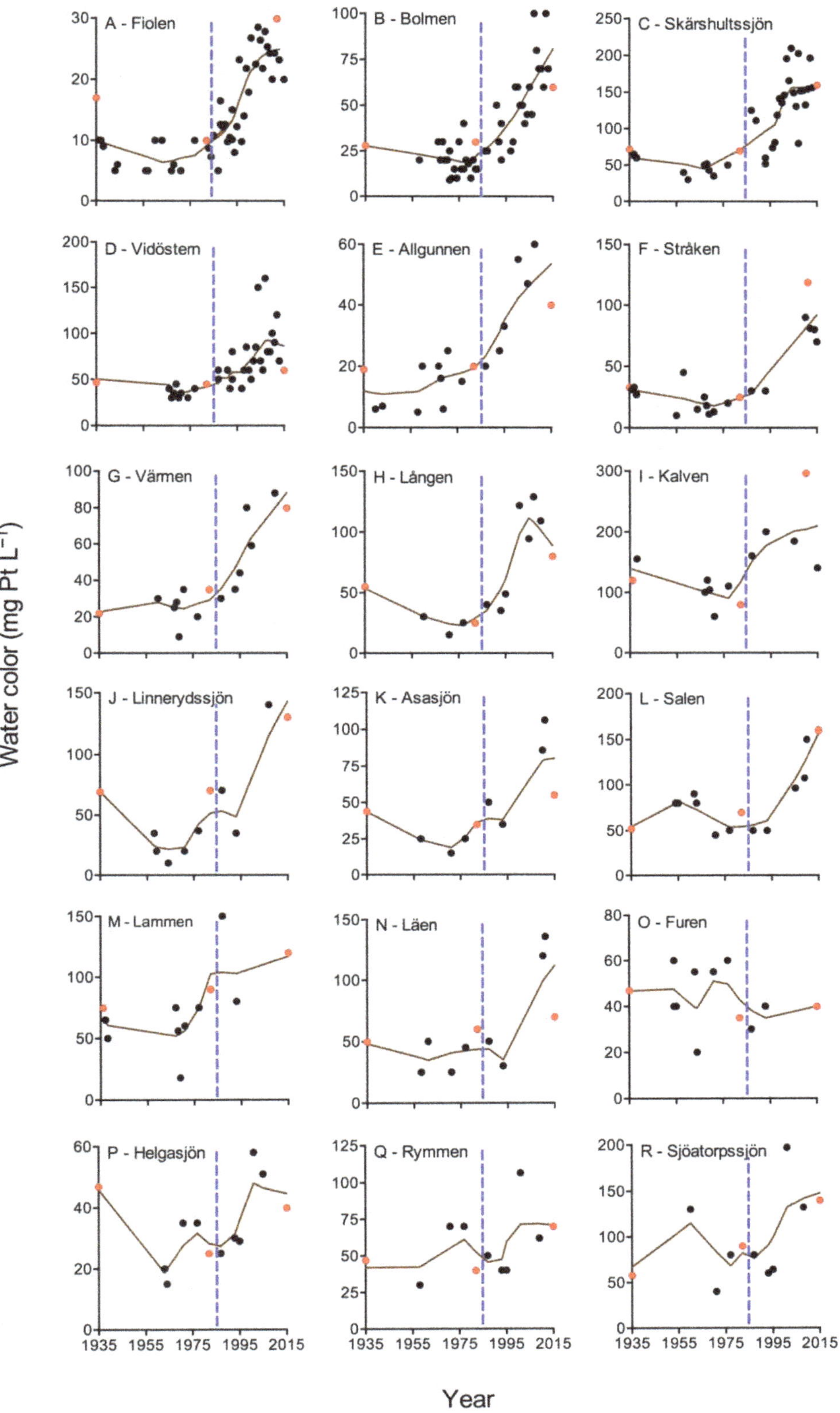

Fig. 3. Historical data on water color in 18 lakes (A-R). Red circles are measurements from 1935, 1982, and 2015, respectively, i.e., data used in Fig. 2. The brown lines represent Loess smoothing fitted to the data. The dashed blue lines denote the peak of S deposition in 1985.

in 2015 was considerably higher than in 1935 by 88% ± 75% (mean, 1 SD; $t_{49} = 6.156$, $p < 0.001$; Fig. 2C).

Water color and DOC are known to be variable within and between years, and in that respect comparisons between single time points are not necessarily representative of long-term trends. In this unique dataset, I was able to compile long-term time series from a subset of the lakes (Fig. 3), and the results clearly show that the water color in 1982 does not stand out from the overall temporal pattern and that the water color in 2015 even tends to be in the low range compared to data after 2005. Water color in 1935 tended to be in the higher range of measurements from the early period of the dataset, e.g., water color was significantly higher in 1935 than in 1936 ($t_3 = 3.29$, $p < 0.05$) for the four lakes where data exists for both years. Precipitation in the region was unusually low during 1932–1934, and the return to more average levels of precipitation and discharge during 1935 (Andersson 1971; Supporting Information Fig. S1) may have mobilized accumulated organic matter from the catchments and resulted in relatively high water color for that particular year. Thus, to the extent that water color in 1935 and 2015 are not representative for the respective periods, the direction in which they deviate is conservative to the conclusions drawn when comparing 1935, 1982, and 2015: that water color was not lower during peak acidification than before, that water color is higher today than during peak acidification, and that water color today is significantly higher than before peak acidification.

The general pattern that emerges from water color dynamics in the 18 lakes, is that water color has increased strongly during the last few decades to levels well above those in 1935 (Fig. 3). On average, the Loess fits yield water color in 2015 to be 126% ± 93% (mean, 1 SD) higher than in 1935 ($t_{17} = 6.99$, $p < 0.001$). Moreover, the data support significant increases in water color 1935–2015 and 1985–2015 in 11 and 7 lakes, respectively (Table 1). The absence of significant increases in the other lakes can either be due to real divergence in the temporal dynamics in those lakes, or the limited amount of data. While the minimum values of Loess fits all occur after 1935, the data does not support a significant decline in water color between 1935 and 1985 for any of the lakes ($p = 0.11$–0.80). Finally, Loess fits suggest that the increase in water color began well before the peak of S deposition around 1985, and the minimum water color occurred around 1968 (mean, SD = 10, Fig. 3).

Out of the 1225 possible pairs of lakes in this dataset, 385 showed significant positive correlations in water color ($p < 0.05$). Thus, there was some degree of synchronicity in the temporal variation of water color among the 50 lakes, which is also evident from Figs. 2, 3. Common temporal dynamics, which have been observed in other regions (Pace and Cole 2002; Erlandsson et al. 2008), reflect the role of drivers acting on the regional or larger scale, such as S deposition. The steep increase in water color after 1985, when emissions were reduced, is in line with increasing solubility of DOC in response to reduced acidity and ionic strength in catchment soils (Monteith et al. 2007; Ekström et al. 2011; Evans et al. 2012). However, the fact that water color started to increase well before emissions started to decline, that water color today is drastically higher than in 1935, and that the data does not show a decline in water color during the period in which S deposition was rising strongly, is clearly at odds with the recovery-hypothesis. Instead, it manifests that S deposition is not the dominant driver on the centennial scale and that the on-going browning is not necessarily representing a recovery from acidification and a return to a more natural state. Thus, we need to evaluate the role of other drivers.

Another driver that may operate at a regional scale—when influenced by, e.g., human population dynamics, political decisions or legislation—is land-use. Variation in composition of vegetation and soil and management practices affect production and transport of DOC (Wilson and Xenopoulos 2008; Stanley et al. 2012), e.g., the proportion of coniferous boreal forest in the catchment is the best predictor of lake DOC concentration on the global scale (Sobek et al. 2007). Paleolimnological approaches clearly demonstrate that landscape utilization strongly influence lake DOC on the centennial scale (Bragée et al. 2015; Meyer-Jacob et al. 2015). In spite of that, land-use has rarely been

Table 1. Rate of change in water color (% yr^{-1})† during three different periods.

	1935–2015	1985–2015	1935–1985
Fiolen	**2.5*****	**3.5*****	0.0
Bolmen	**3.3*****	**3.8*****	−1.9
Skärshultssjön	**1.9*****	**1.7*****	−1.3
Vidöstern	**1.6*****	**2.5****	0.0
Allgunnen	**3.1*****	**3.5***	1.5
Stråken	2.1	0.0	−1.9
Värmen	**2.9*****	**3.5***	1.0
Lången	**3.0****	3.4	2.4
Kalven	1.1	0.2	−1.4
Linnerydssjön	**3.4***	2.8	0.0
Asasjön	**2.3***	1.7	−0.5
Salen	**1.0**	**3.8**	−0.2
Lammen	**1.2***	−9.7**	0.6
Läen	**1.4***	3.5	0.0
Furen	−0.8	0.0	−0.2
Helgasjön	1.1	1.5	−1.7
Rymmen	0.6	1.3	0.5
Sjöatorpssjön	1.4	2.0	1.1

† Rate of change determined by dividing the Theil Sen slope by the median water color for the period and multiplied by 100. *, **, and *** denote statistical significance at the $p < 0.05$, $p < 0.01$, and $p < 0.001$ level, respectively.

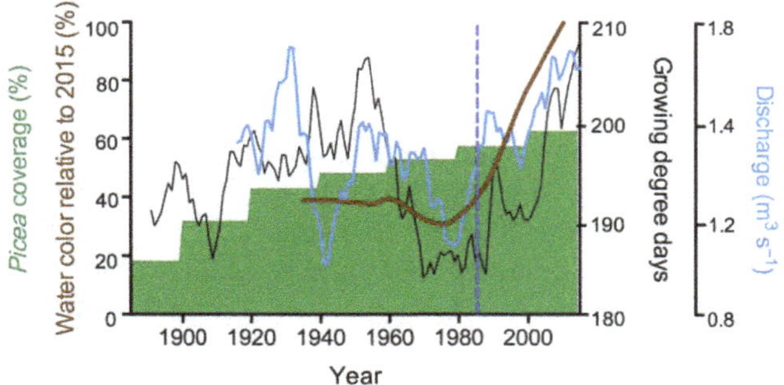

Fig. 4. Historical patterns in water color and the potential drivers land-use, growing degree days, hydrology, and S deposition. The brown line shows the general pattern of water color over time across the 18 lakes displayed in Fig. 3: the change in color for the individual lakes was normalized to that in 2015 (according to the Loess fit) and the average of the relative changes were fit by Loess smoothing. The green area shows spruce coverage in the region (adopted from Fredh et al. 2012). The black and light blue lines show the 10-yr running mean of growing degree days (i.e., days with mean temperature > 5°C) and discharge in one of the lakes (Allgunnen), respectively. The dashed blue line denotes the peak in S deposition in 1985.

included when analyzing drivers of recent browning, since land-use has generally been stable during the last few decades for which monitoring data show browning of waters. While land cover in this region has not changed much during the last few decades, a major transition from agriculture to a modern forest management occurred between 1880 and 1940, when the total forest cover increased dramatically and the species composition changed from deciduous to coniferous dominance (Fredh et al. 2012). Based on lake sediment pollen records, spruce (*Picea*) coverage in the investigated region expanded from 19% in the late 1800s to 63% at present (Fredh et al. 2012; Fig. 4). Shifting land-use from agriculture to forestry induces major changes in soil organic carbon stocks (Guo and Gifford 2002) and thereby also DOC export, but the change is not instant. Exploring long-term effects of afforestation on DOC dynamics, it was found to take >40 yr for organic soil horizons (O horizons) to develop after planting spruce onto agricultural land (Rosenqvist et al. 2010). Moreover, the organic carbon content of the O horizon, which sets the base line for organic carbon export to surface waters (Weyhenmeyer et al. 2012), was increasing strongly with the age of forest stands (15–90 yr; Rosenqvist et al. 2010). Hence, the effect of afforestation on soil processes is gradual, and the influence on downstream surface waters should exhibit a substantial lag after planting, on the order of several decades. Knowing that land-use has been an important modulator of lake DOC on the time scale of hundreds of years (Meyer-Jacob et al. 2015), and the major landscape-wide shift from agricultural activities during the last century, strongly suggests that the recent change in water color and DOC concentrations is to a large extent the result of a delayed and still on-going response to the transition to modern forestry.

Another driver that could potentially contribute to synchronicity, and both short- and long-term variability in DOC, is climate. While some studies invoke different expressions of climate change as causes of browning (Larsen et al. 2011; Weyhenmeyer et al. 2016), others argue that climate has not been an important driver, since, e.g., the effect of a <1°C temperature rise on relevant processes such as decomposition rates could only explain a minor share of observed increases in DOC, and since changes in precipitation and hydrology are not monotonic or consistent on the larger scale. However, other representations of climate change reveal considerable changes. The number of growing degree days have increased by ~20 since the 1970s in this area, and also precipitation and hydrology has changed as indicated by increasing discharge during the same period (Fig. 4). The length of the growing season is a significant predictor of DOC concentrations across space and time (Weyhenmeyer and Karlsson 2009), and climate mediated greening has been shown to contribute to increasing DOC in Norwegian lakes during the past 30 yr (Finstad et al. 2016). Similarly, variability in DOC is influenced by variation in precipitation and flow (Pace and Cole 2002; Erlandsson et al. 2008). For the lakes in this region, the past decades of browning is paralleled by increasing number of growing degree days and discharge (Fig. 4), but the pronounced and extended decline in the same variables from the 50s to the 70s seem not to be paralleled by declining water color. Thus, decadal variability in the length of the growing season and hydrology alone do not explain the long-term pattern for DOC and water color, but they may have accentuated the effect of land-use change, by further enhancing the potential for production and leaching of DOC from afforested catchments during the past few decades.

To fully disentangle the relative importance of land-use, climate change and S deposition to lake DOC and water color, modeling with site specific information on land-use would be the best way forward. The results presented here suggest that the major transition in land-use in this region is likely a major factor driving lake browning, and that this browning has been augmented by climate change. While historical changes in S deposition have not been the *sole* driver, it has most probably played an important role by delaying the browning through enhanced adsorption of organic matter during the period of high deposition. Thus, the high rate at which water color has increased in the past few decades should be highly influenced by reduced S deposition. The data and analysis presented here is specific for this particular region, but it should be noted that the transition to forestry was facilitated by the introduction of artificial fertilizers, which reduced the land area required for cultivating crops and supporting animals that produce manure—a development common to many parts of northern Europe (Myrdal 1997; Morell 2001). Also in North America, regrowth of forest at the expense of agriculture has resulted in enhanced vegetation productivity and carbon accumulation (Caspersen et al. 2000; Birdsey et al. 2006). Nevertheless, the importance of land-use, climate change and acid deposition as drivers of browning is variable both on temporal and spatial scales.

Importantly, appreciating that land-use is a major driver behind browning opens up the potential for local measures to control water color. When Stanley et al. (2012) reviewed effects of land-use on river DOC the focus was on agriculture. Nevertheless, several of the fundamental arguments made, e.g., about land cover controlling terrestrial organic carbon accumulation, which sets the first constraint on aquatic DOC, and about extensive ditching, which promotes rapid routing of DOC from soil to channel, largely apply to managed forestry as well. The importance of the riparian zone as a source of DOC in forested catchments (Bishop et al. 2004), suggests that riparian management, e.g., by selecting for deciduous tree species, may be efficient in controlling DOC in receiving surface waters.

References

Åberg, B., and W. Rodhe. 1942. Über die Milieufaktoren in einigen Südschwedischen Seen. Symb. Bot. Upsalienses, Uppsala.

Andersson, G. 1971. Vattenkemi och långtidsmässiga förändringar i Aneboda-områdets sjöar. Limnologiska Institutionen, Lund Univ.

Battin, T. J., S. Luyssaert, L. A. Kaplan, A. K. Aufdenkampe, A. Richter, and L. J. Tranvik. 2009. The boundless carbon cycle. Nat. Geosci. **2**: 598–600. doi:10.1038/ngeo618

Birdsey, R., K. Pregitzer, and A. Lucier. 2006. Forest carbon management in the United States: 1600–2100. J. Environ. Qual. **35**: 1461–1469. doi:10.2134/jeq2005.0162

Bishop, K., J. Seibert, S. Köhler, and H. Laudon. 2004. Resolving the Double Paradox of rapidly mobilized old water with highly variable responses in runoff chemistry. Hydrol. Process. **18**: 185–189. doi:10.1002/hyp.5209

Björk, S. 1967. Ecologic investigations of *Phragmites communis*. Studies in theoretic and applied limnology. Folia Limnol. Scand. **14**: 248.

Bragée, P., F. Mazier, A. B. Nielsen, P. Rosén, D. Fredh, A. Broström, W. Granéli, and D. Hammarlund. 2015. Historical TOC concentration minima during peak sulfur deposition in two Swedish lakes. Biogeosciences **12**: 307–322. doi:10.5194/bg-12-307-2015

Caspersen, J. P., S. W. Pacala, J. C. Jenkins, G. C. Hurtt. P. R. Moorcroft, and R. A. Birdsey. 2000. Contributions of land-use history to carbon accumulation in U.S. forests. Science **290**: 1148–1151. doi:10.1126/science.290.5494.1148

Chow, A. T., R. A. Dahlgren, and J. A. Harrison. 2007. Watershed sources of disinfection byproduct precursors in the Sacramento and San Joaquin rivers, California. Environ. Sci. Technol. **41**: 7645–7652. doi:10.1021/es070621t

Downing, J. A., J. J. Cole, J. J. Middelburg, R. G. Striegl, C. M. Duarte, P. Kortelainen, Y. T. Prairie, and K. A. Laube. 2008. Sediment organic carbon burial in agriculturally eutrophic impoundments over the last century. Global Biogeochem Cycles **22**: GB1018. doi:10.1029/2006gb002854

Ekström, S. M., E. S. Kritzberg, D. B. Kleja, N. Larsson, P. A. Nilsson, W. Granéli, and B. Bergkvist. 2011. Effect of acid deposition on quantity and quality of dissolved organic matter in soil-water. Environ. Sci. Technol. **45**: 4733–4739. doi:10.1021/es104126f

Erlandsson, M., I. Buffan, J. Fölster, H. Laudon, J. Temnerud, G. A. Weyhenmeyer, and K. Bishop. 2008. Thirty-five years of synchrony in the organic matter concentrations of Swedish rivers explained by variation in flow and sulfate. Glob. Chang. Biol. **14**: 1191–1198. doi:10.1111/j.1365-2486.2008.01551.x

Evans, C. D., and others. 2012. Acidity controls on dissolved organic carbon mobility in organic soils. Glob. Chang. Biol. **18**: 3317–3331. doi:10.1111/j.1365-2486.2012.02794.x

Ferland, M.-E., P. A. del Giorgio, C. R. Teodoru, and Y. T. Prairie. 2012. Long-term C accumulation and total C stocks in boreal lakes in northern Québec. Global Biogeochem. Cycles **26**: GB0E04. doi:10.1029/2011GB004241

Finstad, A. G., T. Andersen, S. Larsen, K. Tominaga, S. Blumentrath, H. A. de Wit, H. Tømmervik, and D. O. Hessen. 2016. From greening to browning: Catchment vegetation development and reduced S-deposition promote organic carbon load on decadal time scales in Nordic lakes. Sci. Rep. **6**: 31944. doi:10.1038/srep31944

Fredh, D., A. Broström, L. Zillén, F. Mazier, M. Rundgren, and P. Lagerås. 2012. Floristic diversity in the transition from traditional to modern land-use in southern Sweden AD 1800–2008. Veg. Hist. Archaeobot. **21**: 439–452. doi:10.1007/s00334-012-0357-z

Guo, L. B, and R. M. Gifford. 2002. Soil carbon stocks and land use change: A meta analysis. Glob. Chang. Biol. **8**: 345–360. doi:10.1046/j.1354-1013.2002.00486.x

Jane, S. F., L. A. Winslow, C. K. Remurcal, and K. C. Rose. 2017. Long-term trends and synchrony in dissolved organic matter characteristics in Wisconsin, USA, lakes: Quality, not quantity, is highly sensitive to climate. J. Geophys. Res. Biogeosci. **122**: 546–561. doi:10.1002/2016JG003630

Karlsson, J., P. Byström, J. Ask, L. Persson, and M. Jansson. 2009. Light limitation of nutrient-poor lake ecosystems. Nature **460**: 506–509. doi:10.1038/nature08179

Kritzberg, E. S., and S. M. Ekström. 2012. Increasing iron concentrations in surface waters - a factor behind brownification? Biogeosciences **9**: 1465–1478. doi:10.5194/bg-9-1465-2012

Larsen, S., T. Andersen, and D. O. Hessen. 2011. Climate change predicted to cause severe increase of organic carbon in lakes. Glob. Chang. Biol. **17**: 1186–1192. doi:10.1111/j.1365-2486.2010.02257.x

Laudon, H., and others. 2012. Cross-regional prediction of long-term trajectory of stream water DOC response to climate change. Geophys. Res. Lett. **39**: L18404. doi: 10.1029/2012GL053033

Lessmark, O. 1997. Sjöar i Kronobergs län 1971–1993. Länsstyrelsen i Kronobergs län.

Malmer, N. 1961. Ecologic studies on the water chemistry of lakes in South Sweden. Bot. Not. **114**: 121–143.

Meyer-Jacob, C., J. Tolu, C. Bigler, H. Yang, and R. Bindler. 2015. Early land use and centennial scale changes in lake-water organic carbon prior to contemporary monitoring. Proc. Natl. Acad. Sci. USA. **112**: 6579–6584. doi:10.1073/pnas.1501505112

Moldan, F., B. J. Cosby, and R. F. Wright. 2013. Modeling past and future acidification of Swedish Lakes. Ambio **42**: 577–586. doi:10.1007/s13280-012-0360-8

Monteith, D. T., and others. 2007. Dissolved organic carbon trends resulting from changes in atmospheric deposition chemistry. Nature **450**: 537–541. doi:10.1038/nature06316

Morell, M. 2001. Jordbruket i industrisamhället. Natur och Kultur.

Myrdal, J. 1997. En agrarhistorisk syntes, p. 302–322. *In* B. M. P. Larsson, M. Morell, and J. Myrdal [eds.], Agrarhistoria. Forumförlag.

Pace, M. L., and J. J. Cole. 2002. Synchronous variation of dissolved organic carbon and color in lakes. Limnol. Oceanogr. **47**: 333–342. doi:10.4319/lo.2002.47.2.0333

Ranåker, L., M. Jönsson, P. A. Nilsson, and C. Brönmark. 2012. Effects of brown and turbid water on piscivore-prey fish interactions along a visibility gradient. Freshw. Biol. **57**: 1761–1768. doi:10.1111/j.1365-2427.2012.02836.x

Rosenqvist, L., D. B. Kleja, and M.-B. Johansson. 2010. Concentrations and fluxes of dissolved organic carbon and nitrogen in a *Picea abies* chronosequence on former arable land in Sweden. For. Ecol. Manage. **259**: 275–285. doi: 10.1016/j.foreco.2009.10.013

Roulet, N., and T. R. Moore. 2006. Environmental chemistry - browning the waters. Nature **444**: 283–284. doi:10.1038/444283a|ISSN 0250–6971

Sobek, S., L. J. Tranvik, Y. T. Prairie, P. Kortelainen, and J. J. Cole. 2007. Patterns and regulation of dissolved organic carbon: An analysis of 7,500 widely distributed lakes. Limnol. Oceanogr. **52**: 1208–1219. doi:10.4319/lo.2007.52.3.1208

Stanley, E. H., S. M. Powers, N. R. Lottig, I. Buffam, and J. T. Crawford. 2012. Contemporary changes in dissolved organic carbon (DOC) in human-dominated rivers: Is there a role for DOC management? Freshw. Biol. **57**: 26–42. doi:10.1111/j.1365-2427.2011.02613.x

Thunmark, S. 1937. Über die Regionale Limnologie von Südschweden. Nordstedt & Söner.

Thunmark, S. 1945. Zur Sociologie des Süsswasserplanktons, einer Methodologisch-ökologische Studie. Folia Limnol. Scand. **3**: 1–66.

Weyhenmeyer, G. A., and J. Karlsson. 2009. Nonlinear response of dissolved organic carbon concentrations in boreal lakes to increasing temperatures. Limnol. Oceanogr. **54**: 2513–2519. doi:10.4319/lo.2009.54.6_part_2.2513

Weyhenmeyer, G. A., M. Fröberg, E. Karltun, M. Khalili, D. Kothawala, J. Temnerud, and L. J. Tranvik. 2012. Selective decay of terrestrial organic carbon during transport from land to sea. Glob. Chang. Biol. **18**: 349–355. doi:10.1111/j.1365-2486.2011.02544.x

Weyhenmeyer, G. A., R. A. Muller, M. Norman, and L. J. Tranvik. 2016. Sensitivity of freshwaters to browning in response to future climate change. Clim. Change **134**: 225–239. doi:10.1007/s10584-015-1514-z

Wilson, H. F., and M. A. Xenopoulos. 2008. Ecosystem and seasonal control of stream dissolved organic carbon along a gradient of land use. Ecosystems **11**: 555–568. doi: 10.1007/s10021-008-9142-3

Wilson, H. F., and M. A. Xenopoulos. 2009. Effects of agricultural land use on the composition of fluvial dissolved organic matter. Nat. Geosci. **2**: 37–41. doi:10.1038/NGEO391

Acknowledgments

I thank M. Skerlep for assistance in the field and laboratory. I thank the Swedish Agricultural University and the Swedish Institute for Meteorology and Hydrology for running the monitoring programs and making data available. I thank IVL Swedish Environmental Research Institute and J. Stadmark for access to and help with EMEP data, and A. Broström and D. Fredh for sharing land-use data. Discussions with S. Björk, H. Linge, C. Brönmark, L.-A. Hansson, and A. Broström improved the manuscript. I am thankful to J. F. Lapierre and one anonymous reviewer for constructive comments. This research was performed with funding from the Swedish Research Council Formas (http://www.formas.se;grant 2015–05450) and through the Strong Research Environment "Managing multiple stressors in the Baltic Sea" (http://www.formas.se;grant 217–2010-126).

8

Phosphorus availability regulates intracellular nucleotides in marine eukaryotic phytoplankton

Elizabeth B. Kujawinski,[1]* *Krista Longnecker*,[1] *Harriet Alexander*,[2,3,a] *Sonya T. Dyhrman*,[2] *Cara L. Fiore*,[1,b] *Sheean T. Haley*,[2] *Winifred M. Johnson*[3]

[1]Department of Marine Chemistry and Geochemistry, Woods Hole Oceanographic Institution, Woods Hole, Massachusetts; [2]Department of Earth and Environmental Science and the Lamont-Doherty Earth Observatory, Columbia University, Palisades, New York; [3]MIT-WHOI Joint Program in Oceanography/Applied Ocean Science and Engineering, Cambridge, Massachusetts and Woods Hole, Massachusetts

Scientific Significance Statement

Phosphorus (P) is a central element in cellular metabolism that can limit primary production of eukaryotic phytoplankton in the ocean. Adaptation to low P is known to drive metabolic restructuring in eukaryotic phytoplankton, but the specific adaptive responses of cellular metabolites are poorly understood. Here we show that three model phytoplankton alter their metabolites under P deficiency, relative to P-replete conditions. We present evidence for a new model of P allocation within cells, where monophosphate nucleotides can act as a flexible P storage pool, allowing rapid and dynamic distribution of P to cellular processes.

Abstract

Marine eukaryotic phytoplankton adapt to low phosphorus (P) in the oceans through a variety of step-wise mechanisms including lipid substitution and decreased nucleic acid content. Here, we examined the impact of low P concentrations on intracellular metabolites whose abundances can be quickly adjusted by cellular regulation within laboratory cultures of three model phytoplankton and in field samples from the Atlantic and Pacific Oceans. We quantified the relative abundances of monophosphate nucleotides and their corresponding nucleosides, using a combination of targeted and untargeted metabolomics methods. Under P-deficient conditions, we observed a marked decrease in adenosine 5′-monophosphate (AMP) with a concomitant increase in adenosine. This shift occurred within all detected pairs of monophosphate nucleotides and nucleosides, and was consistent with previous work showing transcriptional changes in nucleotide synthesis and salvage under P-deficient conditions for model eukaryotes. In the field, we observed AMP-to-adenosine ratios that were similar to those in laboratory culture under P-deficient conditions.

*Correspondence: ekujawinski@whoi.edu

Author Contribution Statement: KL and EK designed and guided the lab and field metabolomics experiments. HA and SD contributed transcriptomics data, HA and KL linked the metabolomics and transcriptomics data. WJ and EK contributed the field data, CF and SH contributed culture data. EK, KL, and SD wrote the paper. All authors contributed editorial comments and approved the final version of the paper.

[a]Present address: Population Health and Reproduction, University of California – Davis, Davis, California
[b]Present address: Biology Department, Appalachian State University, Boone, North Carolina

It is increasingly recognized that phosphorus (P) can limit primary production and carbon export, and shape phytoplankton community composition in the ocean (reviewed by Karl 2014). A number of strategies have been identified for mitigating the metabolic impacts of P deficiency in phytoplankton. These include increased scavenging from the surrounding environment through upregulation of inorganic P transporters and enzymes to access dissolved organic phosphorus (DOP; Ammerman and Azam 1985; Dyhrman et al. 2012) as well as structural changes such as the substitution of P-based lipids with sulfur-based lipids (Van Mooy et al. 2009; Martin et al. 2011). Many of these P deficiency responses show conservation across other functional groups of eukaryotic phytoplankton (reviewed by Dyhrman 2016).

Investigations to date have understandably focused on the largest intracellular P stores such as lipids, nucleic acids, and polyphosphate. However, the extent to which minor P stores are scavenged under low P is largely unknown (Karl 2014) but could prove important to maintenance of P homeostasis. For example, nucleotides carry up to three phosphate groups (e.g., adenosine 5′-mono-, di-, and triphosphate: AMP, ADP, and ATP, respectively), and play critical roles in cellular energy balances and phosphorylation capabilities. While it is reasonable to assume that phosphate-containing nucleotides would be subject to cleavage under P deficiency, few studies have explicitly examined the impact of low P on the dynamics of these critical metabolites. Instead, most studies have tacitly assumed that this pool is impervious to P deficiency due to its critical role in P homeostasis. However, similar investigations with nitrogen (N) deficiency have observed reduced concentrations of N-rich metabolites such as amino acids (Bromke et al. 2013) in N-deficient cultures of the centric diatom *Thalassiosira pseudonana*, concurrent with biochemical indicators of catabolism of the same amino acids (Hockin et al. 2012).

The overall metabolic response of an organism to nutrient limitation can be elucidated with metabolomics, or the study of molecules produced by a cell during metabolism and growth. These data provide unique insights into a cell's physiological state through the description and quantification of primary and secondary metabolites (Poulson-Ellestad et al. 2014). Yet, metabolic profiling studies under nutrient-limiting growth conditions are rare, relative to transcriptome or proteome investigations.

Metabolomics complements gene-based omics techniques by quantifying metabolic intermediates within active biochemical pathways. "Untargeted" metabolomics is used to describe, semi-quantitatively, all metabolites within a cell that can be detected with a chosen analytical method; thus it is appropriate for open-ended exploration of metabolic intermediates (Patti et al. 2012). In contrast, "targeted" metabolomics is used to quantify a pre-determined group of metabolites for precise cellular dynamics and assessment of relative rates of different metabolic pathways (Kido Soule et al. 2015). Metabolomics studies thus offer critical insights into cellular physiology that cannot be attained from other omics approaches.

In this study, we used untargeted and targeted metabolomics to examine the effect of low P on intracellular metabolites, in general, and on nucleotide abundances, specifically. We sought to understand how cells maintained critical P-dependent processes in the face of P deficiency. To answer this question, we quantified the relative abundances of intracellular nucleotides and nucleosides within three model phytoplankton and compared our laboratory-derived patterns to field observations.

Materials and methods

Overview

In this study, we cultured three phytoplankton—a diatom (*Thalassiosira pseudonana*), a prasinophyte (*Micromonas pusilla*), and a pelagophyte (*Aureococcus anophagefferens*)—under P-replete and P-deficient conditions. The chosen conditions facilitated comparison with published transcriptome and proteome data (Wurch et al. 2011*a*; Dyhrman et al. 2012; Whitney and Lomas 2016). We examined the relative abundances of intracellular monophosphate nucleotides (NMPs) and nucleosides in P-replete relative to P-deficient conditions. Metabolites observed in both targeted and untargeted methods showed similar trends across samples both within the diatom experiment (Supporting Information Fig. S1) and in complementary work (Johnson et al. 2016), facilitating comparisons across datasets. We then quantified particle-associated AMP and adenosine in field samples for comparison with our laboratory results. Detailed methods are provided in the Supporting Information.

Laboratory cultures

We cultured axenic *T. pseudonana* (CCMP 1335) in a modified L1 media with two phosphate conditions (Table 1) for two transfers prior to the start of this experiment, with a modified semi-continuous culturing method. The experiment contained two replicate cultures and one cell-free control for each treatment (12 flasks total for each treatment). We sampled one set of flasks destructively 0, 2, 8, and 10 days after inoculation. *T. pseudonana* cell abundances were calculated by converting the TOC concentrations using the value of 5.94 pg C cell^{-1} from Montagnes et al. (1994). The cultures were maintained under a 12 : 12 light : dark regime (84 μmol m^{-2} s^{-1}). We cultured axenic *M. pusilla* (CCMP 1545) under the same conditions as *T. pseudonana*. Samples for metabolomics were removed during exponential growth (day 6 in P-replete, day 4 in P-deficient) and at the onset of stationary growth (day 13 in P-replete and day 5 in P-deficient). We cultured axenic *A. anophagefferens* (CCMP 1984) in a modified L1 media, monitoring growth by relative fluorescence, and harvested cells for metabolite analyses at the onset of stationary phase. Growth conditions for the *T.*

Table 1. Overview of culture experiments.

	T. pseudonana		*M. pusilla*		*A. anophagefferens*	
	P-replete	**P-deficient**	**P-replete**	**P-deficient**	**P-replete**	**P-deficient**
Initial phosphate concentrations (μM)	36	0.4	36	0.4	36	0.4
Growth rate	0.5 day^{-1}	0.03 day^{-1}	0.46 day^{-1}	0.03 day^{-1}	0.47 day^{-1}	0.31 day^{-1}
Maximum cell yield	7.5×10^5 cells mL^{-1}	2.4×10^5 cells mL^{-1}	3.3×10^7 cells mL^{-1}	2.5×10^6 cells mL^{-1}	3.3×10^6 cells mL^{-1}	1.8×10^6 cells mL^{-1}
Maximum TOC concentration	1010 μM	278 μM	861 μM	309 μM	n.d.	n.d.
Number of significantly more abundant features*,†	462	121	n.d.	n.d.	n.d.	n.d.
Number of features (all time-points)‡	4979	4205	n.d.	n.d.	n.d.	n.d.

* Note that untargeted metabolic profiles were not analyzed for *M. pusilla* or *A. anophagefferens*.
† Using *p*-values adjusted for multiple comparisons allowing a False Discovery Rate of 10% (Storey and Tibshirani 2003).
‡ Sum of positive and negative ion modes.
n.d., not determined.

pseudonana transcriptome cultures are described in Dyhrman et al. (2012).

Metabolomics

For all laboratory samples, we captured cells on 0.2 μm Omnipore (Millipore) filters by gentle vacuum filtration, and extracted intracellular metabolites using a method modified from Rabinowitz and Kimball (2007). The extracts for targeted mass spectrometry analysis were dissolved in 95 : 5 (v/v) water : acetonitrile with deuterated biotin as an internal injection standard, and then analyzed with a C_{18} liquid chromatography (LC) column coupled to a Thermo Scientific TSQ Vantage Triple Stage Quadrupole Mass Spectrometer (Kido Soule et al. 2015). All reported concentrations are relative to an external calibration curve. We report fold changes and relative ratios wherever possible, i.e., when the denominator was not zero.

We desalted extracts for untargeted metabolomics prior to LC/MS analysis. We redissolved dried extracts in 0.01 M hydrochloric acid and re-extracted them using a 50 mg/1 cc PPL cartridge following the protocol of Dittmar et al. (2008). The resulting methanol extracts were dried, dissolved in 95 : 5 water : acetonitrile with deuterated biotin and analyzed in negative and positive ion modes with LC preseparation coupled to a 7-Tesla Fourier-transform ion cyclotron resonance mass spectrometer (FT-ICR MS). Parallel to the FT acquisition, we collected four data dependent fragmentation (MS/MS) scans at nominal mass resolution in the ion trap (LTQ). In this dataset, a mass spectral feature, or a putative metabolite, is defined as a unique combination of a mass-to-charge (*m/z*) ratio and a retention time along a LC gradient. Although untargeted metabolomics data are not fully quantitative, relative changes in feature (or metabolite) abundance can be discerned and statistically evaluated.

Comparison to previously published transcriptomes

We used the P-replete and P-deficient transcriptomes of *T. pseudonana* from Dyhrman et al. (2012) to survey the transcriptional responses related to purine and pyrimidine metabolism. We identified sequence tags with 100% identity to the *T. pseudonana* v3 genome (http://genome.jgi.doe.gov/Thaps3/Thaps3.home.html) (Armbrust et al. 2004) and assessed the statistical significance in differential expression of tags between P-replete and P-deficient conditions using Analysis of Sequence Counts (Wu et al. 2010). We predicted potential functionality of genes linked with metabolites with the Kyoto Encyclopedia of Genes and Genomes annotations from the *T. pseudonana* v3 genome. We manually searched targets of interest that were absent in the predicted pathways using BLASTX v2.2.27+ (e-value = 10 e −5) (Altschul et al. 1997). We then used Pathview (Luo and Brouwer 2013) to simultaneously plot the transcriptomics and metabolomics data on the biochemical pathways for *T. pseudonana*.

Field samples

Suspended particles from the Atlantic Ocean were collected in April and May 2013, from Uruguay to Barbados (38°S to 10°N). We collected water using Niskin bottles and filtered cells from 4 L onto pre-combusted GF/F glass-fiber filters (nominal pore size: 0.7 μm). Field samples from the Pacific Ocean (Line P) were collected in May 2012 in the same way except that cells from 2 L were filtered onto pre-combusted GF/A glass-fiber filters (nominal pore size: 1.6

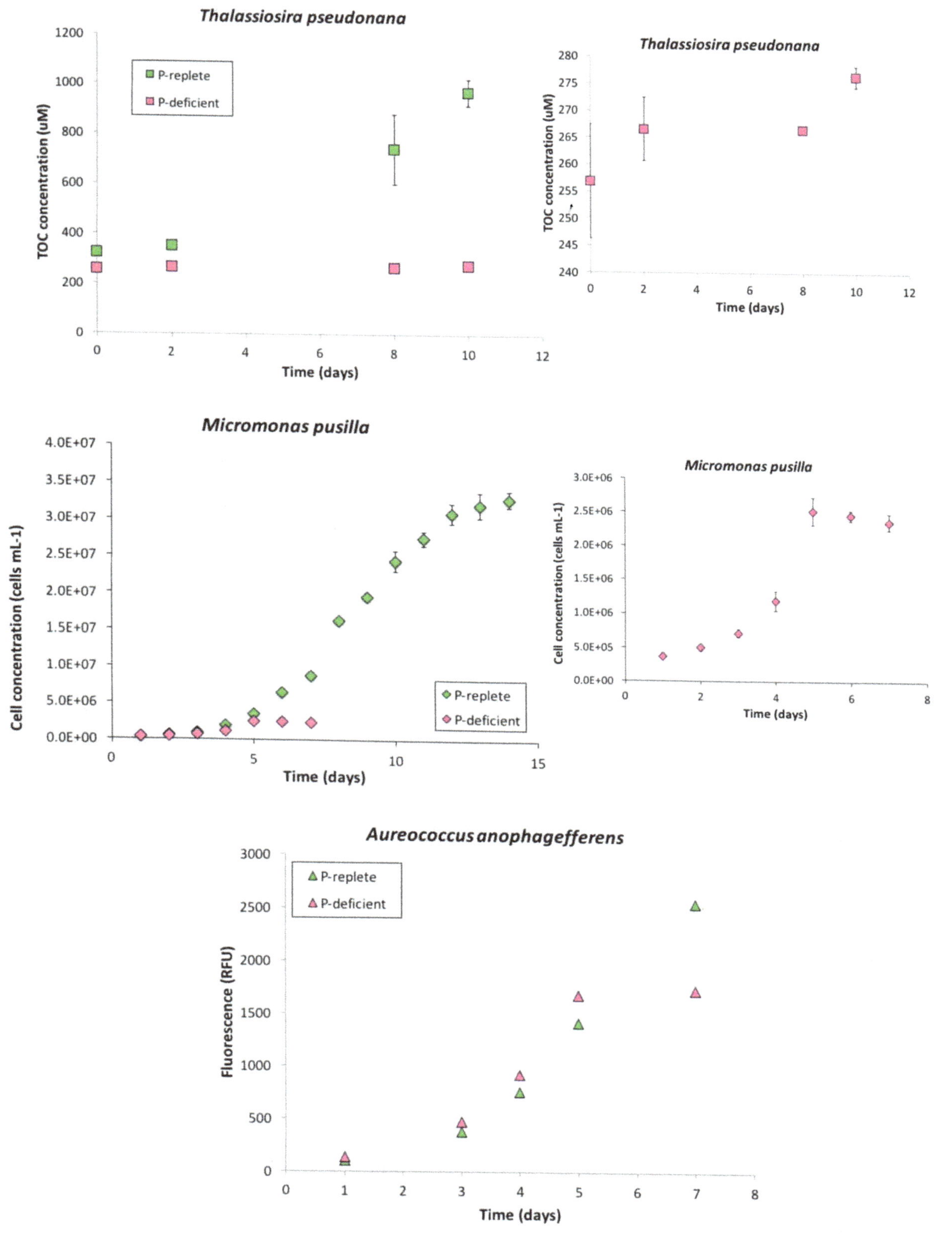

Fig. 1. Growth curves for *T. pseudonana* (top), *M. pusilla* (middle), and *A. anophagefferens* (bottom); for all, pink symbols represent P-deficient treatments and green symbols represent P-replete cultures. For *T. pseudonana*, growth was monitored using changes in total organic carbon concentrations; for *M. pusilla*, growth was monitored using cell counts; and for *A. anophagefferens*, growth was monitored using fluorescence. Insets are provided for the top two cultures to better visualize the low, but positive, growth. Metabolomics samples were collected on days 0, 2, 8, and 10 for *T. pseudonana*, on days 4 and 5 for P-deficient *M. pusilla*, on days 6 and 13 for P-replete *M. pusilla* and on day 7 for *A. anophagefferens*.

μm). We conducted procedural blanks for metabolomics analysis (using Milli-Q water) at the beginning and end of both field campaigns. We stored all filters at −80°C until extraction using the protocols described above. Salinities and fluorescence values are the average values (between 4.5 and 5.5 m) for each station collected on the downward CTD cast. Fluorescence values were then normalized to the highest value measured at these depths for the cruise.

Results

P deficiency affects growth rate and broad distribution of intracellular metabolites

In this study, we compared the metabolite response in three phytoplankton species under two P conditions, with the intention of imposing P deficiency with our lower P treatment. Consequently, we use the term "P deficiency" to conservatively describe any sample from a low-P treatment. In all cultures, P deficiency had a significant impact on phytoplankton growth (Table 1; Fig. 1). For *M. pusilla*, cell counts during the acclimation period confirmed growth under P deficiency (Supporting Information Fig. S2). We lack similar data for *T. pseudonana* but visual inspection of cultures showed growth prior to each transfer and carbon measurements during the experiment confirmed low, but positive, growth. However, these data cannot constrain whether *T. pseudonana* was in stationary or exponential growth at our different sampling points. In all three phytoplankton experiments, P deficiency affected the intracellular metabolite abundances across the targeted and untargeted metabolomics datasets (Table 1). We detected thousands of features at least once in the intracellular untargeted data from each treatment (Table 1), after blank correction. In the diatom culture, we detected the bulk of the features (4019) in both treatments, consistent with earlier metabolomics studies with these methods (Johnson et al. 2016) and expectations that intracellular features represent core metabolic pathways (Dettmer et al. 2007). After normalization to diatom biomass, 462 features exhibited significantly higher abundances at all time-points under P-replete conditions; similarly, 121 features exhibited significantly higher abundances under P-deficient conditions.

Nucleotide concentrations are affected by P deficiency

Of the significantly modulated features, we focused on intermediates within the purine and pyrimidine metabolic pathways (Supporting Information Table S1; Fig. S2). Combined, the targeted and untargeted metabolomics data revealed notable shifts between nucleosides and NMPs under the two growth conditions (Fig. 2). These molecules serve as precursors to the di- and tri-phosphorylated nucleotides (e.g., ADP, ATP; not measured in this study) and are the primary products of RNA degradation.

Under P-replete conditions, *T. pseudonana* contained detectable intracellular concentrations of nucleotides such as

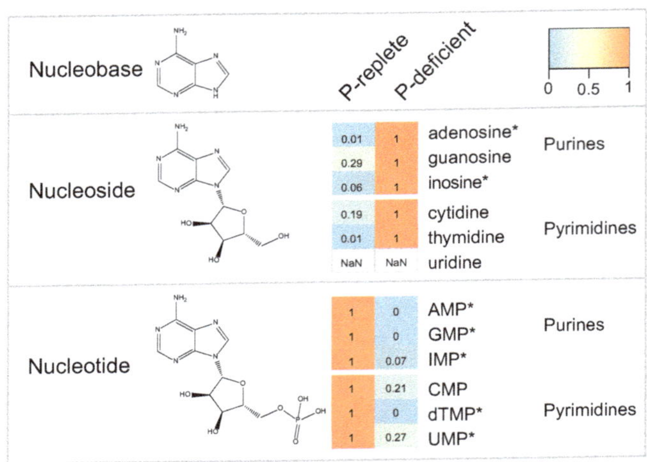

Fig. 2. Relative concentrations of nucleosides and nucleotides in phosphorus (P)-replete and P-deficient conditions for the diatom *T. pseudonana*. The average concentration across all sampling points for each treatment is presented after normalization to the maximum value of each metabolite across the sample set (numerical values in each block). This normalization enables the comparison of targeted and untargeted datasets. The color bar provides the scale used for the colors, where orange represents the maximum value and blue represents the minimum value. Example structures are provided for the nucleobase adenine, the nucleoside adenosine and the nucleotide, adenosine 5′-monophosphate (AMP). (*) indicates statistically significant differences between the two treatments. Individual data from each sampling point is available in Supporting Information Table S2.

AMP (Fig. 2; Supporting Information Fig. S3) and inosine 5′-monophosphate (IMP; Fig. 2). In contrast, we did not detect these P-containing molecules in the intracellular metabolite pool under P-deficient conditions. Instead, we observed elevated nucleosides at all time points under P deficiency. For example, adenosine and inosine were elevated and positively correlated to one another (Spearman's $r = 0.9076$, p-value $\ll 0.0001$). Similarly, the NMPs were correlated across the dataset (for AMP and IMP: Spearman's $r = 0.9166$, p-value $\ll 0.0001$). We observed a parallel pattern in the pyrimidine pathway (Supporting Information Fig. S2B), where the pyrimidine-based NMPs were all elevated relative to their nucleoside counterparts such as cytidine (Supporting Information Fig. S4) in P-replete *T. pseudonana* (Fig. 2). Not all nucleoside and NMP concentrations had statistically significant differences between treatments (Fig. 2), but all NMP-nucleoside pairs exhibited similar overall patterns across all time-points (Supporting Information Table S2). We estimated the minimum concentration of NMPs in a P-deficient cell would be 0.0005 amol, well below our detection limit (Supporting Information Calculation S1). Thus P-deficient cells in our experiments could reasonably contain the necessary, albeit undetectable, NMPs for nucleic acid synthesis during growth.

A conservative comparison between the *T. pseudonana* metabolic profiles and transcriptomes (Dyhrman et al. 2012)

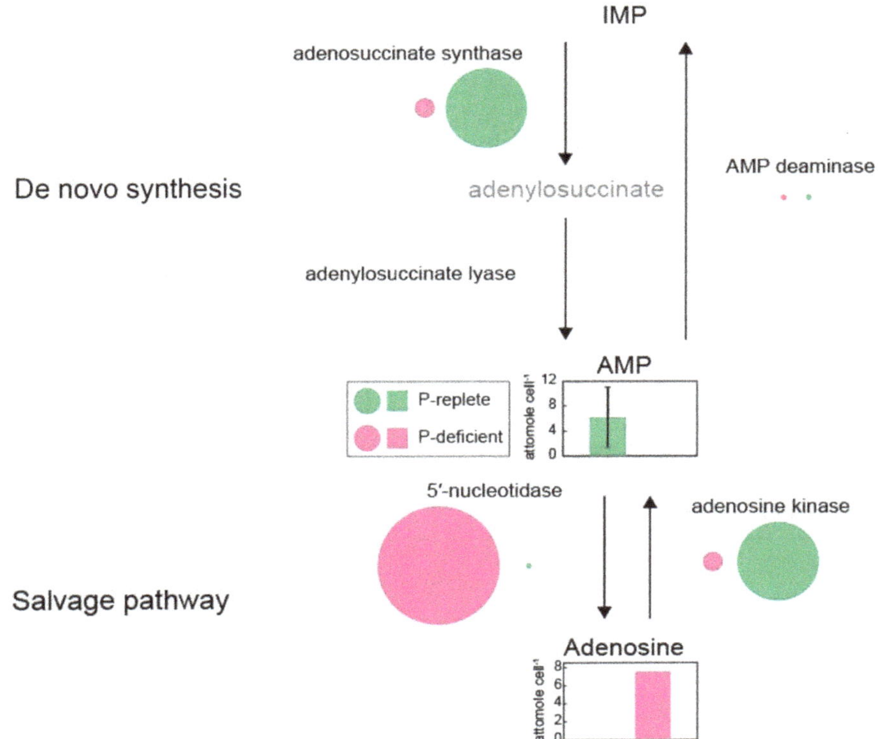

Fig. 3. Relationship between abundance patterns of metabolites and transcripts for a model purine in *T. pseudonana*. Both the de novo synthesis of purines and the salvage pathways are shown. The bar graphs show the metabolite concentrations, relative to an external standard curve and averaged ($n = 8$) across the treatments in amol cell^{-1}; the error bars represent one standard error and are too small to be seen for adenosine. The circles are scaled to denote the transcripts per million for each gene (Dyhrman et al. 2012). The arrows show biochemical pathways linking metabolites. IMP, inosine 5′-monophosphate; AMP, adenosine 5′-monophosphate.

generated from similar treatments but different experiments, revealed relationships between transcripts and metabolites in the purine and pyrimidine pathways (Supporting Information Table S3; Fig. S2). Multiple genes in the *T. pseudonana* purine metabolic pathway had significantly different numbers of transcripts under P-replete compared to P-deficient conditions (Fig. 3; Supporting Information Fig. S2A; Table S3). One of the most highly expressed genes under P-deficient conditions was a phosphohydrolytic 5′-nucleotidase that cleaves phosphate from NMPs to generate the nucleosides. In addition, there were significantly higher numbers of transcripts under P-replete conditions for an adenosuccinate synthase, which catalyzes the conversion of IMP to AMP, and adenosine kinase, which controls production of AMP from adenine (Fig. 3). These patterns suggest that diatoms transcriptionally control nucleotide production under P-replete conditions through the de novo biosynthesis and salvage pathways, and scavenge P from nucleotides under P-deficient conditions.

Nucleotide concentrations in other phytoplankton are affected by P deficiency

We examined the generality of our observations in other eukaryotic phytoplankton by quantifying AMP and adenosine in cultures of *M. pusilla* and *A. anophagefferens* under P-replete and P-deficient conditions (Table 2; Fig. 4). Similar to *T. pseudonana*, the relative concentrations of AMP in *A. anophagefferens* and *M. pusilla* increased under P-replete conditions, while adenosine content increased under P-deficient conditions. Both species upregulate a 5′-nucleotidase gene (Wurch et al. 2011b; Whitney and Lomas 2016) under P deficiency, parallel to the response seen in *T. pseudonana*. In *A. anophagefferens*, the 5′-nucleotidase protein product is also up regulated (Wurch et al. 2011a). In cultures of *M. pusilla*, a cosmopolitan prasinophyte, we observed the same shifts in AMP and adenosine but over a narrower range, suggesting either intrinsic variability in the magnitude of the metabolite restructuring between taxa or variability as a function of relative levels of P deficiency. In summary, despite unavoidable variability in culture conditions across laboratories, P deficiency consistently resulted in a relative decrease in AMP, an increase in adenosine, and an increase in transcripts for a 5′-nucleotidase that could dephosphorylate AMP to produce adenosine.

Liberation of P from NMP dephosphorylation provides a significant quantity of P under P deficiency

Although critical to central metabolism, nucleotide-bound P has typically been viewed as an insignificant component of cellular P relative to P-containing biopolymers

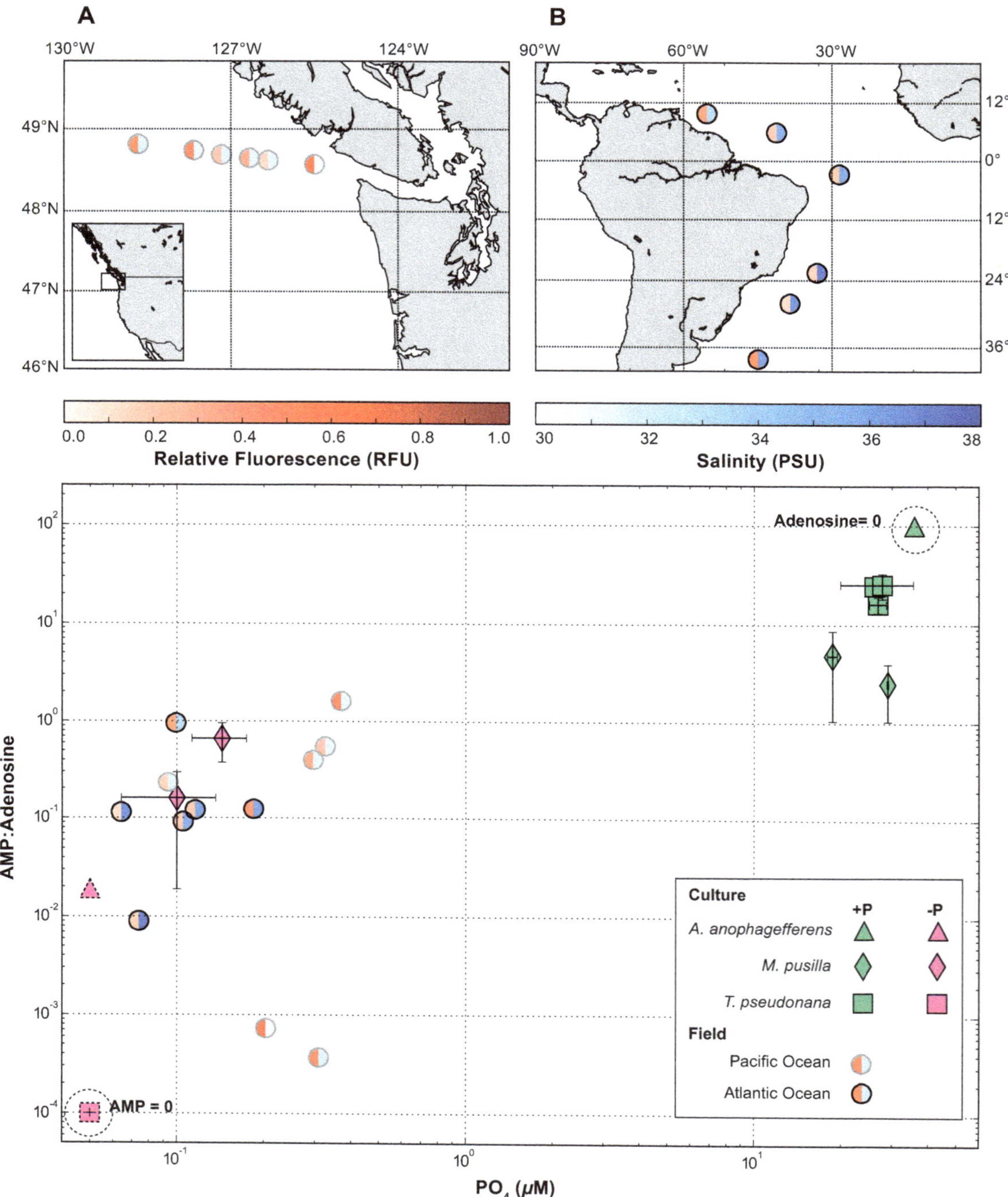

Fig. 4. A comparison of AMP to adenosine ratios from cultures and field sites as a function of the phosphate concentration. (**A**) Field sample locations (gray-outlined circles) collected on 1.6-μm GF/A filters from 5 m depth in the Pacific Ocean. Half of each circle shows the salinity (practical salinity units (PSU): right) and the relative fluorescence (RFU: left), as measured during the downward CTD cast at 5 m. The scales are provided for each measurement in the colorbars. (**B**) Field sample locations (black-outlined circles) collected on 0.7-μm GF/F filters from 5 m depth in the Atlantic Ocean. Same colors as in (**A**). (**C**) AMP to adenosine ratios in field (circles) and culture experiments for *A. anophagefferens* (triangles), *M. pusilla* (diamonds), and *T. pseudonana* (squares) grown under P-replete (green) and P-deficient (pink) conditions. Dashed outlines indicate estimated phosphate concentrations at the method detection limit. A log–log scale is presented to allow all data to be visualized appropriately on one figure. Where available, the mean and one standard deviation of biological replicates at a time-point are presented. Zero concentration values for adenosine and AMP are highlighted.

Table 2. Ratios of AMP to adenosine in all datasets, calculated as moles AMP divided by moles adenosine.

		P treatment	AMP: adenosine range	Mean (± SD)
Cultures	*T. pseudonana*	P-replete	14–32	23 (±6.5) *n*=5
		P-deficient	0	0 *n*=6
	M. pusilla	P-replete	1.5–8.8	3.9 (±3.1) *n*=5
		P-deficient	0–1	0.25 (±0.39) *n*=6
	A. anophagefferens	P-replete	∞ (adenosine b.d.)	n/a
		P-deficient	0.019	n/a
Field	Atlantic Ocean*	0.06–0.18 μM	0.01–0.96	0.26 (±0.39) *n*=6
	Pacific Ocean*	0.09–0.37 μM	0.0–1.6	0.47 (±0.61) *n*=6

* The range of dissolved inorganic P concentrations across stations.
b.d., below detection. n/a, not applicable.

such as nucleic acids. If we assume that P is liberated by NMP de-phosphorylation only once within a growth cycle, de-phosphorylation of the observed NMPs liberates ~ 0.08–0.4% of total cellular P (depending on P cell quota; Supporting Information Calculation S2). However, we posit that the flux through a minimal standing concentration of NMPs will need to be high to maintain P homeostasis in P-deficient cells, thus liberating P many times over a growth cycle. We calculate the total amount of P moving through this cycle by first defining the amount of P needed to replicate cell biomass with the observed doubling time (23 days). If we use 0.67 fmol P as our estimate for P cellular quota (Supporting Information Calculation S2), a cell must assimilate 0.02 fmol P day^{-1} to acquire the necessary P. In comparison, the empirical relationship presented in Perry (1976) estimates that P-deficient diatoms assimilate P at a rate of 0.2 fmol day^{-1}. If we conservatively assume no recycling of nucleic acids, the difference between the assimilation rate predicted by Perry (1976) and the minimal assimilation rate calculated here represents the amount of phosphate liberated through this pathway, or (0.2–0.02 fmol) day^{-1} * 23 days = 4 fmol P. We examined the sensitivity of this calculation to growth rate as well as cellular P quota. For the higher cell quota of 2 fmol P and the observed growth rate, the amount of P available would be 2.5 fmol P. Using a higher growth rate (doubling time = 10 days; Dyhrman et al. 2012), the amount of P available would 3 fmol P, assuming the lower P-deficient cell P quota. Only when we combine the higher cell quota with the higher growth rate do we see negligible liberation of P via this mechanism. Given the uncertainties in P-deficient cell quotas and growth rates for P-deficient cells, we estimate that the total P released through rapid NMP dephosphorylation could be the same order of magnitude as the structural P-sparing strategies of lipid substitution (3 fmol; Martin et al. 2011).

Field AMP and adenosine concentrations are similar to laboratory observations for P deficiency

If this is a widespread adaptation to P deficiency in the eukaryotic phytoplankton, then the nucleotide to nucleoside ratio should be low in oligotrophic (low P) environments when eukaryotic phytoplankton are present. We quantified adenosine and AMP using targeted metabolomics in particulate samples collected from 5-m depth in the Atlantic and Pacific Oceans. We then compared AMP to adenosine ratios in field and culture samples as a function of inorganic P concentrations (Fig. 4). In the *T. pseudonana* culture, AMP: adenosine is representative of the shift observed in all nucleotides between the two P conditions. In the field, this ratio is a more tractable measurement than the adenylate charge (ATP + ½ ADP)/(ATP + ADP + AMP) due to the challenge of quenching metabolism quickly enough (within ~ 30 s) to accurately preserve this intracellular ratio (Villas-Boas and Bruheim 2007). Sample filters contained eukaryotes such as diatoms, haptophytes, dinoflagellates, and picoeukaryotes, as well as cyanobacteria and particle-associated bacteria. Even with this increase in microbial diversity in filter samples, we observed AMP-to-adenosine ratios that were quite similar to those observed for our laboratory cultures under P-deficient conditions. Higher field AMP: adenosine values could be due to many factors, including species-derived variability in AMP: adenosine and broad differences in nutrient distributions. Indeed, the highest ratio in the Atlantic Ocean was sampled in the Amazon River plume, which was characterized by high nutrients, low salinities and high surface chlorophyll fluorescence relative to other sampling sites (Fig. 4). Further work over a range of P environments would help confirm the consistency of the AMP to adenosine ratios as a function of exogenous P concentration or physiological metrics of P deficiency.

Discussion

The combination of transcripts and metabolites studied here point to a new view of P allocation within P-deficient eukaryotic phytoplankton—specifically, that NMPs serve as a dynamic and flexible pool of P for critical energy balance and P homeostasis. We observed uniformly depleted concentrations of NMPs, relative to nucleosides, under P-deficient growth in laboratory cultures of three model phytoplankton. These metabolic shifts appear to be transcriptionally controlled, and the results imply that nucleotides are critical to P homeostasis. Dyhrman et al. (2012) previously attributed the strong pattern of 5′-nucleotidase induction in *T. pseudonana* to the utilization of exogenous DOP. However, the metabolite profiles in this study suggest that the 5′-nucleotidase is acting on intracellular substrates, at least in part. The prevalence of this metabolic shift is further supported by previous studies in marine eukaryotic phytoplankton, where P-deficient conditions were associated with the induction of transcripts and proteins encoding 5′-nucleotidases (Wurch et al. 2011*a,b*; Dyhrman et al. 2012; Forster et al. 2014; Whitney and Lomas 2016) as well as enzyme activities with an affinity for nucleotides (Dyhrman and Palenik, 2003; Xu et al. 2013), and enzymes of the purine and pyrimidine pathways (Yang et al. 2014; Feng et al. 2015). Given that variations in physiology can occur between experiments with the same species, the degree of synergy in these disparate data types from separate experiments is striking, and may underscore that phosphate scavenging from NMPs is not a transient response, but one that is a strong and sustained reflection of a low P environment.

The P-modulated changes in the pyrimidine and purine pathways observed herein may be present even more broadly in microbial eukaryotes, as the accumulation of nucleosides was also observed in studies with yeast, where P-deficient growth led to higher cell-normalized concentrations of adenosine, inosine, and other nucleosides (Boer et al. 2010; Ljungdahl and Daignan-Fornier 2012). Further, a nucleotidase that scavenges the phosphate group from NMPs derived from RNA degradation is central to yeast's survival under P deficiency (Xu et al. 2013). We calculated that movement of P through a vanishingly small NMP pool could provide a quantitatively significant source of P to prioritize the most critical cellular processes. Indeed, the conversion between ADP and ATP is the central energy currency of the cell and these reactions must be maintained, even in the face of P deficiency. Studies in yeast, a model eukaryote, show that ATP levels are affected by P deficiency, but are never fully depleted (Boer et al. 2010). In summary, P deficiency induces shifts between NMPs and nucleosides among disparate eukaryotic taxa, suggesting that scavenging of P from NMPs may be common and an evolutionarily conserved feature of P homeostasis in microbial eukaryotes. Although 5′-nucleotidases in marine cyanobacteria and heterotrophic bacteria are not generally upregulated as a function of P deficiency (Ammerman and Azam 1985; Tammienen 1989), recent metabolic modeling work implicates P recycling mechanisms such as this in the response of the cyanobacterium *Prochlorococcus marinus* to P deficiency (Casey et al. 2016).

The environmental shift in the particulate nucleotide to nucleoside ratios is likely to concomitantly impact dissolved organic matter, as some fraction of these and other cellular metabolites will be exuded by the cell during growth, viral lysis or microzooplankton grazing (Moran et al. 2016). Dissolved concentrations of triphosphate nucleotides (e.g., ATP) have been measured in seawater for decades (Azam and Hodson 1977; Björkman and Karl 2001) as a small fraction of the DOP standing stock (Björkman et al. 2000). In contrast, few measurements of AMP are available (Karl and Holm-Hansen 1978) and thus its contribution to DOP is unknown. In general, nucleotides (mono-, di-, and tri-phosphate) are highly bioavailable to both marine prokaryotic and eukaryotic microbes, particularly when phosphate concentrations are low (Azam and Hodson 1977; Nawrocki and Karl 1989; Björkman and Karl 1994). The metabolomics data presented here suggests, however, that these highly labile molecules are depleted in eukaryotic phytoplankton under low P and thus would not enter the dissolved organic matter pool to an appreciable extent.

Low-P environments in both the Atlantic and the Pacific exhibit similar AMP-to-adenosine ratios to those in culture, suggesting that scavenging of nucleotide phosphate is occurring in eukaryotic phytoplankton field populations. Incubation studies are needed to assess the temporal sensitivity of this ratio to P concentrations, the flux of P through NMPs, and the reversibility of low AMP to adenosine ratios under P addition. Despite these uncertainties, the similarity of field ratios with those in our culture experiments indicates that NMPs may play a more important role in environmental P homeostasis in marine eukaryotic phytoplankton than was previously appreciated. When P is low, cells need a flexible storage pool from which to distribute P quickly among cellular phosphorylation processes such as energy balance and signal transduction. NMPs could provide such a pool of P, allowing eukaryotic cells to use the NMP pool as their wallet, living from paycheck to paycheck, redistributing their P currency as needed to maintain cellular processes.

References

Altschul, S. F., T. L. Madden, A. A. Schäffer, J. Zhang, Z. Zhang, W. Miller, and D. J. Lipman. 1997. Gapped BLAST and PSI-BLAST: A new generation of protein database search programs. Nucleic Acids Res. **25**: 3389–3402. doi: 10.1093/nar/25.17.3389

Ammerman, J. W., and F. Azam. 1985. Bacterial 5'-nucleotidase in aquatic ecosystems: A novel mechanism of phosphorus regeneration. Science **227**: 1338–1340. doi: 10.1126/science.227.4692.1338

Armbrust, E. V., and others. 2004. The genome of the diatom Thalassiosira pseudonana: Ecology, evolution, and metabolism. Science **306**: 79–86. doi:10.1126/science.1101156

Azam, F., and R. E. Hodson. 1977. Dissolved ATP in the sea and its utilisation by marine bacteria. Nature **267**: 696–698. doi:10.1038/267696a0

Björkman, K. M., and D. M. Karl. 1994. Bioavailability of inorganic and organic phosphorus compounds to natural assemblages of microorganisms in Hawaiian coastal waters. Mar. Ecol. Prog. Ser. **111**: 265–273. doi:10.3354/meps111265

Björkman, K. M., A. L. Thomson-Bulldis, and D. M. Karl. 2000. Phosphorus dynamics in the North Pacific subtropical gyre. Aquat. Microb. Ecol. **22**: 185–198. doi: 10.3354/ame022185

Björkman, K. M., and D. M. Karl. 2001. A novel method for the measurement of dissolved adenosine and guanosine triphosphate in aquatic habitats: Applications to microbial ecology. J. Microbiol. Methods **47**: 159–167. doi: 10.1016/S0167-7012(01)00301-3

Boer, V. M., C. A. Crutchfield, P. H. Bradley, D. Botstein, and J. D. Rabinowitz. 2010. Growth-limiting intracellular metabolites in yeast growing under diverse nutrient limitations. Mol. Biol. Cell **21**: 198–211. doi:10.1091/mbc.E09-07-0597

Bromke, M. A., P. Giavalisco, L. Willmitzer, and H. Hesse. 2013. Metabolic analysis of adaptation to short-term changes in culture conditions of the marine diatom *Thalassiosira pseudonana*. PLoS One **8**: e67340. doi:10.1371/journal.pone.0067340

Casey, J. R., A. Mardinoglu, J. Nielsen, and D. M. Karl. 2016. Adaptive evolution of phosphorus metabolism in *Prochlorococcus*. mSystems **1**: e00065–00016. doi:10.1128/mSystems.00065-16

Dettmer, K., P. A. Aronov, and B. D Hammock. 2007. Mass spectrometry-based metabolomics. Mass Spectrom. Rev. **26**: 51–78. doi:10.1002/mas.20108

Dittmar, T., B. Koch, N. Hertkorn, and G. Kattner. 2008. A simple and efficient method for the solid-phase extraction of dissolved organic matter (SPE-DOM) from seawater. Limnol. Oceanogr. Methods **6**: 230–235. doi: 10.4319/lom.2008.6.230

Dyhrman, S. T. 2016. Nutrients and their acquisition: Phosphorus physiology in microalgae, p. 155–183. *In* J. Beardall, M. Borowitzka, and J. Raven [eds.], Developments in applied phycology. Springer.

Dyhrman, S. T., and B. Palenik. 2003. Characterization of ectoenzyme activity and phosphate-regulated proteins in the coccolithophorid *Emiliana huxleyi*. J. Plankton Res. **25**: 1215–1225. doi:10.1093/plankt/fbg086

Dyhrman, S. T., and others. 2012. The transcriptome and proteome of the diatom *Thalassiosira pseudonana* reveal a diverse phosphorus stress response. PLoS One **7**: e33768. doi:10.1371/journal.pone.0033768

Feng, T. -Y., and others. 2015. Examination of metabolic responses to phosphorus limitation via proteomic analyses in the marine diatom *Phaeodactylum tricornutum*. Sci. Rep. **5**: 10373. doi:10.1038/srep10373

Forster, R. R., and others. 2014. Extensive liquid meltwater storage in firn within the Greenland ice sheet. Nat. Geosci. **7**: 95–98.

Hockin, N. L., T. Mock, F. Mulholland, S. Kopriva, and G. Malin. 2012. The response of diatom central carbon metabolism to nitrogen starvation is different from that of green algae and higher plants. Plant Physiol. **158**: 299–312. doi:10.1104/pp.111.184333

Johnson, W. M., M. C. Kido Soule, and E. B. Kujawinski. 2016. Evidence for quorum sensing and differential metabolite production in response to DMSP by a marine heterotrophic bacterium. ISME J. **10**: 2304–2316. doi:10.1038/ismej.2016.6

Karl, D. M. 2014. Microbially mediated transformations of phosphorus in the sea: New views of an old cycle. Ann. Rev. Mar. Sci. **6**: 279–337. doi:10.1146/annurev-marine-010213-135046

Karl, D. M., and O. Holm-Hansen. 1978. Methodology and measurement of adenylate energy charge ratios in environmental samples. Mar. Biol. **48**: 185–197. doi:10.1007/BF00395018

Kido Soule, M. C., K. Longnecker, W. M. Johnson, and E. B. Kujawinski. 2015. Environmental metabolomics: Analytical strategies. Mar. Chem. **177**: 374–387. doi:10.1016/j.marchem.2015.06.029

Ljungdahl, P. O., and B. Daignan-Fornier. 2012. Regulation of amino acid, nucleotide, and phosphate metabolism in *Saccharomyces cerevisiae*. Genetics **190**: 885–929. doi: 10.1534/genetics.111.133306

Luo, W., and C. Brouwer. 2013. Pathview: An R/Bioconductor package for pathway-based data integration and visualization. Bioinformatics **29**: 1830–1831. doi:10.1093/bioinformatics/btt285

Martin, P., B. A. S. Van Mooy, A. Heithoff, and S. T. Dyhrman. 2011. Phosphorus supply drives rapid turnover of membrane phospholipids in the diatom *Thalassiosira pseudonana*. ISME J. **5**: 1057–1060. doi:10.1038/ismej.2010.192

Montagnes, D. J. S., J. A. Berges, P. J. Harrison, and F. J. R. Taylor. 1994. Estimating carbon, nitrogen, protein, and chlorophyll *a* from volume in marine phytoplankton. Limnol. Oceanogr. **39**: 1044–1060. doi:10.4319/lo.1994.39.5.1044

Moran, M. A., and others. 2016. Deciphering ocean carbon in a changing world. Proc. Natl. Acad. Sci. USA **113**: 3143–3151. doi:10.1073/pnas.1514645113

Nawrocki, M. P., and D. M. Karl. 1989. Dissolved ATP turnover in the Bransfield Strait, Antarctica during a spring bloom. Mar. Ecol. Prog. Ser. **57**: 35–44. doi:10.3354/meps057035

Patti, G. J., O. Yanes, and G. Siuzdak. 2012. Metabolomics: The apogee of the omics trilogy. Nat. Rev. Mol. Cell Biol. **13**: 263–269. doi:10.1038/nrm3314

Perry, M. J. 1976. Phosphate utilization by an oceanic diatom in phosphorus-limited chemostat culture and in the oligotrophic waters of the central North Pacific. Limnol. Oceanogr. **21**: 88–107. doi:10.4319/lo.1976.21.1.0088

Poulson-Ellestad, K. L., C. M. Jones, J. Roy, M. R. Viant, F. M. Fernández, J. Kubanek, and B. L. Nunn. 2014. Metabolomics and proteomics reveal impacts of chemically mediated competition on marine plankton. Proc. Natl. Acad. Sci. USA **111**: 9009–9014. doi:10.1073/pnas.140213011

Rabinowitz, J. D., and E. Kimball. 2007. Acidic acetonitrile for cellular metabolome extraction from *Escherichia coli*. Anal. Chem. **79**: 6167–6173. doi:10.1021/ac070470c

Storey, J. D., and R. Tibshirani. 2003. Statistical significance for genomewide studies. Proc. Natl. Acad. Sci. USA **100**: 9440–9445. doi:10.1073/pnas.1530509100

Tammienen, T. 1989. Dissolved organic phosphorus regeneration by bacterioplankton: 5′-nucleotidase activity and subsequent phosphate uptake in a mesocosm enrichment experiment. Mar. Ecol. Prog. Ser. **58**: 89–100. doi:10.3354/meps058089

Van Mooy, B. A. S., and others. 2009. Phytoplankton in the ocean use non-phosphorus lipids in response to phosphorus scarcity. Nature **458**: 69–72. doi:10.1038/nature07659

Villas-Boas, S. G., and P. Bruheim. 2007. Cold glycerol-saline: The promising quenching solution for accurate intracellular metabolite analysis of microbial cells. Anal. Biochem. **370**: 87–97. doi:10.1016/j.ab.2007.06.028

Whitney, L. P., and M. W. Lomas. 2016. Growth on ATP elicits a P-stress response in the picoeukaryote *Micromonas pusilla*. PLoS One **11**: e0155158. doi:10.1371/journal.pone.0155158

Wu, Z., B. D. Jenkins, T. A. Rynearson, S. T. Dyhrman, M. A. Saito, M. Mercier, and L. P. Whitney. 2010. Empirical bayes analysis of sequencing-based transcriptional profiling without replicates. BMC Bioinform **11**: 564. doi:10.1186/1471-2105-11-564

Wurch, L. L., E. M. Bertrand, M. A. Saito, B. A. S. Van Mooy, and S. T. Dyhrman 2011a. Proteome changes driven by phosphorus deficiency and recovery in the brown tide-forming alga *Aureococcus anophagefferens*. PLoS One **6**: e28949. doi:10.1371/journal.pone.0028949

Wurch, L. L., S. T. Haley, E. D. Orchard, C. J. Gobler, and S. T. Dyhrman. 2011b. Nutrient-regulated transcriptional response in the brown tide-forming alga *Aureococcus anophagefferens*. Environ. Microbiol. **13**: 468–481. doi:10.1111/j.1462-2920.2010.02351.x

Xu, Y. -F., and others. 2013. Nucleotide degradation and ribose salvage in yeast. Mol. Syst. Biol. **9**: 665. doi:10.1038/msb.2013.21

Yang, Z.-K., J.-W. Zheng, Y.-F. Niu, W.-D. Yang, J.-S. Liu, and H.-Y. Li. 2014. Systems-level analysis of the metabolic responses of the diatom *Phaeodactylum tricornutum* to phosphorus stress. Environ. Microbiol. **16**: 1793–1807. doi:10.1111/1462-2920.12411

Acknowledgments

We thank C. Breier, G. Swarr, E. Melamund, and G. Hennon for assistance with laboratory work and data analysis. We thank J. Jennings (Oregon State Univ.) and the Armbrust laboratory (Univ. of Washington) for the Atlantic and Pacific inorganic P concentrations, respectively. We thank M. Kido Soule and the WHOI FT-MS Users' Facility for metabolomics analysis. We thank the scientific parties, captains and crews of the R/V *Knorr* and the R/V *Thomas Thompson* for assistance in acquiring the Atlantic and Pacific field samples, respectively. The cruises were funded by the National Science Foundation (Atlantic Ocean: OCE-1154320 to EBK and KL; Pacific Ocean: OCE-1205233 to EV Armbrust). This research was funded by the Gordon and Betty Moore Foundation (Grant 3304 to EBK), the Simons Foundation (SCOPE award ID 329108 to STD), and the National Science Foundation Chemical and Biological Oceanography Programs (OCE-01316036 and OCE-0723667 to STD). HA was supported by a WHOI Ocean Life Institute Fellowship and WMJ was supported by a National Defense Science and Engineering Fellowship.

9

Finding patches in a heterogeneous aquatic environment: pH-taxis by the dispersal stage of choanoflagellates

Gastón L. Miño,[1,2] *M. A. R. Koehl,*[3] *Nicole King,*[4] *Roman Stocker*[1,5]*

[1]Ralph M. Parsons Laboratory, Department of Civil and Environmental Engineering, Massachusetts Institute of Technology, Cambridge, Massachusetts; [2]Laboratorio de Microscopía Aplicada a Estudios Moleculares y Celulares (LAMAE), Centro de Investigaciones y Transferencia de Entre Ríos (CITER), Facultad de Ingeniería, Universidad Nacional de Entre Ríos (FIUNER), Oro Verde, Argentina; [3]Department of Integrative Biology, University of California, Berkeley, California; [4]Howard Hughes Medical Institute and the Department of Molecular and Cell Biology, University of California, Berkeley, California; [5]Institute of Environmental Engineering, Department of Civil, Environmental, and Geomatic Engineering, ETH Zurich, Zurich, Switzerland

Scientific Significance Statement

Microbial eukaryotes that feed on bacteria are critical links in aquatic food webs. Because the aquatic environment is often patchy at the microscale, feeding of microbial eukaryotes can depend strongly on their ability to find patches of concentrated prey. Choanoflagellates provide a system in which feeding performance can be studied in the same species for different life forms—slow swimmers, fast swimmers, and colonies—yet, it has remained unknown whether choanoflagellates are capable of directed motion ("taxis") to find prey patches. Here, we report for the first time that choanoflagellates are capable of taxis toward pH 6–7, which is characteristic of the acidification caused by concentrated patches of bacteria. This behavior is limited to the dispersal form of choanoflagellates—fast swimmers—and may allow them to exploit prey-rich microhabitats.

Abstract

Microbial eukaryotes that feed on bacteria are critical links in aquatic food webs. We used the choanoflagellate *Salpingoeca rosetta*, whose life history includes fast- and slow-swimming unicellular forms and multicellular colonies, to study performance of these different forms at finding and accumulating in patches of prey. Video microscopy of behavior in microfluidic experiments showed that only unicellular fast swimmers exhibited taxis into patches of attractant. Of the chemical cues tested, only pH affected behavior: fast swimmers moved toward patches of water that had pH in the range of 6–7 from water with higher or lower pH. Since bacteria can lower the pH of the water in their vicinity, we suggest that seeking regions with a moderately more acidic pH than average seawater helps the fast-swimming dispersal stage of choanoflagellates find particles or surfaces rich in prey.

*Correspondence: romanstocker@ethz.ch

Author Contribution Statement: All authors designed the research and wrote the paper. GLM performed the main experiments and the data analysis; MARK, NK, and RS conducted pilot experiments; MARK and NK provided biological advice for experimental procedures; RS provided advice for microfluidic approaches.

Heterotrophic microbial eukaryotes that feed on bacteria are important components of food webs in marine (Azam et al. 1983; Worden et al. 2015) and freshwater systems (Tikhonenkov and Mazei 2008). The feeding performance of microbial eukaryotes depends not only on the rate at which they capture bacteria from the surrounding water, but also on their ability to find patches of prey. Choanoflagellates provide a system in which the consequences of different life forms on feeding performance can be studied in the same species. The life history of the best-studied choanoflagellate, *Salpingoeca rosetta*, includes at least three unicellular forms ("slow swimmers," "fast swimmers," "thecate cells") and two multicellular forms ("rosettes," "chains") (Dayel et al. 2011). While the ecological relevance of these life stages is not yet well understood, the fast swimmers are typically thought to be a dispersal form (Leadbeater 1983, 2015; Dayel et al. 2011). Yet, whether dispersal is used in conjunction with chemical sensing, and more generally whether choanoflagellates are capable of taxis to locate prey, has remained unknown.

The aquatic environments in which choanoflagellates live are far from homogeneous, with a plethora of nutrient-rich "hotspots" producing microscale heterogeneity (*see* Stocker 2012), including sinking marine particles (Kiørboe and Jackson 2001; Stocker et al. 2008), plumes of dissolved organic matter around phytoplankton cells (Smriga et al. 2016), and nutrient-rich filaments produced by turbulence (Crimaldi and Koseff 2001; Koehl et al. 2007; Taylor and Stocker 2012). Many microorganisms, such as bacteria, are able to utilize these nutrient hotspots by chemotaxis (Stocker et al. 2008). We hypothesize that choanoflagellates may also be capable of chemotaxis to exploit hotspots where bacterial density is high. The specific questions addressed by this study are: (1) Do choanoflagellates show taxis toward patches of high bacterial density?, (2) If so, which of their life stages show the response?, (3) What is the chemical cue for their taxis? Here, we address these questions by using quantitative image analysis of choanoflagellates in microfluidic experiments.

Methods

Growth media and solutions

Artificial seawater (ASW; 32.9 g L^{-1} of Tropic Marine sea salts in Milli-Q water) was prepared with pH of 8.0 ± 0.1 and salinity of 32 parts per thousand (King et al. 2009). Cereal grass media (CGM; 5 g L^{-1} of cereal grass pellets from Basic Sciences Supplies infused in ASW) was diluted in ASW (10% vol/vol) and used to grow rosettes (ATCC PRA-366) and a rosette-free culture (Dayel et al. 2011). Seawater complete medium (SWC; 24 g L^{-1} of Tropic Marine sea salts, 5 g L^{-1} peptone, 3 g L^{-1} yeast extract, 3 mL glycerol in Milli-Q water) was diluted in ASW (5% vol/vol) to grow *S. rosetta* fed with *Echinicola pacifica* (SrEpac, ATCC PRA-390) (Levin and King 2013).

2X phosphate buffered saline (PBS; diluted from a 10X stock, pH = 7.2 ± 0.1; Sigma-Aldrich) was used to wash bacteria and as a chemoattractant in the taxis experiments. Choanoflagellate-conditioned medium (CM) was obtained by centrifuging the choanoflagellate culture, then filtering with a 0.2 μm syringe filter (VWR, Radnor). To match the concentration of potassium in PBS, KCl was added to ASW to reach a potassium concentration of $\sim$ 10 mM (final pH = 8.2). Similarly, NaH_2PO_4 was added to ASW to match the concentration of PO_4 in PBS ($\sim$ 20 mM) and NaOH was used to maintain the pH at 8. Addition of NaOH and HCl was used to increase or decrease the pH of ASW, respectively (Supporting Information Fig. S2a).

Choanoflagellate cultures

Different cultures of *S. rosetta* were used to test taxis by different cell types (Dayel et al. 2011).

i. Rosettes: A monoxenic strain (ATCC PRA-366, *S. rosetta* and *Algoriphagus machipongonensis*) containing a high concentration of rosettes was maintained by diluting 1 mL of choanoflagellate suspension into 9 mL of fresh 10% CGM every 5–7 d at room temperature.
ii. Fast and slow swimmers: Two protocols were used to control the ratio of fast to slow swimmers in culture (Supporting Information Fig. S1):

a. ATCC PRA-390 (SrEpac) cultures produce high concentrations of slow swimmers (Levin et al. 2014) and they were maintained by diluting 1 mL of culture into 9 mL fresh 5% SWC medium every 4–5 d at room temperature. Experiments were performed after the SrEpac cultures reached stationary phase and cell concentrations were $(4 \pm 1) \times 10^6$ cells mL^{-1}. Choanoflagellate concentrations were determined using a hemocytometer (Bright-Line, Hausser Scientific).
b. Rosette-free cultures (Dayel et al. 2011) produce a higher percentage of fast swimmers from thecate cells. A flask containing thecate cells was washed twice with ASW and refilled with 15 mL of 10% CGM 36–48 h before experiments were conducted. In order to reach concentrations of $(1.5 \pm 0.5) \times 10^6$ cells mL^{-1}, 10 mL of culture was centrifuged for 10 min at 1000 $\times$ g and the cell pellet was resuspended in 1 mL of CM. Thecate cell cultures were maintained by diluting 1 mL of a suspension of fast swimmers in 9 mL of 10% CGM every 4–5 d and incubated at room temperature.

Microinjector assay to measure taxis

With standard soft lithography, we fabricated microinjector channels (Seymour et al. 2008). This device consists of a 3 mm wide, 0.1 mm deep channel that injects the chemoattractant into a larger channel (Fig. 1a,b). Imposing a flow allows the formation of two regions within the channel: a central band (1/5 of the channel width) containing the

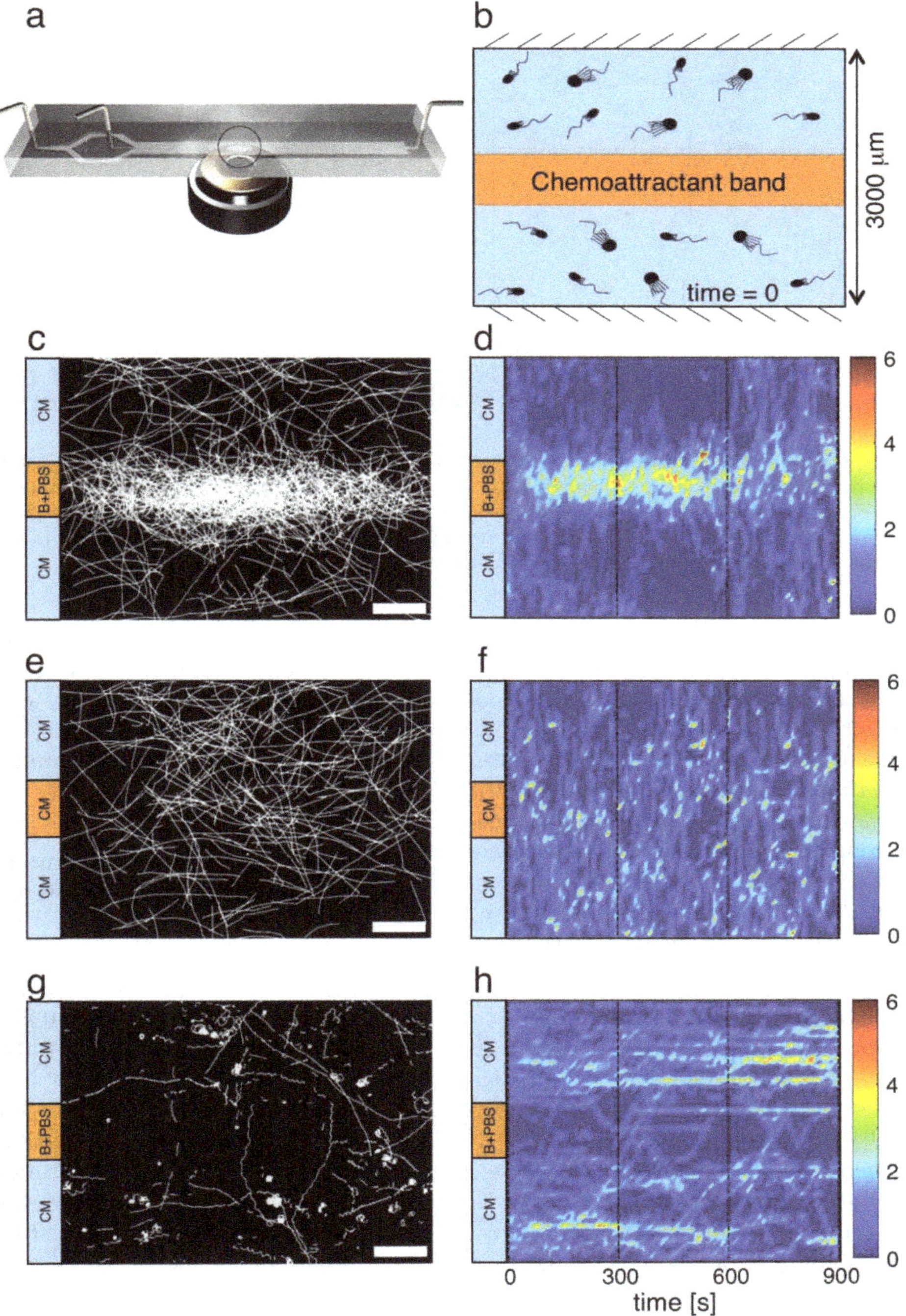

Fig. 1. Chemotaxis by the choanoflagellate *S. rosetta.* (a) Schematic view of the microfluidic chemotaxis assay, and (b) zoomed-in view of the observation region. The device consists of a microfluidic injector through which chemoattractant was injected as a 600-μm wide band (through the middle inlet in panel a), into a channel otherwise filled with a suspension of choanoflagellate cells (injected through the left-most inlet in panel (a)). In panel (b), cells are not drawn to scale (actual length $\sim 5\ \mu$m) and diagonal lines denote the sidewalls of the channel. (c) Trajectories of unicellular choanoflagellates exposed to a band of *A. machipongonensis* bacteria suspended in PBS ("B + PBS"), observed over 900 s after cells were exposed to the band. (d) Normalized distribution (color scale) across the width of the channel (vertical axis) of the cells from the experiment in panel (c), plotted as a function of time after exposure to the central band (in this case, PBS). Note the intense accumulation of choanoflagellates in the central band during the first 300 s, showing strong chemotaxis. At longer times, the chemoattractant band diffuses and cell accumulation ceases. (e) Control experiments showing trajectories of unicellular choanoflagellates exposed to a band of conditioned medium ("CM," prepared by filtration of a choanoflagellate culture; *see* Methods), as in the rest of the chamber, observed over 900 s as in panel (c). (f) Normalized distribution across the width of the channel (vertical axis) of the cells from the experiment in panel (e), plotted as a function of time. Note the absence of accumulation in the center, indicative of a lack of chemotaxis, for this control case. (g) Trajectories of rosettes exposed to a band of *A. machipongonensis* bacteria suspended in PBS, observed over 900 s after cells were exposed to the band. (h) Normalized distribution across the width of the channel (vertical axis) of the cells from the experiment in panel (g), plotted as a function of time. Note the absence of accumulation in the center, indicative of a lack of chemotaxis. Scale bar = 600 μm.

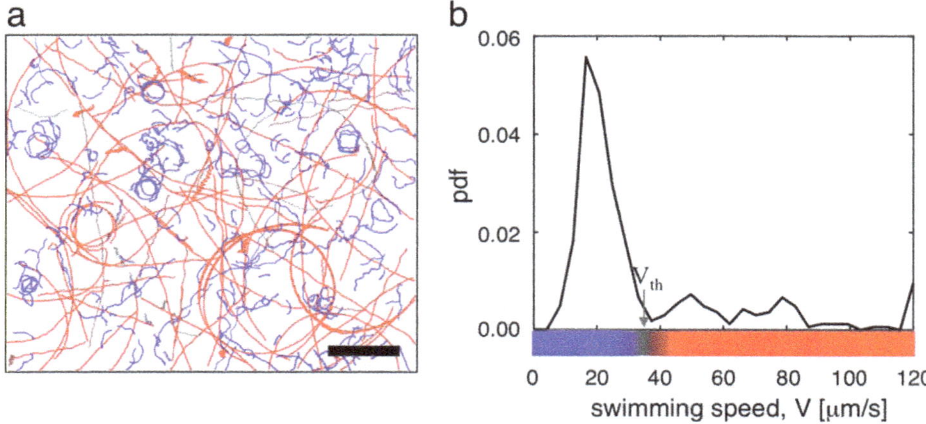

Fig. 2. Diversity of swimming speeds and trajectories in cultures of unicellular *S. rosetta*. (a) Trajectories of choanoflagellates swimming close to the bottom surface of the culture flask, collected at the beginning of stationary phase ($t = 120$ h), over 30 s. Each trajectory is colored by its mean speed (*see* x-axis of panel (b) for color-coding). Note that slow swimmers (blue) followed looping trajectories, whereas faster (red) cells follow slightly curved trajectories that approach linear. Scale bar = 100 μm. (b) Probability density function (pdf) of the swimming speed (a measure of the distribution of swimming speeds within the population) for the same culture ($n = 407$ trajectories).

chemical to be tested, and the region extending on both sides of the band containing the choanoflagellates.

Taxis experiments were performed in triplicate with choanoflagellate concentrations of $\sim 10^6$ cells mL^{-1}. In experiments where bacteria were tested as potential chemoattractants, overnight cultures of *A. machipongonensis* were grown in 10% CGM, centrifuged and washed twice in CM. Moderate variations in the initial cell concentration among experiments are unlikely to have affected results, because even at cell concentrations of $\sim 10^7$ cells mL^{-1} individual cells are ~ 45 μm (i.e., many body lengths) from each other, making cell-to-cell contact unlikely.

Imaging and motility characterization

An inverted microscope (Ti-Eclipse, Nikon) was used with a Neo sCMOS digital camera (Andor Technology) to image cell responses to chemoattractants. Image analysis and cell tracking were performed with ImageJ and TrackMate (NIH) and MATLAB. For chemotaxis characterization, image sequences were acquired at 10 frames s^{-1} for 300–900 s at 10 mm from the tip of the injector using dark-field microscopy and a 4× objective (field of view = 3000 × 4200 μm^2). For motility characterization, image sequences were acquired at 20 frames s^{-1} for 30 s using bright-field microscopy and a 20× objective, near (but not directly on) the bottom of the culture flask. Spatiotemporal diagrams were computed by dividing the channel width into 60 bins and for each bin (area = 50 × 4200 μm^2) the choanoflagellate numbers were averaged over 10-s windows.

The probability distribution function (pdf) of the swimming speed (i.e., the relative occurrence of different speeds within the population) indicated the presence of two types of swimmers, fast and slow (Fig. 2; $n = 407$ trajectories), corresponding to mean speeds of 19.1 ± 7.1 μm s^{-1} and 56.3 ± 41.8 μm s^{-1} (mean ± sd, determined by fitting the pdf with the superposition of two Gaussians). The transition between the two groups occurs at 30–40 μm s^{-1} and we therefore used a threshold of $V_{th} = 35$ μm s^{-1} to compute the fraction of slow and fast swimmers in each sample.

Results

S. rosetta performs directional migration

The chemotaxis experiments were set up in a microfluidic device that generates a narrow (~ 600 μm wide) band of a potential attractant along the midline of a channel 3000 μm wide and 100 μm deep (Seymour et al. 2008). This "band" is established by flowing the hypothesized attractant through a central microinjector and by flowing a suspension of choanoflagellates on either side of the band (Fig. 1a,b; Methods). At time zero, the flow is stopped, allowing the lateral diffusion of the attractant and the creation of attractant gradients to which choanoflagellates can respond by swimming toward the band (attraction) or away from it (repulsion). Chemotaxis was quantified by live imaging and image analysis, recording the positions of choanoflagellates at 10 frames s^{-1}.

Motivated by the fact that bacteria represent a primary nutrient source for choanoflagellates, we first tested whether a suspension of bacteria elicited directional migration by *S. rosetta*. We used a concentrated suspension of *A. machipongonensis*, washed and resuspended in PBS solution, then injected as the central band in the microdevice (Fig. 1b). Unicellular choanoflagellates exhibited strong directional migration into the bacterial suspension (Fig. 1c,d; Supporting Information Movie SV1), accumulating in the central band and indicating that unicellular swimmers of *S. rosetta* are capable of taxis. In contrast, rosettes showed no taxis in response to bacteria suspended in PBS (Fig. 1g,h).

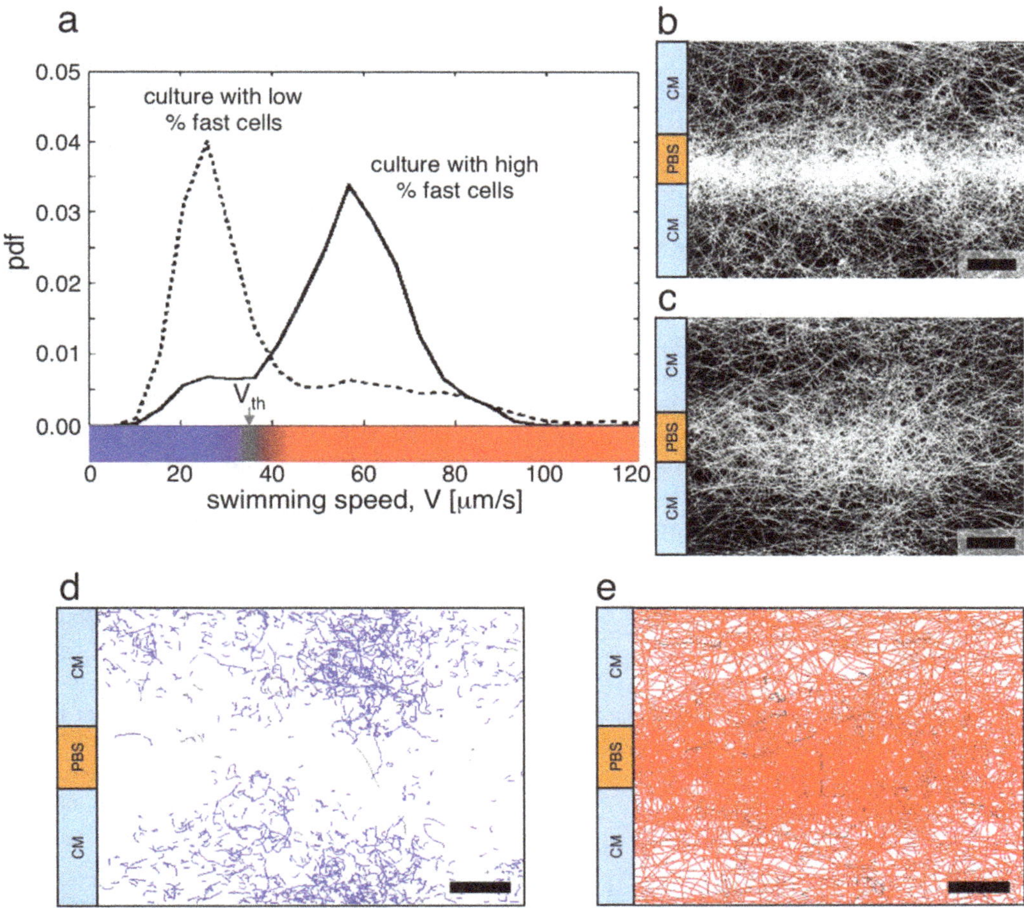

Fig. 3. Taxis is performed by unicellular fast swimmers, but not by slow swimmers. (a) Swimming speed distribution (probability density function, pdf) of individual cells in two cultures with different proportions of fast-swimming choanoflagellate cells, obtained with two different growth protocols (Methods). The arrow indicates the speed threshold, $V_{th} = 35\ \mu m\ s^{-1}$, used to separate fast (red) from slow (blue) cells. In the first culture (dashed line), 11% of cells were fast swimmers, whereas in the second culture (solid line) the fast swimmers represented 87% of all cells. (b) Taxis experiment toward PBS performed with the choanoflagellate culture containing 87% fast swimmers (corresponding to the solid line in panel (a)), observed over 900 s after cells were exposed to the band. (c) Same as (b), but for the choanoflagellate culture containing 11% of fast swimmers (corresponding to the dashed line in panel (a)). Note the reduced accumulation of trajectories in the central band, relative to case (b). (d,e) When the trajectories of slow swimmers and fast swimmers in a culture composed of 11% fast swimmers (dashed line in panel (a)) are plotted separately, it becomes clear that slow swimmers ($V < V_{th}$; panel (d); in blue) fail to accumulate in the PBS band, whereas fast swimmers ($V > V_{th}$; panel (e); in red) swim toward and accumulate in the PBS band. Scale bar = 600 μm in panels (b–e).

To determine which component of the bacterial suspension induced the observed taxis response, we exposed unicellular swimmers to a band of *A. machipongonensis* washed and resuspended in the CM (*see* Methods). No taxis was observed (Supporting Information Fig. S3a,b), suggesting that the bacteria themselves were not the source of the cue. In contrast, when we exposed planktonic, unicellular *S. rosetta* to a band of PBS lacking bacteria, we observed strong taxis, resulting in accumulation of cells within a few tens of seconds (Supporting Information Fig. S4a,b). As a further control, we verified that CM did not cause taxis (Fig. 1e,f). Rosettes did not respond to PBS, to CM, or to bacteria in CM (Supporting Information Fig. S5). Together, these experiments revealed that PBS is sufficient as a source of attraction for unicellular *S. rosetta*.

To quantify the taxis response, we computed the chemotaxis index, $I_C = \mathrm{Cells_{IN}}/\mathrm{Cells_{OUT}}$ (Stocker et al. 2008), which measures the ratio of the cell concentration within the central 600 μm band ($\mathrm{Cells_{IN}}$) to that outside the band ($\mathrm{Cells_{OUT}}$). Values of $I_C > 1$ denote accumulation of cells in the central band (attraction). The chemotaxis index (Supporting Information Fig. S4c) confirms the strength of the response to PBS, with the cells inside the central band reaching a concentration 5.4-fold higher than the average concentration in the channel (mean of I_C between 200 s and 400 s; Supporting Information Fig. S4c). For comparison, the chemotaxis index of *Escherichia coli* did not exceed 2 under nearly optimal conditions (Stocker et al. 2008; Garren et al. 2014).

Fast swimmers, but not slow swimmers or rosettes, perform taxis

To determine which form(s) of unicellular *S. rosetta* perform taxis, we measured swimming speeds to distinguishing fast- from slow-swimming cells (Dayel et al. 2011). The distribution of swimming speeds within different *S. rosetta* cultures (Fig. 2; Methods) was found to display two peaks (Fig. 2b), one above and one below a speed of $V_{th} = 35$ μm s^{-1}, which we chose as the operational threshold speed to separate slow and fast swimmers. The considerable difference in the swimming patterns of cells swimming slower and faster than this threshold (Figs. 2a, 3d,e) further corroborates the distinction of two speed classes.

Two observations revealed that only fast swimmers performed taxis in response to PBS. First, a comparison of the accumulation of cells from two cultures, one with a high and one with a low percentage of fast swimmers (Fig. 3a; Supporting Information Fig. S1; Methods), revealed a considerably stronger accumulation in the former than the latter (Fig. 3b,c). Second, a comparison of the accumulation of slow ($V < V_{th}$) and fast swimmers ($V > V_{th}$) from the same experiment performed in post-processing revealed that slow swimmers did not migrate into the PBS band, even when given ample time (900 s), whereas fast swimmers did (Fig. 3d,e).

The observed taxis is a response to pH gradients and its direction is pH-dependent

A systematic search revealed that pH gradients were responsible for the taxis to PBS. We reached this conclusion by modifying the concentration of particular substances in ASW and using those solutions in the central band, while choanoflagellates suspended in CM were introduced into the rest of the channel. No accumulation was observed in response to ASW containing the same concentration of K^+ (10 mM) or PO_4^{3-} (20 mM) as in PBS (Supporting Information Fig. S3c–f), ruling out K^+ and PO_4^{3-}. The pH of the choanoflagellate cultures (8.0; Methods) was higher than the pH of PBS (7.0), so we next investigated pH as the cue. When the pH of ASW (8.0) was adjusted by addition of hydrochloric acid (HCl) to 7.0 and used as the candidate attractant, *S. rosetta* accumulated in the central band to levels comparable to those elicited by PBS (Supporting Information Fig. S4). This is also apparent in the values of the chemotaxis index, which in both cases reached values considerably above 1, over a comparable time (100–200 s) (Supporting Information Fig. S4a,c). Although we cannot rule out that additional components of PBS also elicited taxis, these results indicate that pH differences are sufficient to elicit taxis. Such directional motion of cells along pH gradients has previously been observed for bacteria (Seymour and Doetsch 1973; Tsang et al. 1973; Croxen et al. 2006; Zhuang et al. 2015), *Dictyostelium discoideum* (Bonner et al. 1985), and fungal and algal zoospores (Morris et al. 1995; Govorunova and Sineshchekov 2005).

The response of *S. rosetta* to different values of pH (from 5.2 to 9.0) was investigated by running experiments in which we changed the pH of the ASW band (Supporting Information Fig. S2a). Cells moved away from a band with pH = 5.2 and accumulated at the sidewalls (Fig. 4c), suggesting repulsion from very low pH. When the pH of the central band was between 6.0 and 7.0, cells swam from the CM (pH = 8.0) to the band (Fig. 4e,f). When the pH was 8.0 both outside the band (CM) and within the band (either CM, Fig. 4b; or ASW with pH = 8.0, Fig. 4g), no accumulation was observed. Moreover, when the pH of the central band was higher (pH = 9.0) than that of CM, no accumulation was observed (Fig. 4h). Together, these experiments indicate that *S. rosetta* fast swimmers swim towards a preferential pH range of 6–7, considerably lower than the typical pH of seawater (pH = 8.18 ± 0.15 [Marion et al. 2011]). Additional analysis revealed that the observed accumulations are a true case of taxis, rather than being caused for example by modulation in swimming speed, and are consistent with a strategy to seek the source of the pH gradient, rather than optimal physiological pH conditions per se.

The cell accumulation resulted from directional swimming, rather than from a local reduction in swimming speed, which would also cause accumulation (Schnitzer 1993). To show this, we measured choanoflagellate swimming speeds at different pH values, and found no change in speed even after 3 h (Fig. 5a, Supporting Information Fig. S2b). Instead, we directly observed directional swimming into the band. Cells swam toward the central band of PBS from both sides of the channel during the first ~ 100 s (shown by blue in the area above the PBS band and by red in the area below it in Fig. 5b). The timescale of 100 s is consistent with the lifetime of the pH gradients. Using the diffusion coefficient of H^+ ions in seawater at 25°C ($D = 9.31 \times 10^{-9}$ m^2 s^{-1} [Yuan-Hui and Gregory 1974]), we solved the one-dimensional diffusion equation across the width of the microchannel to predict the evolution of the pH profile over time (Fig. 5c) (without taking into account the buffer capacity of ASW). This computation showed that after a timescale in the order of 100 s, pH gradients became small across the channel. Thus, the timescale of directional, inward swimming corresponds to the timescale over which pH gradients persisted.

Discussion

We hypothesized that seeking regions with a more acidic pH than average seawater helps choanoflagellates find patches rich in bacterial prey. Bacterial metabolic products can acidify the water when they grow in high concentrations (Anderson et al. 1920; Solé et al. 2000) and bacterial

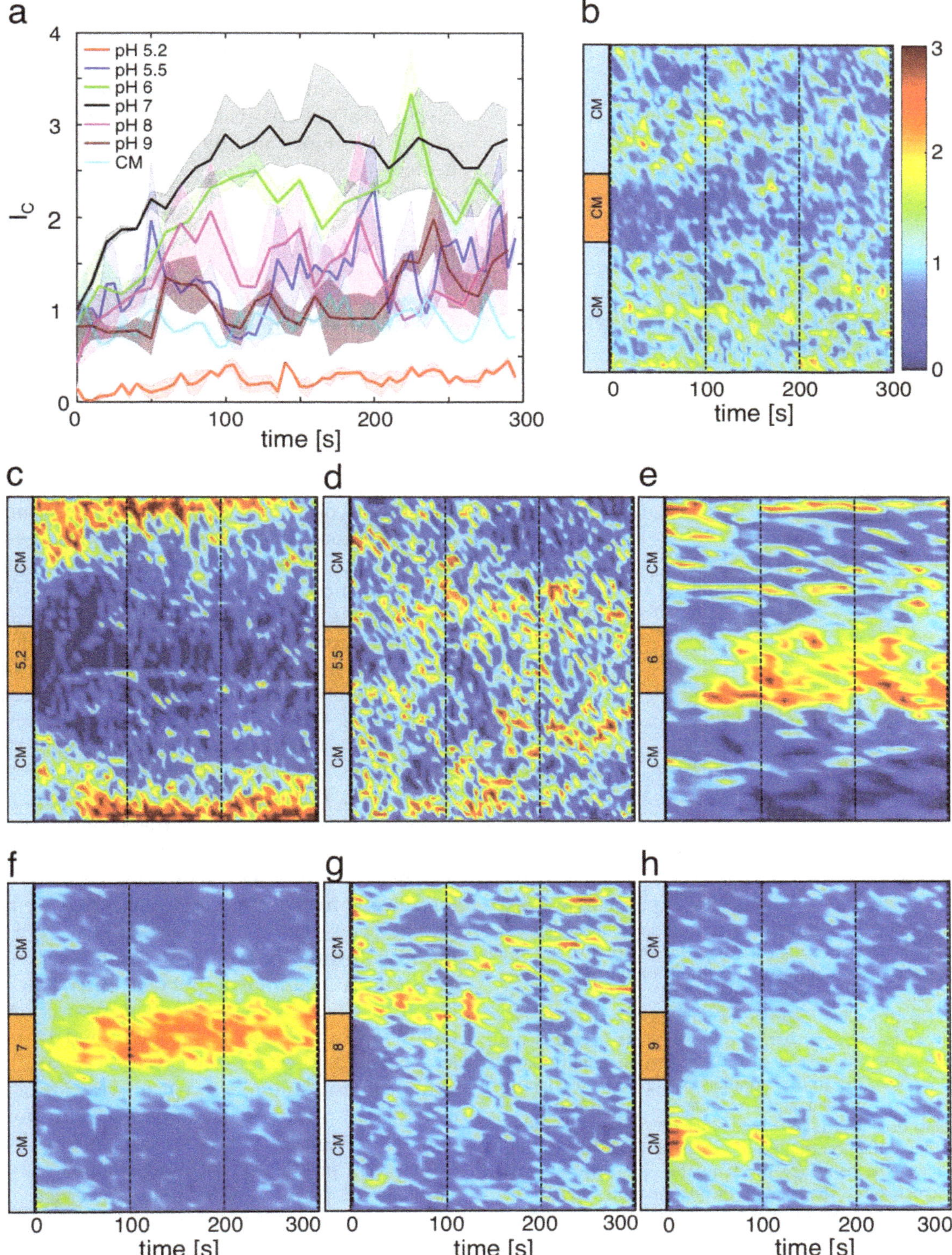

Fig. 4. Unicellular *S. rosetta* perform pH-taxis, a response to gradients in pH. (a) Chemotaxis index (I_C) plotted as a function of time from the onset of an experiment. The index I_C, a measure of the strength of taxis, is the ratio of the number of cells inside the central band (600 μm wide) and the number of cells in the rest of the channel width. $I_C = 1$ denotes no taxis (uniform cell distribution), $I_C > 1$ denotes positive taxis (attraction), and $I_C < 1$ denotes negative taxis (repulsion). Lines and shading represent the mean and SE of three replicate experiments. (b) Distribution of choanoflagellate cells over the channel width (vertical axis) in a control experiment with CM (pH = 8) in the central band. (c–h) Distribution of unicellular choanoflagellates over the channel width relative to a central band of ASW amended to pH 5.2 (c), pH 5.5 (d), pH 6 (e), pH 7 (f), pH 8 (g), and pH 9 (h). Note the strong positive taxis response to pH = 7, the somewhat weaker positive response for pH = 6, the negative response (repulsion) for pH = 5.2, and the absence of a response for other pH values. pH-adjusted ASW (rather than CM) was used in the experiments in panels (c–h) to precisely control pH, since CM may contain additional metabolites that affect pH.

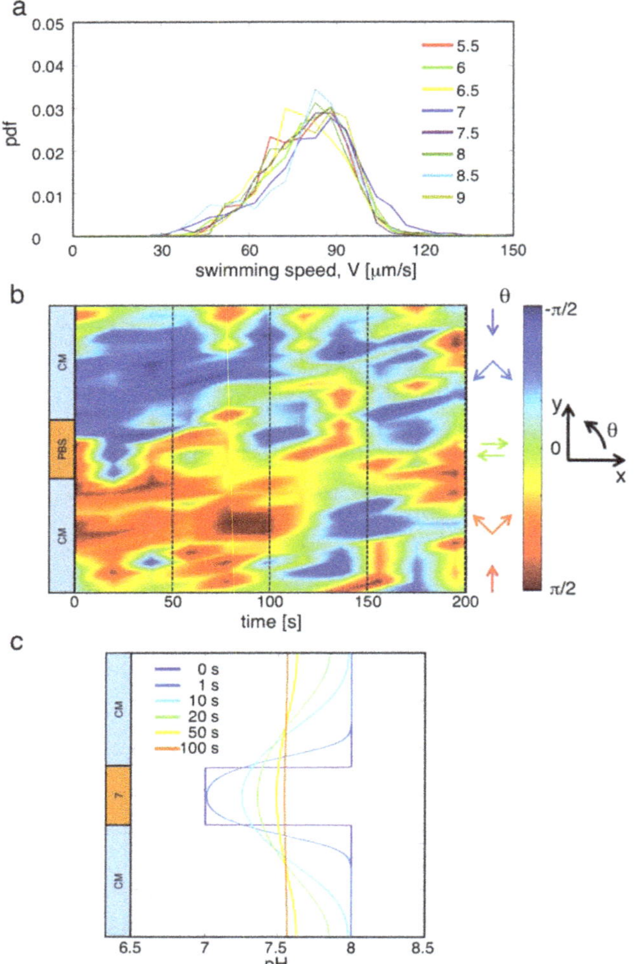

Fig. 5. The response of unicellular *S. rosetta* to pH is a true case of taxis. (a) *S. rosetta* shows no change in swimming speed when grown at a range of different pH values. The distribution of swimming speeds (probability density function, pdf) within a population obtained for fast swimmers in ASW with different pH values. (b) Swimming direction as a function of the position of cells across the width of the channel, plotted as a function of time. The central band contained PBS (corresponding to the experiment in Fig. 3b). Color denotes swimming orientation, as indicated by arrows next to the scale bar. The angle of the instantaneous swimming directions of all the cells at each time were measured relative to the along-channel direction, so that an angle of $\pi/2$ (red) corresponds to swimming upward in the plot and an angle of $-\pi/2$ (blue) corresponds to swimming downward. Over the initial 100 s there was directional swimming into the central band: upward from below and downward from above the band. (c) The temporal evolution of the pH gradient in the microchannel. The solution of the one-dimensional diffusion equation for H^+ ions across the channel width is shown at different times after the injection of the central band.

biofilms have lower pH than the surrounding water (Liermann et al. 2000; Hidalgo et al. 2009). The observation that low values of pH cause repulsion suggests that intermediate acidic conditions ($5 < \mathrm{pH} < 7$) are most indicative of bacterial patches, which is consistent with observations of pH in this range in bacterial biofilms (Hidalgo et al. 2009). In the environment, more extreme acidification may also be prevented by the diluting effect of fluid flow and diffusion, both on benthic surfaces and on marine particles. We thus suggest that the fast-swimming form of *S. rosetta* can pursue pH cues to move rapidly into microenvironments rich in bacterial prey.

Feeding performance of bacterivorous eukaryotes depends on two processes: finding prey-rich microhabitats and capturing prey. Our discovery that only the fast-swimming stage of *S. rosetta* uses taxis is consistent with the hypothesis that fast swimmers are dispersers that can find new prey patches. Rapid swimming is often a feature of chemotactic marine microorganisms (Stocker and Seymour 2012). However, *S. rosetta* fast swimmers have reduced collars and no doubt capture bacterial prey at a lower rate than the long-collared slow swimmers and rosette colonies. We suggest that there is a trade-off between the ability to swim rapidly to find new prey patches and the ability to capture bacteria at a high rate. Larger collars may themselves reduce swimming speed, as they cause more drag. These hypotheses suggest a division of labor between different life cycle stages in choanoflagellates.

In laboratory cultures of fast swimmers, it takes several hours for the long-collared thecate cells and slow swimmers to be produced (Dayel et al. 2011). Therefore, having separate searching and feeding life stages would not enable choanoflagellates to take advantage of ephemeral microscale hotspots in the water column that dissipate in minutes (Stocker 2012; Stocker and Seymour 2012). However, choanoflagellates settle onto marine snow particles and detritus (reviewed in Leadbeater 2015), and *S. rosetta* live in shallow estuaries where dense consortia of bacteria form on suspended particles and benthic surfaces. Having separate searching and feeding stages could be an effective strategy for utilizing particles and their wakes that are rich in bacteria (Kiørboe and Jackson 2001; Stocker et al. 2008), or for foraging near benthic surfaces that continuously shed filaments of resources in turbulent ambient water flow (Crimaldi and Koseff 2001; Koehl et al. 2007). Fast swimmers can attach rapidly to surfaces using filopodia, and then develop into long-collared thecate cells (Dayel et al. 2011). If fast swimmers can find and attach to particles or benthic surfaces, then the long-collared thecate cells would be in the prey-rich boundary layer of water next to these surfaces, and subsequent production of long-collared, slowly-swimming forms would place them in the resource-rich plumes shed by particles and substrata.

Directional responses to pH gradients may in general serve a range of functions, though in few cases has direct evidence for the functional role of pH-taxis been reported. Bacteria have been suggested to migrate away from regions of very low pH to avoid physiologically damaging conditions (Seymour and Doetsch 1973) or overcrowding (Tsang et al. 1973). Slime molds may migrate toward low pH because acidic environments enhance fruiting (Bonner et al. 1985) and fungal zoospores may accumulate in a pH range that enhances encystment (Morris et al. 1995). While we cannot completely rule out that pH taxis in choanoflagellates serves a similar role, of allowing cells to reach physiologically optimal conditions, its occurrence only in the rapid swimming stage of choanoflagellates suggests a prey-seeking function of pH taxis.

Although *S. rosetta* were collected in a shallow muddy estuary, little else is known about their ecology. The colonization by *S. rosetta* of marine snow should be investigated, but studies of particles suggest that it would be feasible for fast swimming choanoflagellates to find them. The success of a chemotactic strategy for finding sinking marine particles depends on the competition between the timescale of availability of the chemical signal and the timescale of the chemotactic response. However, this does not simply reduce to a comparison of the sinking speed of particles (often $\sim$ 100–1000 μm s^{-1}) and the swimming speed of the organisms ($\sim$ 50 μm s^{-1}), because of the boundary layer around the particles. Indeed, considerable attachment to sinking particles was found in a model of bacteria swimming at speeds comparable to those observed here for *S. rosetta* (Kiørboe and Jackson 2001). The same model showed that these swimming speeds are sufficient to move from particle to particle under typical particle concentrations in the ocean. We thus propose that the chemotactic responses described here will be relevant under natural conditions and enable fast-swimming choanoflagellates to exploit pH gradients associated with biofilms, enabling the choanoflagellates to colonize food-rich marine particles and surfaces.

Studies of bacteria and of microscopic larvae using chemical cues to find benthic surfaces suggest that choanoflagellate fast swimmers might also be able to do so. For example, in still water, pathogenic bacteria that swim at speeds comparable to those of *S. rosetta* can successfully use chemotaxis to coral exudates to reach and infect their coral hosts (Garren et al. 2014). Even in turbulent flow across benthic surfaces that continuously release chemical cues, the frequency of encounters with filaments of cue is very high near the surfaces, and the behavior of weakly swimming microorganisms can bias how they are transported to the surfaces by the ambient flow (Koehl et al. 2007; Koehl and Cooper 2015). These hypotheses could be tested in experiments in which choanoflagellates are exposed to biofilms, preferably grown under controlled conditions, for example in microfluidic devices, and both the pH conditions (e.g., through optodes) and the choanoflagellate distribution are measured.

References

Anderson, A. J., E. B. Fred, and W. H. Peterson. 1920. The relation between the number of bacteria and acid production in the fermentation of xylose. J. Infect. Dis. **27**: 281–292. doi:10.1093/infdis/27.4.281

Azam, F., T. Fenchel, J. G. Field, J. S. Gray, L. A. Meyerreil, and F. Thingstad. 1983. The ecological role of water-column microbes in the sea. Mar. Ecol. Prog. Ser. **10**: 257–263. doi:10.3354/meps010257

Bonner, J. T., A. Hay, D. G. John, and H. B. Suthers. 1985. pH affects fruiting and slug orientation in *Dictyostelium discoideum*. J. Embryol. Exp. Morphol. **87**: 207–213.

Crimaldi, J. P., and J. R. Koseff. 2001. High-resolution measurements of the spatial and temporal scalar structure of a turbulent plume. Exp. Fluids **31**: 90–102. doi:10.1007/s003480000263

Croxen, M. A., G. Sisson, R. Melano, and P. S. Hoffman. 2006. The *Helicobacter pylori* chemotaxis receptor TlpB (HP0103) is required for pH taxis and for colonization of the gastric mucosa. J. Bacteriol. **188**: 2656–2665. doi:10.1128/JB.188.7.2656-2665.2006

Dayel, M. J., R. A. Alegado, S. R. Fairclough, T. C. Levin, S. A. Nichols, K. McDonald, and N. King. 2011. Cell differentiation and morphogenesis in the colony-forming choanoflagellate. Dev. Biol. **357**: 73. doi:10.1016/j.ydbio.2011.06.003

Garren, M., K. Son, J. B. Raina, R. Rusconi, F. Menolascina, O. H. Shapiro, J. Tout, D. G. Bourne, J. R. Seymour, and R. Stocker. 2014. A bacterial pathogen uses dimethylsulfoniopropionate as a cue to target heat-stressed corals. ISME J. **8**: 999–1007. doi:10.1038/ismej.2013.210

Govorunova, E. G., and O. A. Sineshchekov. 2005. Chemotaxis in the green flagellate alga *Chlamydomonas*. Biochemistry **70**: 717–725. doi:10.1007/s10541-005-0176-2

Hidalgo, G., A. Burns, E. Herz, A. G. Hay, P. L. Houston, U. Wiesner, and L. W. Lion. 2009. Functional tomographic fluorescence imaging of pH microenvironments in microbial biofilms by use of silica nanoparticle sensors. Appl. Environ. Microbiol. **75**: 7426–7435. doi:10.1128/AEM.01220-09

King, N., S. L. Young, M. Abedin, M. Carr, and B. S. Leadbeater. 2009. The choanoflagellates: Heterotrophic nanoflagellates and sister group of the metazoa. Cold Spring Harb. Protoc. **4**: 1–5. doi:10.1101/pdb.emo116

Kiørboe, T., and G. A. Jackson. 2001. Marine snow, organic solute plumes, and optimal chemosensory behavior of bacteria. Limnol. Oceanogr. **46**: 1309–1318. doi:10.4319/lo.2001.46.6.1309

Koehl, M. A., J. A. Strother, M. A. Reidenbach, J. R. Koseff, M. G. Hadfield. 2007. Individual-based model of larval transport to coral reefs in turbulent, wave-driven flow: Effects of behavioral responses to dissolved settlement cues. Mar. Ecol. Prog. Ser. **335**: 1–18. doi:10.3354/meps335001

Koehl, M. A., and T. Cooper. 2015. Swimming in an unsteady world. Integr. Comp. Biol. **55**: 683–697. doi:10.1093/icb/icv092

Leadbeater, B. S. 1983. Life-history and ultrastructure of a new marine species of *Proterospongia* (Choanoflagellida). J.

Mar. Biol. Assoc. U.K. **63**: 135–160. doi:10.1017/S0025315400049857

Leadbeater, B. S. 2015. The choanoflagellates: Evolution, biology and ecology. Cambridge Univ. Press.

Levin, T. C., and N. King. 2013. Evidence for sex and recombination in the choanoflagellate *Salpingoeca rosetta*. Curr. Biol. **23**: 2176–2180. doi:10.1016/j.cub.2013.08.061

Levin, T. C., A. J. Greaney, L. Wetzel, and N. King. 2014. The *rosetteless* gene controls development in the choanoflagellate *S. rosetta*. eLife **3**: e04070. doi:10.7554/eLife.04070

Liermann, L. J., A. S. Barnes, B. E. Kalinowski, X. Zhou, and S. L. Brantley. 2000. Microenvironments of pH in biofilms grown on dissolving silicate surfaces. Chem. Geol. **171**: 1–16. doi:10.1016/S0009-2541(00)00202-3

Marion, G. M., F. J. Millero, M. F. Camões, P. Spitzer, R. Feistel, C.-T. Chen. 2011. pH of seawater. Mar. Chem. **126**: 89–96. doi:10.1016/j.marchem.2011.04.002

Morris, B. M., B. Reid, and N. A. Gow. 1995. Tactic response of zoospores of the fungus *Phytophthora palmivora* to solutions of different pH in relation to plant infection. Microbiology **141**: 1231–1237. doi:10.1099/13500872-141-5-1231

Schnitzer, M. J. 1993. Theory of continuum random walks and application to chemotaxis. Phys. Rev. E. **48**: 2553–2568. doi:10.1103/PhysRevE.48.2553

Seymour, F. W., and R. N. Doetsch. 1973. Chemotactic responses by motile bacteria. J. Gen. Microbiol. **78**: 287–296. doi:10.1099/00221287-78-2-287

Seymour, J. R., T. Ahmed, Marcos, R. Stocker. 2008. A microfluidic chemotaxis assay to study microbial behavior in diffusing nutrient patches. Limnol. Oceanogr.: Methods **6**: 477–488. doi:10.4319/lom.2008.6.477

Smriga, S., V. I. Fernandez, J. G. Mitchel, and R. Stocker. 2016. Chemotaxis toward phytoplankton drives organic matter partitioning among marine bacteria. Proc. Natl. Acad. Sci. USA. **113**: 1576–1581. doi:10.1073/pnas.1512307113

Solé, M., N. Rius, and J. G. Lorén, 2000. Rapid extracellular acidification induced by glucose metabolism in non-proliferating cells of *Serratia marcescens*. Int. Microbiol. **3**: 39–43.

Stocker, R. 2012. Marine microbes see a sea of gradients. Science **338**: 628–633. doi:10.1126/science.1208929

Stocker, R., J. R. Seymour, A. Samadani, D. E. Hunt, and M. F. Polz. 2008. Rapid chemotactic response enables marine bacteria to exploit ephemeral microscale nutrient patches. Proc. Natl. Acad. Sci. USA **105**: 4209–4214. doi:10.1073/pnas.0709765105

Stocker, R., and J. R. Seymour. 2012. Ecology and physics of bacterial chemotaxis in the ocean. Microbiol. Mol. Biol Rev. **76**: 792–812. doi:10.1128/MMBR.00029-12

Taylor, J. R., and R. Stocker. 2012. Trade-offs of chemotactic foraging in turbulent water. Science **338**: 675–679. doi:10.1126/science.1219417

Tikhonenkov, D. V., and Y. A. Mazei. 2008. Heterotrophic flagellate biodiversity and community structure in freshwater streams. Inland Water Biol. **1**: 129–133. doi:10.1134/S1995082908020041

Tsang, N., R. Macnab, and D. E. Koshland. 1973. Common mechanism for repellents and attractants in bacterial chemotaxis. Science **181**: 60–63. doi:10.1126/science.181.4094.60

Worden, A. Z., M. J. Follows, S. J. Giovannoni, S. Wilken, A. E. Zimmerman, and P. J. Keeling. 2015. Rethinking the marine carbon cycle: Factoring in the multifarious lifestyles of microbes. Science **347**: 1257594-1–1257594-10. doi:10.1126/science.1257594

Yuan-Hui, L., and Gregory S. 1974. Diffusion of ions in sea water and in deep-sea sediments. Geochim. Cosmochim. Ac. **38**: 703–714. doi:10.1016/0016-7037(74)90145-8

Zhuang, J., R. W. Carlsen, and M. Sitti. 2015. pH-taxis of biohybrid microsystems. Sci. Rep. **5**: 11403. doi:10.1038/srep11403

Acknowledgments

We gratefully acknowledge Laura Wetzel, Rosanna A. Alegado, and Karna Gowda for help with early experiments; Ben Larson and Eric Ferro for helpful feedback on the manuscript; Rebecca Schilling for technical support; Javier Sparacino, Adolfo J. Banchio, Andy Chang, and Veronica I. Marconi for useful discussions; and the National Science Foundation for grant IOS-1147215 to M. K., N. K. and R. S.

10

Ice duration drives winter nitrate accumulation in north temperate lakes

S. M. Powers,[1]* *S. G. Labou*,[1] *H. M. Baulch*,[2] *R. J. Hunt*,[3] *N. R. Lottig*,[4] *S. E. Hampton*,[1] *E. H. Stanley*[5]
[1]Center for Environmental Research, Education and Outreach, Washington State University, Pullman, Washington; [2]School of Environment and Sustainability, and Global Institute for Water Security, University of Saskatchewan, Saskatoon, Saskatchewan, Canada; [3]US Geological Survey, Wisconsin Water Science Center, Middleton, Wisconsin; [4]Trout Lake Station, Center for Limnology, University of Wisconsin–Madison, Boulder Junction, Wisconsin; [5]Center for Limnology, University of Wisconsin-Madison, Madison, Wisconsin

Scientific Significance Statement

Many freshwater lakes freeze during winter, yet little is known about the rates and controls on ecological processes under ice, such as nitrification, or how these processes may be affected by ongoing and future changes in ice cover. To address these knowledge gaps, we analyzed three decades of winter chemistry data related to the nitrogen cycle in five seasonally ice-covered north temperate lakes. We found that nitrate accumulated during winter as a result of internal lake processes and was strongly related to the number of days since ice-on. Consequently, winters with shorter ice duration may result in reduced total nitrification and lower peak nitrate concentrations, with potentially important implications for annual lake biogeochemical budgets, primary productivity, and biological communities.

Abstract

The duration of winter ice cover on lakes varies substantially with climate variability, and has decreased over the last several decades in many temperate lakes. However, little is known of how changes in seasonal ice cover may affect biogeochemical processes under ice. We examined winter nitrogen (N) dynamics under ice using a 30+ yr dataset from five oligotrophic/mesotrophic north temperate lakes to determine how changes in inorganic N species varied with ice duration. Nitrate accumulated during winter and was strongly related to the number of days since ice-on. Exogenous inputs accounted for less than 3% of nitrate accumulation in four of the five lakes, suggesting a paramount role of nitrification in regulating N transformation and the timing of chemical conditions under ice. Winter nitrate accumulation rates ranged from 0.15 μg N L^{-1} d^{-1} to 2.7 μg N L^{-1} d^{-1} (0.011–0.19 μM d^{-1}), and the mean for intermediate depths was 0.94 μg N L^{-1} d^{-1} (0.067 μM d^{-1}). Given that winters with shorter ice duration (< 120 d) have become more frequent in these lakes since the late 1990s, peak winter nitrate concentrations and cumulative nitrate production under ice may be declining. As ice extent and duration change, the physical and chemical conditions supporting life will shift. This research suggests we may expect changes in the form and amount of inorganic N, and altered dissolved nitrogen : phosphorus ratios, in lakes during winters with shorter ice duration.

*Correspondence: steve.powers@wsu.edu

Author Contribution Statement: SMP led the manuscript effort. SMP and EHS came up with the research questions and designed the study approach. SMP, SEH, and SGL contributed to statistical analyses. SMP and SGL contributed figures. EHS, NRL, and RJH contributed data. SMP, EHS, HMB, NRL, and SEH interpreted results. All authors contributed to writing of the paper.

Most of the world's naturally formed freshwater lakes are located above 45° North and freeze during winter (Verpoorter et al. 2014). Effects of ice on physical properties of lakes include a convergence of bottom temperatures toward 4°C in freshwater systems, altered mixing dynamics, loss of gas exchange, and winter darkness (McKnight et al. 2000; Kirillin et al. 2012). Despite these fundamental changes, the influence of ice dynamics on ecology and biogeochemical cycling of seasonally frozen lakes are not well understood, often inferred indirectly by comparing lake conditions before and after ice is present (Weyhenmeyer 2009; Hampton et al. 2015). Broad scale declines in ice duration over the past century in the northern hemisphere (Magnuson et al. 2000; Benson et al. 2012) underscore the need to address this winter blind spot (Powers and Hampton 2016; Hampton et al. 2017) and better understand ecosystem responses to varying winter conditions (Özkundakci et al. 2016) and ice duration (Denfeld et al. 2016).

Ice cover results in substantial changes to the physicochemical environment and biogeochemical processes important to nitrogen (N) cycling (Bertilsson et al. 2012). N mineralization and nitrification—the microbial oxidation of ammonium to nitrite and nitrate—are central processes in N cycling that occur in diverse environments. Much understanding about these transformations comes from studies in soils, followed by coastal marine systems (Heiss and Fulweiler 2016), estuaries (Wankel et al. 2011; Beman 2014), permanently frozen Antarctic lakes (Voytek et al. 1999), and open water conditions in streams and freshwater lakes (Hall and Jeffries 1984; Bernhardt et al. 2002). As in marine environments (Konneke et al. 2005), Archaea are often the dominant ammonium oxidizers in lakes (Small et al. 2013; Hayden and Bemen 2014; Yang et al. 2017). The relative abundance of marine Archaea can increase during extreme cold under sea ice (Church et al. 2003), and nitrification is known to be a major influence on N speciation in permanently frozen lakes (Voytek et al. 1999). Consequently, nitrification may occur broadly in cold freshwater systems ($<$ 5°C) and at ecosystem-relevant rates, providing an ammonium sink and nitrate source that could affect other biogeochemical fluxes as well as many taxa that preferentially use ammonium over nitrate (Glibert et al. 2016). The expectation of significant nitrification during seasonal ice cover has been supported by isotopic evidence in subarctic lakes (Gu 2012), but to date there have been relatively few measurements of rates and controls on nitrification in freshwater lakes (Small et al. 2013), especially beneath ice.

Previous lake studies have reported nitrate accumulation during winter. Due to major exogenous N sources, it has often been difficult to elucidate the role of in-lake (endogenous) processes in generating this winter nitrate. For example, George et al. (2010) report winter nitrate concentrations $> 500\ \mu g$ N/L in eight lakes of the UK, Finland, Estonia, and Sweden, where drainage from surrounding agricultural land was identified as a primary source of the N. Similarly, Knowles and Lean (1987) reported rates of nitrate accumulation up to 13 $\mu g\ L^{-1}\ d^{-1}$ (as N) using in situ water column measurements in Lake St. George, Ontario, a small (140 ha) kettle lake with contiguous shoreline development and likely exogenous N sources. Pettersson et al. (2003) report nitrate accumulation under ice in Lake Erken, Sweden, located in a moderately agricultural catchment. In contrast, ice-covered lakes in forested landscapes often have relatively low exogenous inputs of nitrate, potentially offering a simplified window on the N cycle during conditions not well represented in prior aquatic studies.

Nitrification has often been examined under warmer or more saline conditions than those typically observed during seasonal ice cover in freshwater lakes. When oxidized nitrogen species accumulate under lake ice, one simple constraint on the concentration is the number of days that ice has been present. Controls on nitrification in such lakes may also include: levels of connectivity between sediment microbes, N pools, and water column oxygen sources; exposure to excess light (French et al. 2012), ammonium (Chen et al. 2010), acidity (Jeschke et al. 2013), or anthropogenic pollutants such as metals or pharmaceuticals (Kraigher and Mandic-Mulec 2011; Rosi-Marshall and Royer 2012); competition for ammonium or nitrite by heterotrophic microbes or algae (Agbeti and Smol 1995), which may be elevated in lakes with high sediment-water contact or primary productivity. In unpolluted freshwaters, nitrite is thought to occur at trace concentrations and nitrite oxidation is not commonly viewed as a rate-limiting step of nitrification, although nitrite may sometimes accumulate to toxic concentrations in agricultural streams or groundwater (Stanley and Maxted 2008; Chen et al. 2010). While these factors can influence spatio-temporal variability in gross or net N transformation rates, understanding of total winter production of nitrate and ammonium could be enhanced through ecosystem-level studies that employ long-term winter data. Further, a simple predictor—the number of days since ice-on—may facilitate understanding of time-integrated process rates and the role of seasonal ice duration in N cycling.

The goal of this work was to examine dissolved inorganic N (DIN) dynamics and speciation during winter in a set of seasonally ice-covered north temperate lakes, focusing on the link between DIN and ice duration. We asked: (1) How rapidly and in what form (NH_4-N or NO_3-N) does DIN accumulate under ice? and (2) How are these rates related to ice duration? To address these questions and interpret potential roles of underlying mechanisms (e.g., nitrification, transfers between organic and inorganic N pools), we investigated temporal and vertical patterns of DIN, along with supporting measurements using a 30-yr dataset from five north temperate U.S.A. lakes.

Methods

We used 30+ yr of winter limnological data from five North Temperate Lakes Long-Term Ecological Research (NTL-LTER) study lakes (Allequash, Big Muskellunge (Big Musky), Crystal, Sparkling, and Trout Lakes; https://lter.limnology.wisc.edu) in northern Wisconsin, U.S.A. Maximum lake depth ranged from 8 m to 35 m. Lake depth, ice duration, winter data availability, and mean winter conditions are summarized in Table 1. Lakes were sampled 2–3 times each winter and N fractions were determined using consistent field and laboratory methodologies. Samples were collected at three depths at the deepest point of each lake (typically close to center of lake) through a 20-cm bore hole: surface (0–5 cm below the bottom of the ice), deep (1 m above lake bottom), and middle of the water column. Expanded methods are in the Supporting Information.

Relationships between concentration and days since ice-on

We examined winter NO_3-N and NH_4-N at three depths, and supporting physico-chemistry including total dissolved phosphorus (TDP, μg/L) and dissolved inorganic carbon (DIC, mg/L). With days since ice-on as our time variable, we used linear mixed effects models (Bates et al. 2015) to estimate lake-specific and depth-specific rates of change (Supporting Information). For Trout Lake only, a fitting window of 1–100 d since ice-on was examined for deep water, due to a hump-shaped curve for NO_3-N. Analyses were conducted using R (R Core Team 2016). Loess fits were used for visualization.

Exogenous N inputs

Fitted rates of N change under ice were compared to rates of exogenous N input from streams and groundwater (Supporting Information). Exogenous N was quantified using three data sources: (1) groundwater NO_3-N and NH_4-N ($n = 12$ wells), typically sampled each year during autumn; (2) daily winter water fluxes to each lake from a data-driven hydrologic model based on daily precipitation and seasonal measurements of hydraulic head (Hunt et al. 2013; Hunt and Walker 2017); and (3) available stream chemistry data, sampled during 2007–2015. We assumed that N fixation and atmospheric N inputs were negligible during ice cover (Sampson and Brezonik 2003).

Results

Across lakes and depths, the ice-on period began with relatively low NO_3-N concentrations, and higher NH_4-N concentrations at the deep sampling depth. In deep water of Allequash, Big Musky, and Sparkling, NH_4-N frequently reached concentrations > 500 μg/L. At middle and surface depths in all lakes, NO_3-N was significantly higher in late winter compared to early winter (Fig. 1), attaining concentrations > 100 μg/L. These patterns at middle and surface

Table 1. Lake features and data availability.

Lake name	Max depth (m)	Mean depth (m)	Lake area (km^2)	Catchment area (km^2)	Ice duration (days)			Sample group	Sample depth (m)	NO_3N data availability			Mean under ice condition						
					Mean	Max	Min			# of obs.	# of years	Mean # obs per year	NO_3N (μg/L)	NH_4N (μg/L)	O_2 (mg/L)	DIC (mg/L)	TDP (μg/L)	Water temp (deg C)	pH
Allequash Lake, north basin	8	2.9	1.12	21.5	144	182	111	Surface	0	83	31	2.7	51	67.6	10.1	11.6	6.25	0.651	7.3
								Middle	4	87	31	2.8	141	68	5.99	13.6	10.4	3.19	7
								Deep	6	85	31	2.7	78.1	326	2.01	16.1	33.4	4.08	7
Big Musky Lake	21.3	7.5	3.96	4.0	138	182	106	Surface	0	84	31	2.7	45	80.7	13.9	5.57	5.56	0.645	7.3
								Middle	10	91	32	2.8	84.1	71.3	9.12	6.52	9.69	3.19	6.9
								Deep	17	88	32	2.8	13.7	610	2.36	10.3	177	3.93	6.7
Crystal Lake	20.4	10.4	0.367	0.86	136	169	105	Surface	0	83	31	2.7	32.4	56.3	14.7	0.761	3.87	0.739	6.3
								Middle	10	90	32	2.8	27.2	46.9	10.4	1.15	3.67	3.62	6
								Deep	17	89	32	2.8	71	271	5.75	3.28	14.9	4.13	5.8
Sparkling Lake	20	10.9	0.64	1.4	135	169	105	Surface	0	84	30	2.8	36	44.9	13.7	8.51	4.02	0.648	7.7
								Middle	10	88	32	2.8	66.4	34.9	9.99	8.99	4.08	3.35	7.2
								Deep	17	89	32	2.8	37.5	967	3.6	14.5	70.2	3.86	6.9
Trout Lake, south basin	35.7	14.6	10.9	47	133	170	77	Surface	0	84	31	2.7	68.1	27.9	12.8	11	3.19	0.743	7.6
								Middle	10	89	31	2.9	60	13.8	12.1	11.3	3.65	1.76	7.6
								Deep	28	90	32	2.8	155	168	6.9	13.2	9.38	2.35	6.9

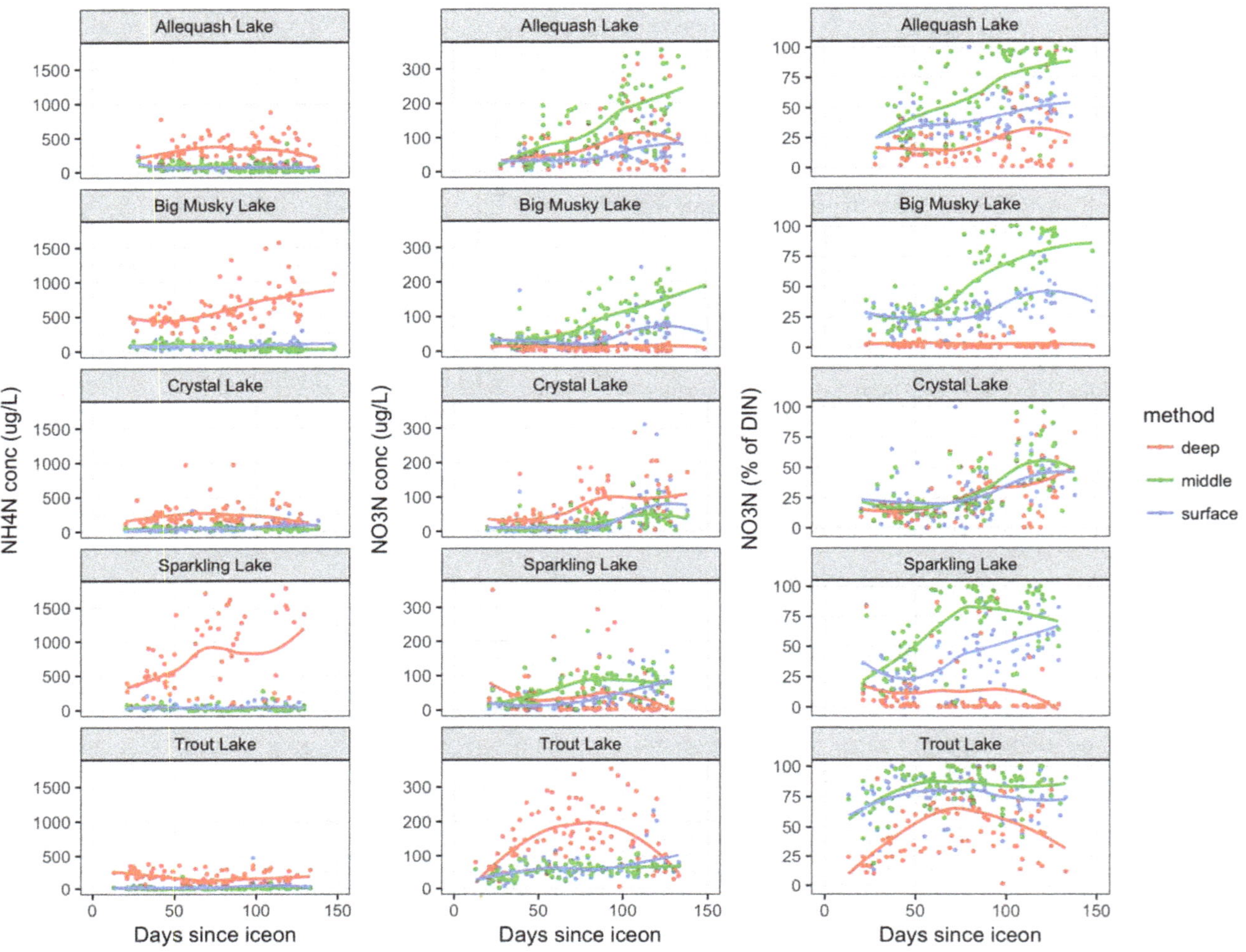

Fig. 1. NH_4-N and NO_3-N observations in deep, middle, and surface water expressed as a function of the number of days since ice-on. Lines are loess fits to aid visualization of winter trends. A few high values exceed the plotted scale.

depths produced a consistent shift in DIN form over winter, from NH_4-N dominance to NO_3-N dominance.

Model-fitted NO_3-N accumulation rates (Supporting Information Table S1) were highest at the middle sampling depth in Allequash, Big Musky, and Sparkling lakes, and at the deep sampling depth in Crystal and Trout (Fig. 2). Trout Lake deep water had the fastest NO_3-N accumulation rate among lakes and depths, at 2.7 μg L^{-1} d^{-1} (0.19 μM d^{-1}) prior to the NO_3-N decline around day 100. The selected NO_3-N model had a random intercept for year, and a heteroscedastic error structure (proportional to days since ice-on), producing a marginal r-squared of 0.79 (Supporting Information). NO_3-N accumulation rates at middle depth ranged from 0.35 μg L^{-1} d^{-1} (0.025 μM d^{-1}) in Crystal to 1.68 μg L^{-1} d^{-1} (0.12 μM d^{-1}) in Allequash, and had a mean of 0.94 μg N L^{-1} d^{-1} (0.067 μM d^{-1}). In deep water, NH_4-N also accumulated at $\geq$2.0 μg L^{-1} d^{-1} (0.14 μM d^{-1}), exceeding NO_3-N accumulation, but in these lakes most benthic surface area occurs above the deep sampling depth (Supporting Information). If middle depth is used to represent the whole lake, corresponding areal NO_3-N accumulation rates ranged from 260 μmol N m^{-2} d^{-1} in Crystal to 790 μmol N m^{-2} d^{-1} in Trout, with rates of 350 in Allequash, 580 in Big Musky, and 660 in Sparkling. The ratio of NO_3-N : DIN accumulation at middle depth was 0.41 in Crystal, and 0.64–0.83 in the other lakes, reflecting the substantial fraction of DIN as NO_3-N. In contrast to DIN, pronounced TDP accumulation was restricted to deep water (Supporting Information). At middle depth, between 30 and 120 d since ice-on, stable TDP combined with DIN accumulation gave predicted shifts in DIN : TDP ratios from 24 to 57 in Allequash, 25 to 42 in Big Musky, 32 to 53 in Crystal, and 39 to 67 in Sparkling; in Trout Lake between 30 and 100 d since ice-on, the predicted shift in DIN : TDP was from 33 to 50.

Exogenous inputs from streams and groundwater were insufficient to account for observed in-lake NO_3-N increases (Supporting Information). In three lakes that lacked perennial streams (Big Musky, Crystal, Sparkling), using the

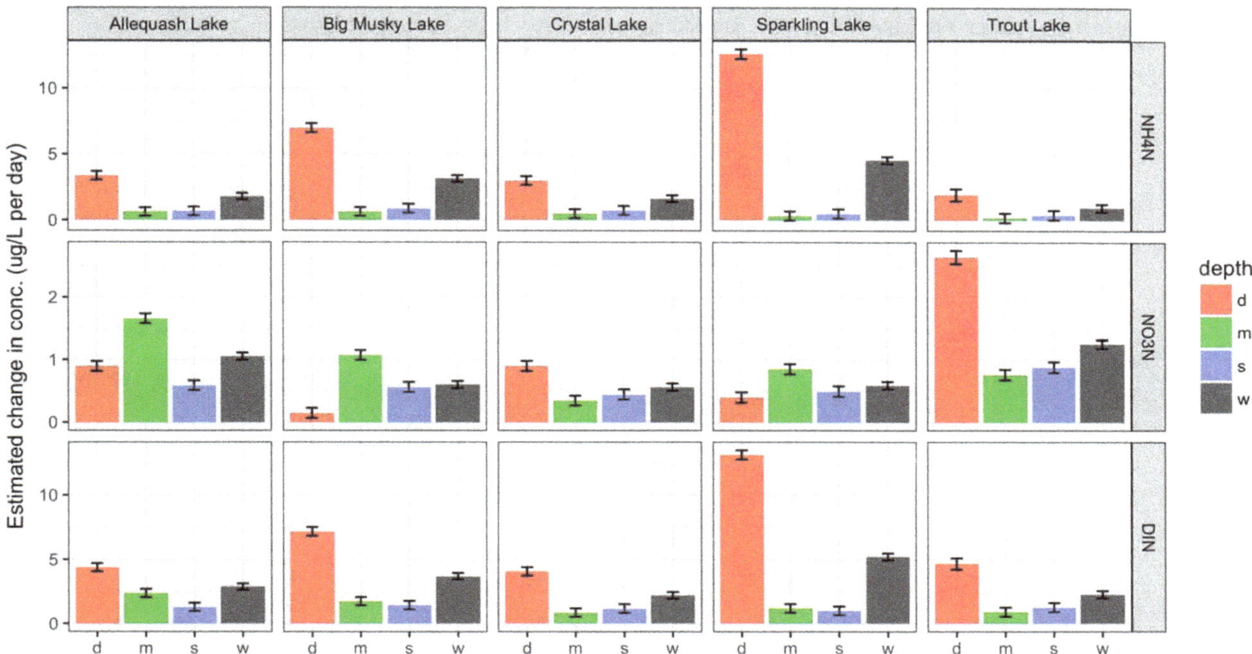

Fig. 2. Fitted winter changes in NH_4-N and NO_3-N under ice in the five lakes. Estimates are from linear mixed effects models. Key: d = deep, m = middle, s = surface. Whole (w) indicates fitted values using data from all three depths. Note different *y*-axis for each row.

median groundwater NO_3-N concentration of 4 μg/L and assuming middle depth is most representative, exogenous inputs accounted for < 0.1% of the water column NO_3-N accumulation rate. Exogenous NO_3-N inputs were similarly negligible in comparison to aggregated (depth-ignorant) NO_3-N accumulation rates across the three depths. In Allequash, mean daily winter water input ranged from 17,000 m^3 d^{-1} to 28,000 m^3 d^{-1} depending on the year, equivalent to 49–113% of the lake volume (mean of 79%) when integrated over the duration of ice cover. However, using a stream and groundwater concentration of 15 μg/L NO_3-N, our calculations suggest exogenous NO_3-N inputs can only account for 3.8–6.0% of the in-lake NO_3-N increase in Allequash, and 0.62–1.1% in Trout. If a more liberal estimate of stream and groundwater NO_3-N concentration is used (35 μg/L, the 75th percentile of groundwater data), exogenous NO_3-N inputs still only account for 8.8–14% of the in-lake increase in Allequash, and 1.5–2.6% in Trout.

Positive correlations between DIN and DIC concentrations (Fig. 3) suggested organic matter decomposition and N mineralization supplied N available for nitrification, particularly in deep and intermediate waters. The highest DIN and DIC concentrations occurred in deep samples, pointing to likely benthic organic sources, with possible additional contributions from NH_4-N that was abiotically released from sediments. Breakdown of dissolved organic N (DON) represents a potential alternative source of water column NH_4-N, or NO_3-N (if DON breakdown is coupled with nitrification). However, there were no consistent, substantial declines in DON concentrations over winter (Supporting Information Fig. S5).

Discussion

While past work has demonstrated that nitrification rates can be inversely related to temperature (Randall and Buth 1984; Bruesewitz et al. 2009), this process may continue in cold winter waters when oxygen is present. There were strong positive relationships between NO_3-N concentration and days since ice-on in all five lakes, indicating NO_3-N accumulation in the water column under ice. External inputs could not account for these pronounced increases, leaving nitrification as a likely explanation for NO_3-N accumulation. In all five lakes, these integrated winter NO_3-N accumulation rates were ≥ 0.75 μg $L^{-1}d^{-1}$ at one or more sampling depths. Over shorter periods of winter, higher rates of NO_3-N accumulation may still be possible in these lakes. Higher NO_3-N accumulation rates have been reported under ice in more eutrophic lakes (Knowles and Lean 1987). Our reported rates reflect net processing, as some N may be consumed by processes such as denitrification or algal uptake, perhaps especially under thin or clear ice (Kerfoot et al. 2008, Katz et al. 2015). The regularity of inorganic N accumulation over these 30+ yr of data suggests the number of days since ice-on can be useful in predicting lake dynamics.

NO_3-N accumulation was more pronounced in middle and deep water compared to the surface, likely because nitrification occurs predominantly in the benthos, as has been

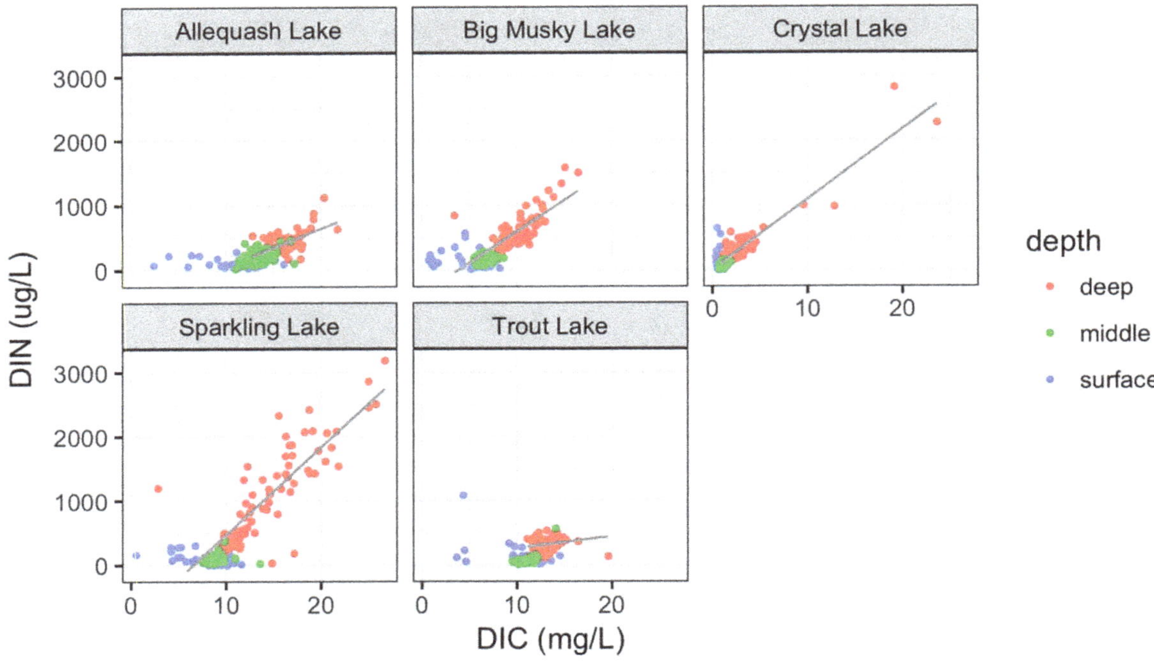

Fig. 3. Positive relationships between DIN and DIC within each lake. Lines are the linear regression fits for the deep water samples (Supporting Information).

widely reported during ice-free conditions (e.g., Pauer and Auer 2000). Prior data on sediment nitrification rates below 5°C are scarce, even among water treatment studies (but *see* Abeysinghe et al. 2002). Benthic mineralization appears to be an important source of DIN in these lakes. This is suggested by elevated NH_4-N and DIC concentrations in deep and middle samples, positive NH_4-N~DIC relationships, and lack of predictable seasonal patterns for DON and suspended particle N (Supporting Information). The DIN accumulation rates at middle and surface depths were sufficient to regularly produce molar DIN : TDP ratios of 40–70 after 100 d of ice cover.

Water column DON turnover may have contributed a fraction of the DIN. However, there were no robust declines in DON concentrations over winter (Supporting Information Fig. S5) as might be expected if DON were the major DIN source. In addition, the hypothesis that water column DON supplies the majority of DIN runs counter to the paradigm of organic matter breakdown as a benthic-dominated process within inland lakes, perhaps especially during low light conditions that limit winter DOM photodegradation (Hampton et al. 2017).

One of the most conspicuous consequences of under-ice N transformation is the marked rise in NO_3-N : NH_4-N ratios over winter (Fig. 1), and its potential effects on autotrophic communities in late winter (Kerfoot et al. 2008) and spring (Gächter et al. 1974). For example, many taxa preferentially use NH_4-N as an inorganic N source, and will be at a disadvantage as NO_3-N becomes a larger fraction of the DIN pool at the expense of NH_4-N (Glibert et al. 2016). As additional examples, altered NO_3-N concentration could influence biogenic production of gases (N_2, N_2O) through coupled nitrification-denitrification, or spring/summer production of cyanobacterial secondary metabolites such as microcystin, which may be favored under lower NO_3-N : NH_4-N (Harris et al. 2016). Weyhenmeyer et al. (2007) indicate that nitrate-depleted summer conditions have become more common in shallow lakes of northern Europe, yet potential linkages between winter nitrification variability and summer NO_3-N remain an area of needed investigation. While P availability often limits productivity, N may co-limit autotrophs across a range of aquatic environments (Elser et al. 2007; Paerl et al. 2016), and can influence metabolic processes of phytoplankton and other organisms at non-limiting concentrations (Glibert et al. 2016).

Much benthic surface area in these lakes falls closer to middle sampling depths than surface or deep depths (Supporting Information), so the middle may be more representative of ecosystem-level process rates. Processes near the ice surface in these lakes are probably less influential on whole lake N transformation rates, as shallow samples represented a relatively thin layer of the water column according to temperature-DO profiles (Supporting Information Fig. S6), and shallow littoral sediments may become biogeochemically inactive if they freeze. Nonetheless, shallower lakes also often have higher surface area : volume ratios and

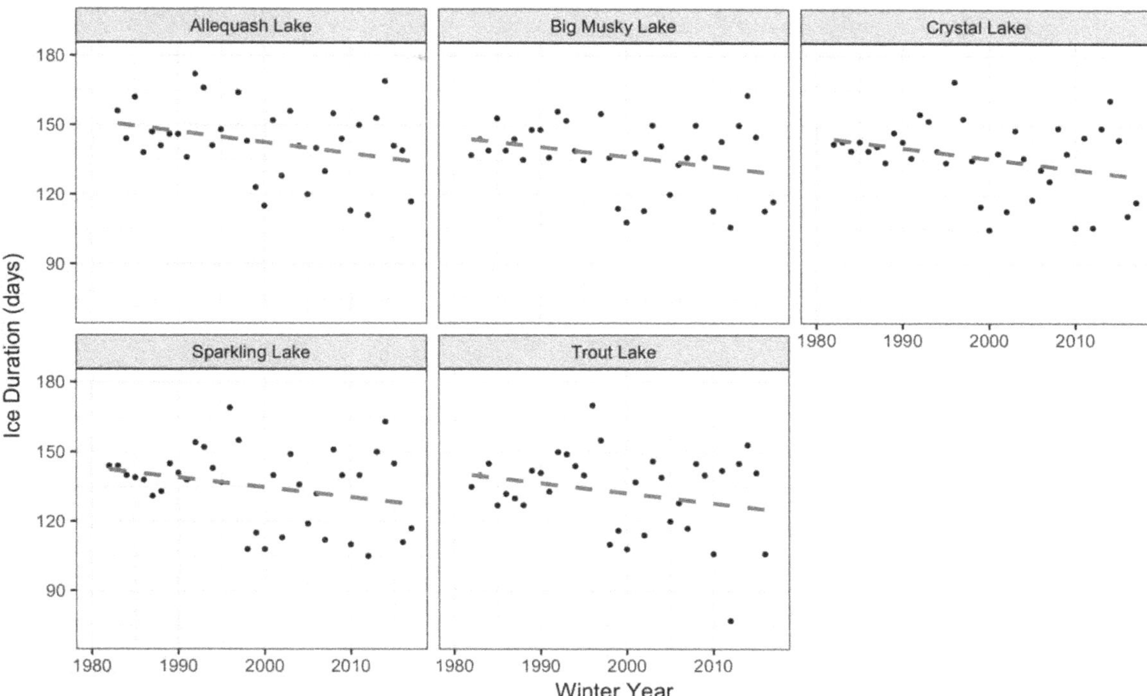

Fig. 4. Long-term patterns of ice duration in the five lakes between 1982 and 2017. Fitted lines are from simple linear regression. Over this period of record, there were marginally significant linear declines in ice duration ($p \leq 0.08$, $r^2 < 0.10$) of 4–5 d per decade in Allequash, Big Musky, and Crystal Lakes. Linear declines in Sparkling or Trout Lakes were not significant ($p > 0.1$), in part due to the disproportionately long ice duration of 2013–2014.

potentially higher sediment-water contact, which could influence N transformation rates. The two shallowest lakes, Allequash and Big Musky, had higher surface area : volume ratios (> 0.13 m^2/m^3) and in turn the highest NO_3-N accumulation rates ($\mu M\ d^{-1}$) at middle depth, yet estimated areal rates were intermediate (350 μ mol m^{-2} d^{-1} and 580 μ mol m^{-2} d^{-1}) to the other lakes (260–790 μ mol m^{-2} d^{-1}). Understanding of the role of morphometry could be aided by research across a more diverse suite of lakes.

Given differences in DIN among depths, changes in NO_3-N concentrations may reflect depth-specific nitrification. However, in three lakes, surface and middle NO_3-N increases coincided with NH_4-N accumulation in deep waters (Allequash, Big Musky, Sparkling), raising the possibility of vertical transfer, perhaps especially during late winter. Below 4.0°C, water density increases as it warms (Wetzel 2001), and recent research has demonstrated the potential role of convective mixing under ice (Bruesewitz et al. 2014; Pernica et al. 2017). Such mixing may deliver oxygen to benthic nitrifiers, perhaps contributing to the high NO_3-N accumulation rate in deep water of Trout Lake (28 m). In larger, deeper lakes, nitrification may occur at depths > 100 m when oxygen is present (Small et al. 2013). If NO_3-N accumulation in surface or deep water arrived through vertical transfer, our reported accumulation rates may still underestimate the contribution of intermediate depths to whole lake nitrification. While nitrification is likely a spatially heterogeneous process, identifying habitats that contribute disproportionately to NO_3-N accumulation is often challenging (Clevinger et al. 2014).

Ice duration data from 1982 to 2017 indicate that winters with shorter ice duration (< 120 d) have become more frequent in these lakes since the late 1990s (Fig. 4), with much interannual variability. Since 1982, there were marginally significant simple linear declines in ice duration ($p \leq 0.08$) of 4–5 d per decade in Allequash, Big Musky, and Crystal Lakes, but no significant declines in Sparkling and Trout ($p > 0.1$). A disproportionately long ice duration in 2013–2014 influenced significance of these regressions, as all lakes had significant declines ($p \leq 0.04$) of $\sim$ 6 d per decade when 2013–2014 was excluded. Other lakes in this region have shown similar declines of 4–9 d per decade over a longer period of record (Jensen et al. 2007). Given these patterns, it is possible that nitrate-rich winters are becoming less frequent in forested lakes experiencing shorter ice duration. The fitted accumulation rates can inform future expectations about peak winter NO_3-N concentration. For example, based on accumulation rates at middle depth, the expected NO_3-N concentration after 90 d of ice cover (corresponding to a short winter) is 21–40% lower compared to 140 d of ice cover, a decrease of 17–96 μg/L depending on the lake. Associated predictions at 90 d give peak DIN : TDP ratios of 30–60, a decrease of 10–20 units.

These predictions are approximate and can still vary across years and lake types. Nonetheless, truncated winter N dynamics are expected with shorter ice duration, and this could influence annual lake biogeochemical budgets, primary productivity, and biological communities.

References

Abeysinghe, D. H., D. G. De Silva, D. A Stahl, and B. E Rittmann. 2002. The effectiveness of bioaugmentation in nitrifying systems stressed by a washout condition and cold temperature. Water Environ. Res. **74**: 187–199. doi:10.2175/106143002X139901

Agbeti, M. D., and J. P. Smol. 1995. Winter limnology - A comparison of physical, chemical and biological characteristics in 2 temperate lakes during ice cover. Hydrobiologia **304**: 221–234. doi:10.1007/BF02329316

Bates, D., M. Mächler, B. Bolker, and S. Walker. 2015. Fitting linear mixed-effects models using lme4. J. Stat. Softw. **67**: 1–48. doi:10.18637/jss.v067.i01

Beman, J. M. 2014. Activity, abundance, and diversity of nitrifying Archaea and denitrifying bacteria in sediments of a subtropical estuary: Bahia del Tbari, Mexico. Estuar. Coast. **37**: 1343–1352. doi:10.1007/s12237-013-9716-y

Benson, B., and others. 2012. Extreme events, trends, and variability in Northern Hemisphere lake-ice phenology (1855–2005). Clim. Change **112**: 299–323. doi:10.1007/s10584-011-0212-8

Bernhardt, E. S., R. O. Hall, and G. E. Likens. 2002. Whole-system estimates of nitrification and nitrate uptake in streams of the Hubbard Brook Experimental Forest. Ecosystems **5**: 419–430. doi:10.1007/s10021-002-0179-4

Bertilsson, S., and others. 2012. The under-ice microbiome of seasonally frozen lakes. Limnol. Oceanogr. **58**: 1998–2012. doi:10.4319/lo.2013.58.6.1998

Bruesewitz, D. A., J. L. Tank, and S. K. Hamilton. 2009. Seasonal effects of zebra mussels on littoral nitrogen transformation rates in Gull Lake, Michigan, USA. Freshw. Biol. **54**: 1427–1443. doi:10.1111/j.1365-2427.2009.02195.x

Bruesewitz, D. A., C. C. Carey, D. C. Richardson, and K. C. Weathers. 2014. Under-ice thermal stratification dynamics of a large, deep lake revealed by high-frequency data. Limnol. Oceanogr. **60**: 347–359. doi:10.1002/lno.10014

Chen, G. Y., X. Y. Cao, C. L. Song, and Y. Y. Zhou. 2010. Adverse effects of ammonia on nitrification process: The case of Chinese shallow freshwater lakes. Water Air Soil Pollut. **210**: 297–306. doi:10.1007/s11270-009-0253-z

Church, M. J., E. F. DeLong, H. W. Ducklow, M. B. Karner, C. M. Preston, and D. M. Karl. 2003. Abundance and distribution of planktonic Archaea and Bacteria in the waters west of the Antarctic Peninsula. Limnol. Oceanogr. **48**: 1893–1902. doi:10.4319/lo.2003.48.5.1893

Clevinger, C. C., R. T. Heath, and D. T. Bade. 2014. Oxygen use by nitrification in the hypolimnion and sediments of Lake Erie. J. Great Lakes Res. **40**: 202–207. doi:10.1016/j.jglr.2013.09.015

Denfeld, B. A., P. Kortelainen, M. Rantakari, S. Sobek, and G. A. Weyhenmeyer. 2016. Regional variability and drivers of below ice CO2 in boreal and subarctic lakes. Ecosystems **19**: 461–476. doi:10.1007/s10021-015-9944-z

Elser, J. J., and others. 2007. Global analysis of nitrogen and phosphorus limitation of primary producers in freshwater, marine, and terrestrial ecosystems. Ecol. Lett. **10**: 1135–1142. doi:10.1111/j.1461-0248.2007.01113.x

French, E., J. A. Kozlowski, M. Mukherjee, G. Bullerjahn, and A. Bollmann. 2012. Ecophysiological characterization of ammonia-oxidizing Archaea and Bacteria from freshwater. Appl. Environ. Microbiol. **78**: 5773–5780. doi:10.1128/AEM.00432-12

Gächter, R., R. A. Vollenweider, and W. A. Glooschenko. 1974. Seasonal variations of temperature and nutrients in the surface waters of Lakes Ontario and Erie. J. Fish. Res. Board Can. **31**: 275–290. doi:10.1139/f74-047

George, G., M. Jarvinen, T. Nogas, T. Blenkner, and K. Moore. 2010. The impacts of the changing climate on the supply and recycling of nitrate, p. 161–178. *In* G. George [ed.], Impact of climate change on European Lakes. Elsevier Academic Press.

Glibert, P. M., and others. 2016. Pluses and minuses of ammonium and nitrate uptake and assimilation by phytoplankton and implications for productivity and community composition, with emphasis on nitrogen-enriched conditions. Limnol. Oceanogr. **61**: 165–197. doi:10.1002/lno.10203

Gu, B. 2012. Stable isotopes as indicators for seasonally dominant nitrogen cycling processes in a subarctic lake. Int. Rev. Hydrobiol. **97**: 233–243. doi:10.1002/iroh.201111466

Hall, G. H., and C. Jeffries. 1984. The contribution of nitrification in the water column and profundal sediments to the total oxygen deficit of the hypolimnion of a mesotrophic lake (Grasmere, English Lake District). Microb. Ecol. **10**: 37–46. doi:10.1007/BF02011593

Hampton, S. E., M. V. Moore, T. Ozersky, E. H. Stanley, C. M. Polashenski, and A. W. E. Galloway. 2015. Heating up a cold subject: Prospects for under-ice plankton research in lakes. J. Plankton Res. **37**: 1–8. doi:10.1093/plankt/fbv002

Hampton, S. E., and others. 2017. Ecology under lake ice. Ecol. Lett. **20**: 98–111. doi:10.1111/ele.12699

Harris, T. D., V. H. Smith, J. L. Graham, D. B. Vander Waal, L. P. Tedesco, and N. Clercin. 2016. Combined effects of nitrogen to phosphorus and nitrate to ammonia ratios on cyanobacterial metabolite concentration in eutrophic Midwestern USA reservoirs. Inland Waters **6**: 199–210. doi:10.5268/IW-6.2938

Heiss, E. M., and R. W. Fulweiler. 2016. Coastal water column ammonium and nitrite oxidation are decoupled in summer. Estuar. Coast. Shelf Sci. **178**: 110–119. doi:10.1016/j.ecss.2016.06.002

Hunt, R. J., J. F. Walker, W. R. Selbig, S. M. Westenbroek, and R. S. Regan. 2013. Simulation of climate-change effects on streamflow, lake water budgets, and stream temperature using GSFLOW and SNTEMP, Trout Lake Watershed, Wisconsin, 118 p., U.S. Geological Survey Scientific Investigations Report 2013–5159. Available from http://pubs.usgs.gov/sir/2013/5159/.

Hunt, R. J., and J. F. Walker. 2017. GSFLOW groundwater-surface water model 2016 update for the Trout Lake Watershed, Wisconsin: U.S. Geological Survey data release. Available from https://doi.org/10.5066/F7M32SZ2.

Jensen, O. P., B. J. Benson, J. J. Magnuson, V. M. Card, M. N. Futter, P. A. Soranno, and K. M. Stewart. 2007. Spatial analysis of ice phenology trends across the Laurentian Great Lakes region during a recent warm period. Limnol. Oceanogr. **52**: 2013–2026. doi:10.4319/lo.2007.52.5.2013

Jeschke, C., C. Falagan, K. Knoller, M. Schultze, and M. Koschorreck. 2013. No nitrification in lakes below pH 3. Environ. Sci. Technol. **47**: 14018–14023. doi:10.1021/es402179v

Katz, S. L., L. R. Izmest'eva, S. E. Hampton, T. Ozersky, K. Shchapov, M. V. Moore, S. V. Shimaraeva, and E. A. Silow. 2015. The "Melosira years" of Lake Baikal: Winter environmental conditions at ice onset predict under-ice algal blooms in spring. Limnol. Oceanogr. **60**: 1950–1964. doi:10.1002/lno.10143

Kerfoot, W. C., J. W. Budd, S. A. Green, J. B. Cotner, B. A. Biddanda, D. J. Schwab, and H. A. Vanderploeg. 2008. Doughnut in the desert: Late-winter production pulse in southern Lake Michigan. Limnol. Oceanogr. **53**: 589–604. doi:10.4319/lo.2008.53.2.0589

Kirillin, G., and others. 2012. Physics of seasonally ice-covered lakes: A review. Aquat. Sci. **74**: 659–682. doi: 10.1007/s00027-012-0279-y

Knowles, R., and D. R. S. Lean. 1987. Nitrification: A significant cause of oxygen depletion under winter ice. Can. J. Fish. Aquat. Sci. **44**: 743–749. doi:10.1139/f87-090

Konneke, M., A. E. Bernhard, J. R. de la Torre, C. B. Walker, J. B. Waterbury, and D. A. Stahl. 2005. Isolation of an autotrophic ammonia-oxidizing marine archaeon. Nature **437**: 543–546. doi:10.1038/nature03911

Kraigher, B., and I. Mandic-Mulec. 2011. Nitrification activity and community structure of nitrite-oxidizing bacteria in the bioreactors operated with addition of pharmaceuticals. J. Hazard. Mater. **188**: 78–84. doi:10.1016/j.jhazmat.2011.01.072

Magnuson, J. J., and others. 2000. Historical trends in lake and river ice cover in the Northern Hemisphere. Science **289**: 1743–1746. doi:10.1126/science.289.5485.1743

McKnight, D. M., B. L. Howes, C. D. Taylor, and D. D. Goehringer. 2000. Phytoplankton dynamics in a stably stratified Antarctic lake during winter darkness. J. Phycol. **36**: 852–861. doi:10.1046/j.1529-8817.2000.00031.x

Özkundakci, D., A. S. Gsell, T. Hintze, H. Täuscher, and R. Adrian. 2016. Winter severity determines functional trait composition of phytoplankton in seasonally ice-covered lakes. Glob. Chang. Biol. **22**: 284–298. doi:10.1111/gcb.13085

Paerl, H. W., and others. 2016. It takes two to tango: When and where dual nutrient (N & P) reductions are needed to protect lakes and downstream ecosystems. Environ. Sci. Technol. **50**: 10805–10813. doi:10.1021/acs.est.6b02575

Pauer, J. J., and M. T. Auer. 2000. Nitrification in the water column and sediment of a hypereutrophic lake and adjoining river system. Water Res. **34**: 1247–1254. doi: 10.1016/S0043-1354(99)00258-4

Pernica, P., R. L. North, and H. M. Baulch. 2017. In the cold light of day: The potential importance of under ice convective mixed layers to primary producers. Inland Waters **7**: 138–150. doi:10.1080/20442041.2017.1296627

Pettersson, K., K. Grust, G. Weyhenmeyer, and T. Blenckner. 2003. Seasonality of chlorophyll and nutrients in Lake Erken – effects of weather conditions. Hydrobiologia **506**: 75–81. doi:10.1023/B:HYDR.0000008582.61851.76

Powers, S. M., and S. E. Hampton. 2016. Winter limnology as a new frontier. Limnol. Oceanogr. Bull. **25**: 103–108. doi:10.1002/lob.10152

R Core Team. 2016. R: A language and environment for statistical computing. R Foundation for Statistical Computing, Vienna, Austria. Available from https://www.R-project.org/.

Randall, C. W., and D. Buth. 1984. Nitrite build-up in activated sludge resulting from temperature effects. J. Water Pollut. Control Fed. **56**: 1039–1044.

Rosi-Marshall, E. J., and T. V. Royer. 2012. Pharmaceutical compounds and ecosystem function: An emerging research challenge for aquatic ecologists. Ecosystems **15**: 867–880. doi:10.1007/s10021-012-9553-z

Rysgaard, S., N. Risgaard-Petersen, P. N. Sloth, K. Jensen, and P. L. Nielsen. 1994. Oxygen regulation of nitrification and denitrification in sediments. Limnol. Oceanogr. **39**: 1643–1652. doi:10.4319/lo.1994.39.7.1643

Sampson, C. J., and P. L. Brezonik. 2003. Responses of nutrients to experimental acidification and recovery in Little Rock Lake, USA. Water Air Soil Pollut. **142**: 39–57. doi:10.1023/A:1022026610079

Small, G. E., G. S. Bullerjahn, R. W. Sterner, B. F. N. Beall, S. Brovold, J. C. Finlay, M. L. McKay, and M. Mukherjee. 2013. Rates and controls of nitrification in a large oligotrophic lake. Limnol. Oceanogr. **58**: 276–286. doi: 10.4319/lo.2013.58.1.0276

Stanley, E. H., and J. T. Maxted. 2008. Changes in the dissolved nitrogen pool across land cover gradients in Wisconsin streams. Ecol. Appl. **18**: 1579–1590. doi:10.1890/07-1379.1

Verpoorter, C., T. Kutser, D. A. Seekell, and L. J. Tranvik. 2014. A global inventory of lakes based on high-resolution satellite imagery. Geophys. Res. Lett. **41**: 6396–6401. doi:10.1002/2014GL060641

Voytek, M. A., J. C. Priscu, and B. B. Ward. 1999. The distribution and relative abundance of ammonia-oxidizing bacteria in lakes of the McMurdo Dry Valley, Antarctica. Hydrobiologia **401**: 113–130. doi:10.1023/A:1003754830988

Wankel, S. D., A. C. Mosier, C. M. Hansel, A. Paytan, and C. A. Francis. 2011. Spatial variability in nitrification rates and ammonia-oxidizing microbial communities in the agriculturally impacted Elkhorn Slough Estuary, California. Appl. Environ. Microbiol. **77**: 269–280. doi:10.1128/AEM.01318-10

Wetzel, R. G. 2001. Limnology: Lake and river ecosystems, 3rd ed. Gulf Professional Publishing.

Weyhenmeyer, G. A., and others. 2007. Nitrate-depleted conditions on the increase in shallow northern European lakes. Limnol. Oceanogr. **52**: 1346–1353. doi:10.4319/lo.2007.52.4.1346

Weyhenmeyer, G. A. 2009. Do warmer winters change variability patterns of physical and chemical lake conditions in Sweden? Aquat. Ecol. **43**: 653–659. doi:10.1007/s10452-009-9284-1

Yang, Y. Y, Y. Dai, N. N. Li, B. X. Li, S. G. Xie, and Y. Liu 2017. Temporal and spatial dynamics of sediment anaerobic ammonium oxidation (anammox) bacteria in freshwater lakes Microbial Ecology **73**: 285–295. doi:10.1007/s00248-016-0872-z

Acknowledgments

We thank the dedicated scientists who have sustained field sampling and laboratory efforts of the NTL-LTER, in particular Pam Montz and Tim Meinke who collected the vast majority of samples used in this study. Funding was provided by the National Science Foundation (NSF DEB #1431428; NSF DEB #1136637), North Temperate Lakes Long-term Ecological Research (NTL-LTER) project (NSF DEB-1440297), and Washington State University. Several people provided generous input on this research following the "Ecology under ice" session at the 2016 ASLO meeting in Santa Fe, New Mexico. Support was provided by the Center for Environmental Research, Education, and Outreach (CEREO) at WSU. Any use of trade, firm, or product names is for descriptive purposes only and does not imply endorsement by the U.S. government.

11

The role of microbial exopolymers in determining the fate of oil and chemical dispersants in the ocean

Antonietta Quigg,[*1,2] Uta Passow,[3] Wei-Chun Chin,[4] Chen Xu,[5] Shawn Doyle,[2] Laura Bretherton,[1] Manoj Kamalanathan,[1] Alicia K. Williams,[1] Jason B. Sylvan,[2] Zoe V. Finkel,[6] Anthony H. Knap,[2] Kathleen A. Schwehr,[5] Saijin Zhang,[5] Luni Sun,[5] Terry L. Wade,[2] Wassim Obeid,[7] Patrick G. Hatcher,[7] Peter H. Santschi*[2,5]

[1]Department of Marine Biology, Texas A & M University at Galveston, Galveston, Texas; [2]Department of Oceanography, Texas A & M University, College Station, Texas; [3]Marine Science Institute, University of California Santa Barbara, Santa Barbara, California; [4]School of Engineering, University of California - Merced, Merced, California; [5]Department of Marine Science, Texas A & M University at Galveston, Galveston, Texas; [6]Environmental Science, Mount Allison University, New Brunswick, Sackville, Canada; [7]Department of Chemistry and Biochemistry, Old Dominion University, Norfolk, Virginia

Scientific Significance Statement

Extracellular polymeric substances (EPS) are a group of chemically heterogeneous polymers released into the environment by microbes (bacteria, archaea, and phytoplankton), often in response to environmental stresses. EPS serve an important role in determining the fate and transport of oil after a spill, but relatively little is known about EPS production in relation to oil and dispersants, especially at molecular and chemical levels. Here, we summarize the scope of our current knowledge and identify major knowledge gaps.

Abstract

The production of extracellular polymeric substances (EPS) by planktonic microbes can influence the fate of oil and chemical dispersants in the ocean through emulsification, degradation, dispersion, aggregation, and/or sedimentation. In turn, microbial community structure and function, including the production and character of EPS, is influenced by the concentration and chemical composition of oil and chemical dispersants. For example, the production of marine oil snow and its sedimentation and flocculent accumulation to the seafloor were observed on an expansive scale after the Deepwater Horizon oil spill in the Northern Gulf of Mexico in 2010, but little is known about the underlying control of these processes. Here, we review what we do know about microbially produced EPS, how oil and chemical dispersant can influence the production rate and chemical and physical properties of EPS, and ultimately the fate of oil in the water column. To improve our response to future oil spills, we need a better understanding of the biological and physiochemical controls of EPS production by microbes under a range of environmental conditions, and in this paper, we provide the key knowledge gaps that need to be filled to do so.

*Correspondence: quigga@tamug.edu

Author Contribution Statement: AQ, UP, PHS conceived the premise for the paper and coordinated its preparation; WCC, CX, KAS, SZ, LS were major contributors to preparation the operational definitions and chemical components of the EPS including the text box and summary figure; LB, MK, AKW, ZVF contributed the phytoplankton sections while SD, JBS focused on the microbial sections; TLW, AHK contributed the oil related information while WO and PGH examined the MOS and EPS in significant chemical detail. The data (Figs. 2, 3 and 4) presented are the outcome of a 5 d mesocosm experiment in which all authors participated. Quigg and Santschi are accountable for the integrity of the data, analysis, and presentation of findings as a whole.

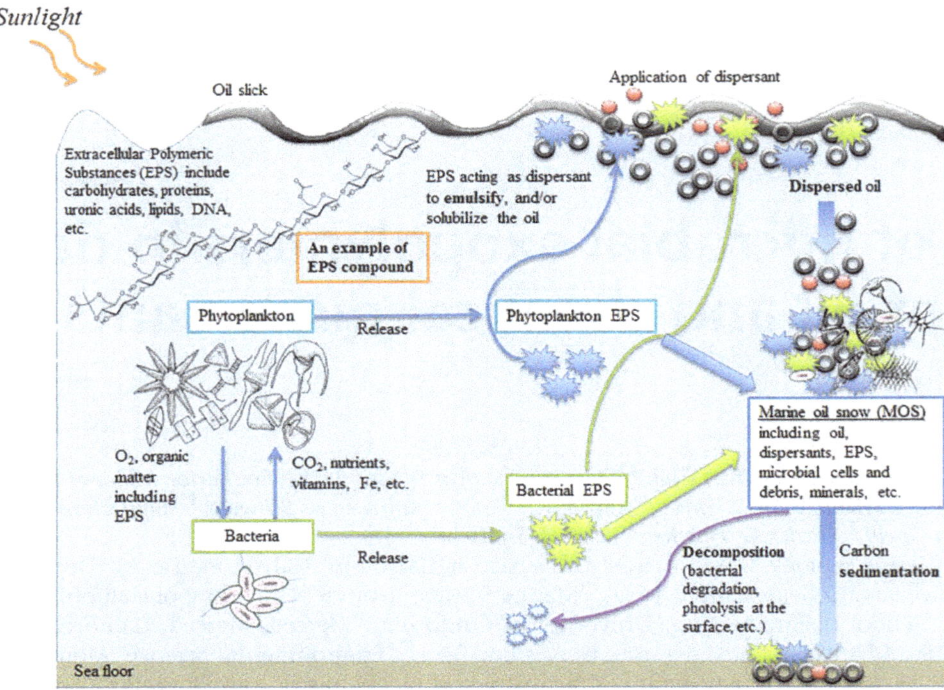

Fig. 1. This summary figure shows the interactions between microbes, EPS, oil, and dispersant. In this figure, we give an example of an EPS molecule (typically polysaccharides, glycoprotein backbones) which is excreted by microbes—bacteria, archaea, or phytoplankton. Factors which affect the production and composition of EPS include but are not limited to nutrients, light, pH, redox, salinity as well as the presence of oil and/or dispersants. Marine oil snow (MOS) forms when compounds such as the EPS, oil, dispersant combine; but it may also contain cells or cellular debris, particulates and other materials floating in the water column. In this capacity, EPS function (1) to **emulsify** and **disperse** petroleum components, and thus enhance their solubility and bioavailability, (2) to provide physical structure allowing the microbial communities to **degrade** oil, and (3) to provide colloidal or particulate traps for petroleum components and subsequent **sedimentation** (the formation of marine oil snow) through chemical coagulation, **aggregation** and cross-linking processes.

Marine microbes are capable of producing high molecular weight exudates, called extracellular polymeric substances (EPS; also: exopolymeric substances, exopolysaccharides) (Fig. 1). The term EPS describes a heterogeneous group of materials existing in a size continuum from dissolved to colloidal phases, and includes gels, which function as particles. EPS may be attached in cell coatings, biofilms or colony matrices, or free floating. These polymeric substances have varying functional roles (e.g., protecting microbes, aiding their attachment), chemical and physical properties (Decho 1990; Hoagland et al. 1993; Decho and Herndl 1995; Leppard 1997; Verdugo 2012), and appearance (Figs. 2–5). The production of EPS by planktonic microbes can influence the fate of oil and chemical dispersants in the ocean through emulsification, degradation, dispersion, aggregation, and/or sedimentation. In turn, microbial community structure and function, including the production and character of EPS, is influenced by the concentration and chemical composition of oil and chemical dispersants. As such, EPS serve an important role in determining the fate and transport of oil after a spill, but relatively little is known about EPS production in relation to oil and dispersants, especially at molecular and chemical levels. Oil spills remain a widespread problem, with an average 96 incidents reported per year over the last decade in the U.S.A. alone (NOAA, https://incidentnews.noaa.gov/raw/index, last accessed 09-22-2016), and marine ecosystems can take many years to show signs of recovery (Peterson et al. 2003). Characterizing the microbial processes that govern EPS production—and ultimately the fate of oil—can contribute to both explaining and predicting the response of marine systems to these catastrophic events.

Transparent exopolymer particles (TEP) are arguably the most abundant and studied form of particulate EPS in the oceans (Passow 2002), in part because they are extremely surface active (Mopper et al. 1995; Zhou et al. 1998) and are the underlying matrix of marine snow (e.g., Alldredge et al. 1993; Passow et al. 1994; Kiørboe 2001; Simon et al. 2002), which drives particle sedimentation (e.g., see Fig. 2). Transport of particles to the bottom of the ocean via marine snow is a key pathway of the carbon cycle (Alldredge and Gotschalk 1988, 1989; Asper et al. 1992; Burd and Jackson 2009; Fu et al. 2014; Daly et al. 2016). The microbial release of EPS in aquatic environments may be thought of as part of an autopoietic system, that is, a self-sustaining community response (Varela et al. 1974;

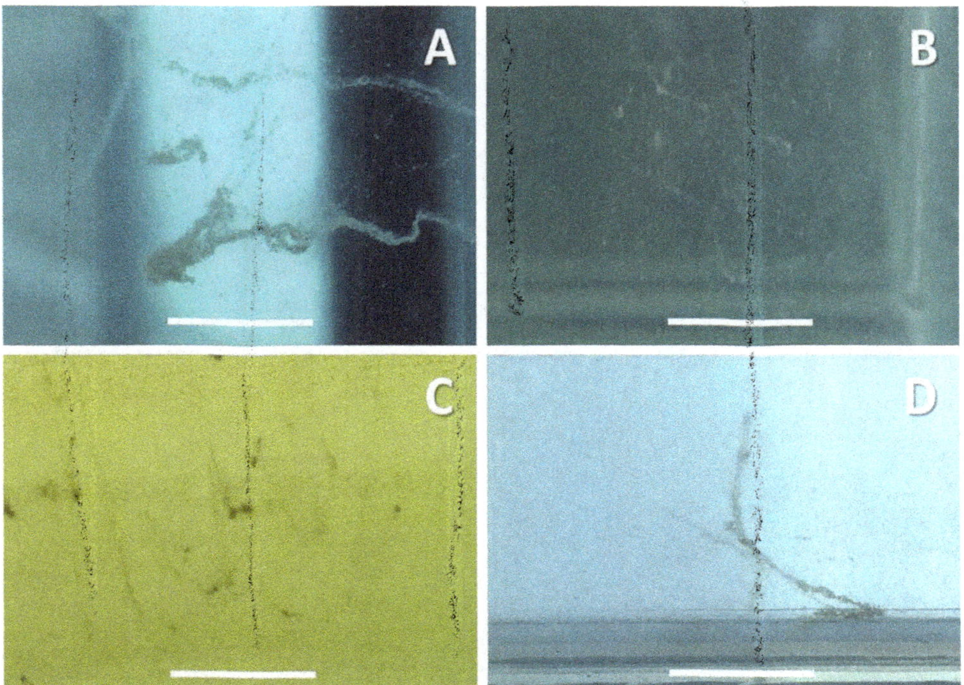

Fig. 2. Mesocosm experiments were performed as part of the ADDOMEx GOMRI funded program with seawater collected from the NGOM and "seeded" with microbial populations collected from Galveston Bay (Texas). Four treatments were set up in triplicate (**A**) Control, (**B**) water accommodated oil fraction, (**C**) water accommodated oil fraction with Corexit 9500, and (**D**) a 10-fold diluted equivalent of (**C**). Bars represent 2 cm.

Maturana and Varela 1980; Santschi et al. 2003). If this is indeed the case, then EPS production is of great ecological importance. The processes, mechanisms, and participants however are poorly known and require further study.

The physical-chemical mechanisms by which EPS enhance aggregation and/or dispersion of oil need to be better understood. Additionally, it is important to investigate how the presence of dispersants affect mechanisms leading to the sedimentation of oil in a ternary system (oil-dispersant-EPS), compared to that in a binary system (oil-EPS) (Fig. 6). EPS have an important role in determining the fate and transport of oil after a spill, but relatively little is known about EPS production in relation to oil and dispersants, especially how the amphiphilicity and surface activity of EPS are controlled at molecular and chemical levels. In this article, we review what we do know about microbially produced EPS, how oil and chemical dispersant can influence the production rate and chemical and physical properties of EPS, and ultimately the fate of oil in the water column. In addition, we discuss this work in the context of the Deep Water Horizon (DwH) oil spill because it is the first such event in which the importance of marine oil snow (MOS), a biological process, as a transport pathway for oil to the seafloor, was observable. The scale of this production of MOS clearly has an important impact of the ecosystem, its alterations and/or recovery, and will do for some time to come.

Background on the study of marine snow and oil spills

Interest in the significance of marine snow can be traced back to observations from the 1800s when EPS-rich marine snow was considered a potentially important food source for deep-sea animals (see review by Silver 2015). Pioneering studies of "suspended, mucous-rich particles, containing abundant microbial populations" however did not begin until the 1940s and 1950s (Silver 2015). Silver et al. (1978) was the first to quantify, in situ, both the abundance and microbial community composition of marine snow particles, revealing them to be a "hot spot" of microbial activity (see also Fig. 3). Later work found that a majority of carbon cycling in the sub-euphotic zone occurs within marine snow, which have elevated bacterial production and exoenzyme activity compared to free-living microbes in the water column (Cho and Azam 1988; Smith et al. 1992; Ziervogel et al. 2012; Arnosti et al. 2016). More recently, EPS have been shown to be a major reservoir of organic carbon (ca. 10–25%) in the ocean, with an estimated global pool of $\sim$ 70 Gigatons, or more than two times the total amount of living biomass (Hansell and Carlson 2001; Verdugo et al. 2004; Verdugo and Santschi 2010). In addition, there has been a renewed interest in better understanding marine snow because of its perceived role in influencing the fate of both oil and dispersants in marine environments.

For example, during the Deepwater Horizon (DwH) oil spill in the Northern Gulf of Mexico (NGOM), different types

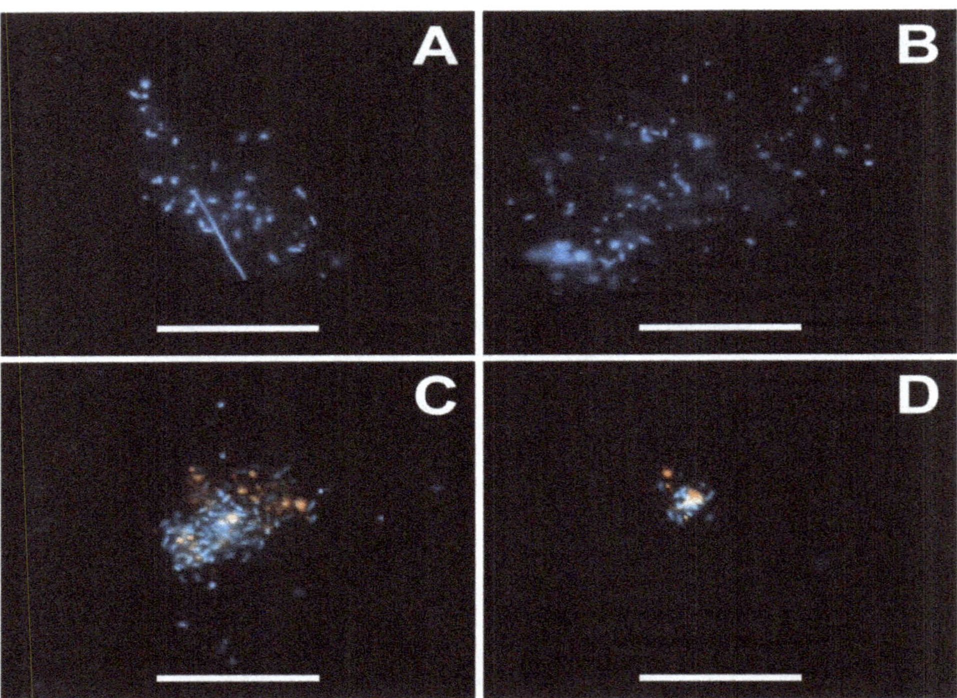

Fig. 3. Mesocosm experiments were performed (as part of the ADDOMEx GOMRI funded program) with seawater collected from the NGOM and "seeded" with microbial populations collected from Galveston Bay (Texas). False-colored photographs of micro-scale microbial aggregates collected from four different mesocosm treatments after 48 h. (**A**) Control, (**B**) water accommodated oil fraction, (**C**) water accommodated oil fraction with Corexit 9500, and (**D**) a 10-fold diluted equivalent of (**C**). Water samples were fixed with formalin (2% v/v final concentration), stained with DAPI (45 μM final concentration), and visualized on a Zeiss Axio Imager 2 microscope. Bars represent 20 μm. The DAPI-stained cells appear blue, while oil /Corexit appears orange. In (**B**), no oil droplets or globules were observed.

of marine snow formed that incorporated oil (Passow et al. 2012; Ziervogel et al. 2012; Daly et al. 2016). It has been suggested that the formation of this MOS (Passow et al. 2012; Daly et al. 2016; Passow and Hetland 2016) ultimately resulted in 4–31% of the DwH oil being returned back to the seafloor or deposited on nearby corals as a loose flocculent (White et al. 2012; Valentine et al. 2014; Chanton et al. 2015, Passow and Ziervogel in press). Mucus-rich MOS differed from aggregates containing oil, in that it formed in the absence of particles other than bacteria. Like a biofilm, MOS consists of a matrix of bacterial colonized EPS, whereas aggregates include diverse particles (e.g., algae) embedded in an EPS matrix. In this way, EPS arguably served not only as a vehicle for the transport of oil to depth via EPS-rich marine snow, but also acted as a natural dispersant. Its amphiphilic properties—possessing both hydrophilic (*water-loving,* polar) and lipophilic (*fat-loving, non-polar*) characteristics—increased the interfacial area between oil and microbes and thus enhanced its biodegradation (e.g., Kappell et al. 2014; Ron and Rosenberg 2002; Head et al. 2006; Quiroz et al. 2006; Ding et al. 2008, 2009; McGenity et al. 2012; Gutierrez et al. 2013; Daly et al. 2016; Joye et al. 2016; Passow and Ziervogel in press).

Microbes (bacteria, archaea, and phytoplankton) that are capable of conducting oil degradation are ubiquitously found in marine waters, but typically represent only a small fraction of the pre-spill communities (Head et al. 2006; Valentine et al. 2014; Baelum et al. 2012). The release of petroleum hydrocarbons, which includes polycyclic aromatic hydrocarbons (PAHs) and other petrocarbons, triggers a complex cascade of microbial responses. No single species dominates, but instead microbial consortia develop (MacNaughton et al. 1999; Kappell et al. 2014; Head et al. 2006; Baelum et al. 2012; Joye et al. 2016), with EPS forming the matrix of microbial aggregates and are functionally comparable to biofilms. Further, tiny oil droplets trapped in the EPS matrix are available to the microbial communities in the aggregates (Gutierrez et al. 2013; Fig. 3B–D). During the DwH oil spill which lasted 3 months, the bacterial community response was dominated by a large bloom of hydrocarbon-degrading Gammaproteobacteria. This bloom included abundant populations of alkane-degrading bacteria from the order *Oceanospirillales* and obligate PAH degraders of the genus *Cycloclasticus* (Kappell et al. 2014; Hazen et al. 2010; Baelum et al. 2012; Mason et al. 2012; Dubinsky et al. 2013; Rivers et al. 2013; Joye et al. 2016). Other microbial groups found in the post-spill community included various members of the metabolically versatile *Roseobacter* clade as well as moderately psychrophilic members of *Colwellia* with

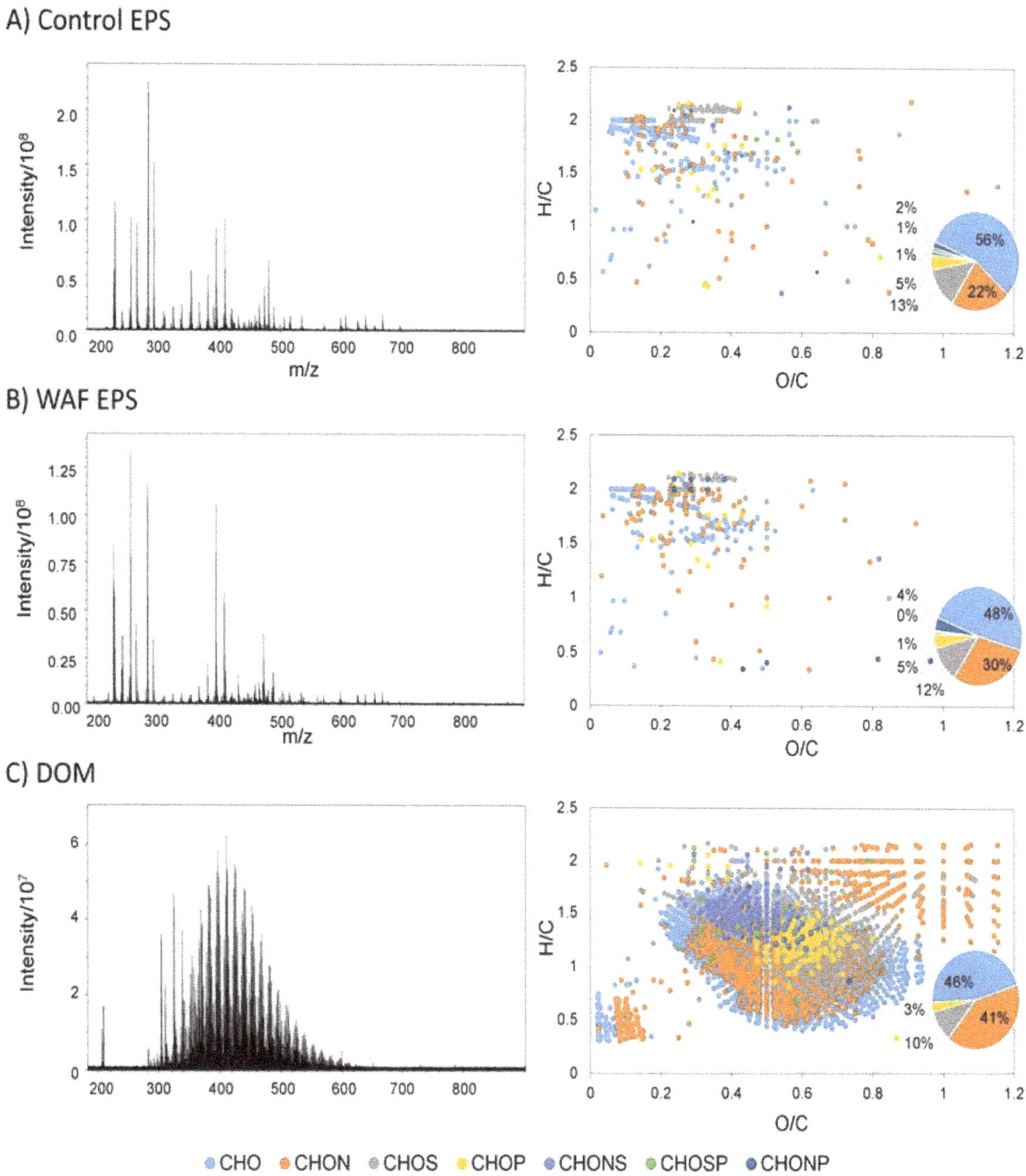

Fig. 4. Mesocosm experiments were performed as part of the ADDOMEx GOMRI funded program with seawater collected from the NGOM and "seeded" with microbial populations collected from Galveston Bay (Texas). Of the treatments, negative mode Electrospray Ionization Mass Spectrometry Coupled Fourier Transform Ion Cyclotron Resonance Mass Spectrometry (ESI-FTICR-MS) spectra (left) and their associated Van Krevelen diagrams (right) of (**A**) pyridine extract of marine snow collected from a control mesocosm tank, (**B**) pyridine extract of marine snow collected from a WAF mesocosm tank, and (**C**) bulk DOM from a control mesocosm tank. Pie chart inserts denote percentages of molecular formulas containing CHO, CHON, CHOS, CHONS, CHOSP, and CHONP.

the ability to degrade a wide range of hydrocarbon compounds (Prince et al. 2010; Redmond and Valentine 2012; Arnosti et al. 2016; Yang et al 2016). Archaeal communities were dominated by *Euryarchaeota* or *Thaumarchaeota* consistent with non-spill conditions, but the relationships between archaea and oil remain difficult to resolve and require further study (Redmond and Valentine 2012). Less is known about the role of eukaryotic autotrophs in the degradation of oil, but some species, like the diatom *Skeletonema* sp. were abundant in sediment traps after the spill

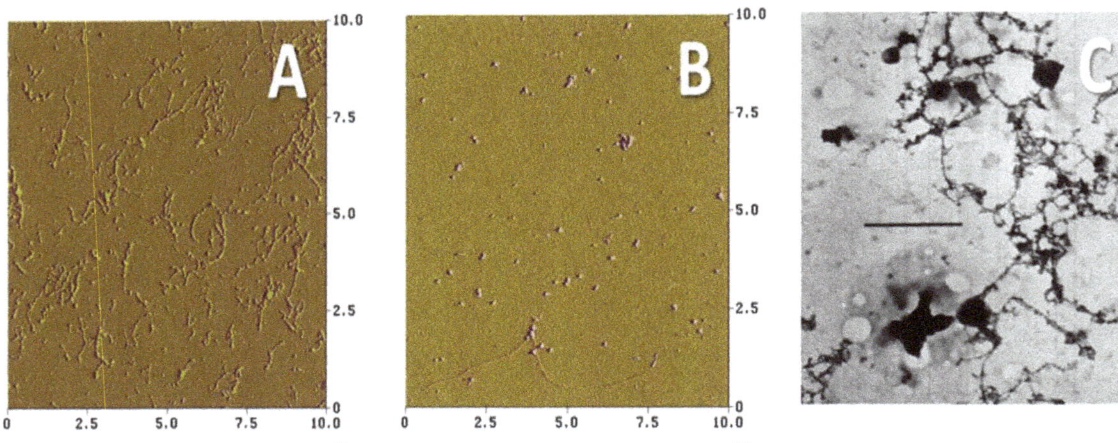

Fig. 5. Fibrils in colloidal organic matter (1–3 nm thickness and 100s to 1000s of nm in length) from the Middle Atlantic Bight, taken at 2 m and 2600 m water depth, which had modern radiocarbon ages when composed of 100% carbohydrate (Santschi et al. 1998; with permission from Wiley Rightslink® by Copyright Clearance Center); **(A)** and **(B)** imaged by Atomic Force Microscopy, after mounting on mica, with image size of 10 × 10 μm; **(C)** imaged by Transmission Electron Microscopy, after staining with Ru oxychloride, and mounted on hydrophilic nanoplast resin. Image bar is 500 μm. All images show the "spiderweb" fibrillar structures appearing as pearls on necklaces; fibrillar forms of colloids were only found in surface and bottom waters, but not in mid-depth waters.

(Yan et al. 2016), and are commonly associated with oil residues in sediments (Parsons et al. 2014).

Extracellular polymeric substances

Operational definitions of EPS

EPS, extracellular polymeric substances, are "operationally" defined based upon their characteristics, size(s), and methods of quantification (Table 1). Exopolymeric substances released by cells (e.g., polysaccharides, proteins) are typically nanofibers (Leppard 1997; Santschi et al. 1998) which self-assemble into colloidal sized nanogels (Chin et al. 1998; Verdugo et al. 2008; Orellana and Leck 2015) that are highly organized crystalline-ridged threads, with widths of 1–2 nm and lengths of several μm (Fig. 5). Nanogels form larger gels (micro-gels) and porous networks (Passow 2000; Verdugo 2007; Verdugo and Santschi 2010). Transparent exopolymer particles, TEP, are an operationally defined group of particles that form from dissolved precursors (see text box) and have the properties of gels (Bar-Zeev et al. 2015). In the ocean, self-assembly of gel particles (nano-or microgels) from colloidal EPS (e.g., Verdugo et al. 2004; Verdugo and Santschi 2010) is rapid, i.e., it only takes hours to days (Chin et al. 1998; Ding et al. 2008, 2009), and is dependent on temperature, salinity, and pH (Chen et al. 2015) as well as ultraviolet (UV) radiation (Orellana and Verdugo 2003) (see also Fig. 6). TEP, which are retained on a polycarbonate filter (usually $> 0.4 - 100$'s μm's; but per definition exclude coatings, or colony matrixes) are stained with Alcian blue, a dye specific for acidic polysaccharides. The amount of dye bound to TEP may be determined colorimetrically and TEP concentrations expressed as gum xanthan equivalents (Alldredge et al. 1993; Passow and Alldredge 1995). When calibrated with alginic acid, the stained particles have also been called acid polysaccharide particles (APS), with results dependent on reference compound (Hung et al. 2003; Santschi et al. 2003). TEP precursors (Thornton 2002) or specifically colloidal TEP (cTEP, 0.05 μm < particle size < 0.4 μm; Villacorte et al. 2009), may be distinguished separately (see text box). Coomassie stained particles (CSP) are proteinaceous particles made visible with the dye Coomassie blue, and exhibit a different dynamic than TEP (Long and Azam 1996; Nagasaki et al. 2004; Verdugo et al. 2008; Cisternas-Novoa et al. 2015). Marine snow (Figs. 2, 5), defined as composite particles > 0.5 mm, consists of a gel matrix and solid particles like cells (bacteria, phytoplankton), feces (zooplankton), detritus and/or minerals (Alldredge and Silver 1988, 1989; Asper et al. 1992; Diercks and Asper 1997; Pilskaln et al. 1998; Armstrong et al. 2009). When EPS concentrations are high, marine snow may be distinctly different in appearance from diatom aggregates or fecal matter. Figure 2 provides examples of some of the differences possible. It is not well understood how the differently defined forms and types of EPS are related to each other (see also Bar-Zeev et al. 2015). This makes the task of identifying their relationships and interactions extremely complex and challenging. Below we make an effort to do so.

Chemical composition of EPS

Chemically, EPS consist largely of acidic polysaccharides and proteins that occur in the form of glycoproteins, proteoglycans, glycolipids, uronic acids and other macromolecules including DNA and enzymes (Azam 1998; Verdugo et al. 2004; Verdugo 2007; Zhang et al. 2008; Xu et al. 2011*a,b*; Zhang and Santschi 2009; Verdugo and Santschi 2010). In the colloidal (but not marine snow sized) fraction of EPS, only 6–18% of carbohydrates are acidic polysaccharides

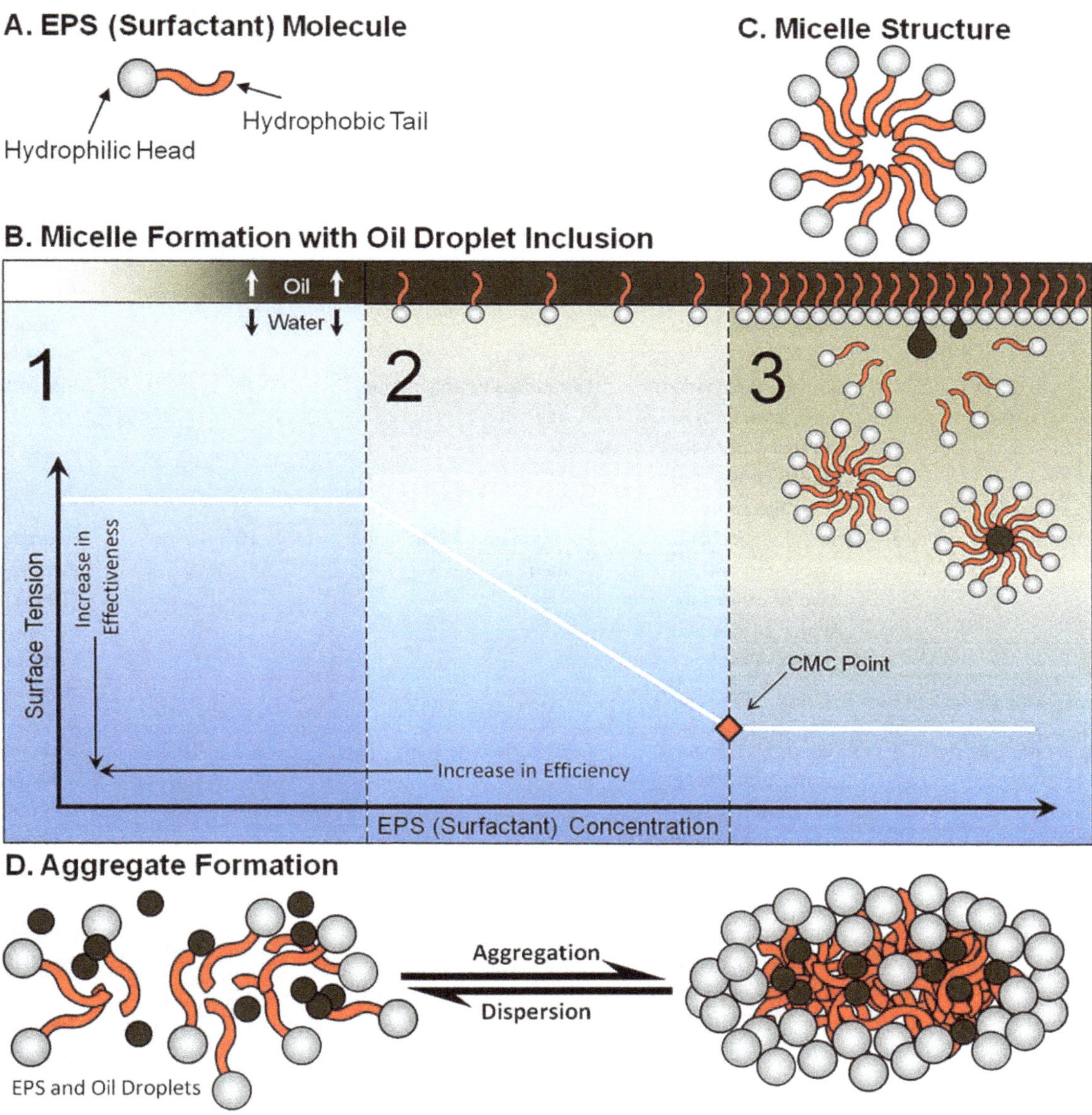

Fig. 6. (**A**) shows the amphiphilic nature of an EPS molecule as a surfactant with a hydrophilic head group and a hydrophobic tail. (**B**) Demonstrates three stages of interfacial tension (IFT) as a function of EPS surfactant concentration, where (**Phase 1**) is the highest IFT at the interface between oil/water and little or no EPS; (**Phase 2**) an increasing concentration of EPS molecules are diffusing to the interface and are thus decreasing the IFT between the oil and water interface; (**Phase 3**) At the onset of this stage, the interfacial surface has become saturated with EPS molecules and the critical micellar concentration (CMC point) is reached, where micelles, (**C**), form. (**C**) is a representation of a micelle with the EPS biosurfactants oriented in a fashion that solubilizes and entrains oil, organic matter, and hydrophobic portions of EPS molecules while still trapping water within, emulsifying and dispersing the trapped material or, (**D**), aggregating into networks of gels.

(including uronic acids; Guo et al. 2002), whereas acidic polysaccharides dominate the < 0.45 μm fraction, and specifically the colloidal fraction which can be captured by a 1 kDa ultra filtering membrane (Zhang et al. 2008; Xu et al. 2009). Total carbohydrates (neutral plus charged) can make up 40% (or less) of the total particulate organic carbon content of EPS (Guo et al. 2002; Hung et al. 2003; Santschi et al. 2003; Xu et al. 2011*a,b*). Of the polysaccharides associated with marine algae and bacteria in the surface waters in the NGOM after the DwH spill, pullulan [α(1,6)-linked maltotriose(glucose)], laminarin [β(1,3-glucose)], xylan (xylose), fucoidan (sulfated fucose), arabinogalactan (arabinose and galactose), and chondroitin sulfate (sulfated N-acetylgalactoseamine and glucuronicacid) were measured in aggregate-oil water-ambient water combinations by Arnosti et al. (2016). In culture, phytoplankton have been found to produce highly diverse and complex sugars, primarily pentoses, hexoses, 6-deoxyhexoses, *O*-methylated sugars, and

Table 1. Operationally defined types of exopolymeric substances with corresponding size, characteristics, and methods using to identify them (primary, but not only methods), and their known abundances in the ocean.

Type (EPS)	Size	Characteristics /composition	Identification method	Abundance*	References
Nanofibers	< 1 μm		AFM		Leppard (1997), Santschi et al. (1998)
Nanogels	< 1 μm		DLS; AFM		Chin et al. (1998), Verdugo (2012)
Microgels	1–10 μm	Proteins, polysaccharides, lipids, DNA	DLS; flow cytometer	$\sim 3 \times 10^6 – 10^{12}$/L	Chin et al. (1998), Verdugo et al. (2008), Verdugo (2012)
Colloidal TEP	> 0.05 μm and < 0.4 μm	Acidic polysaccharide, associated with diverse other substance classes and trace elements	Filtration and Alcian blue stain		Villacorte et al. (2009)
TEP	Operationally defined > 0.2 or 0.4 μm	Acidic polysaccharide, associated with diverse other substance classes and trace elements	Filtration and Alcian blue stain	$\sim 10^3 – 10^9$/L	Alldredge et al. (1993), Passow and Alldredge (1995), Passow (2000,2002), Verdugo (2007), Verdugo and Santschi (2010)
CSP	Operationally defined	Proteins, presumably associated with diverse other substance classes and trace elements	Coomassie blue stain	$\sim 2–3 \times 10^7$/L	Long and Azam (1996), Nagasaki et al. (2004), Verdugo et al. (2008), Cisternas-Novoa et al. (2015)
Marine snow	> 0.5 mm	Composite particles (algae, minerals, feces, detritus. . .) encased in an EPS matrix:	Macroscopically visible, cameras, optical microscopy	$\sim 5 \times 10^{-4} – 5 \times 10^2$/L	Alldredge and Silver (1988), Santschi et al. (1999)
MOS	> 0.5 mm	Oil associated marine snow	Macroscopically visible, cameras, optical microscopy		Ziervogel et al. (2012), Passow et al. (2012), Passow (2016), Daly et al. (2016), Passow and Ziervogel (2016)

*Abundance numbers are intended to be "representative" as they are known to be highly variable.

aminohexoses (e.g., Hoagland et al. 1993; Chiovitti et al. 2003). It is because of their high polysaccharide content, mainly cross-linked with divalent Ca^{2+} cations, that these substances have polymeric characteristics, with their extracellular occurrence leading to their designation as exopolymeric substances.

There is some preliminary evidence that the protein/carbohydrate ratio of EPS exerts control on their hydrophobicity, surface activity and aggregate formation (Stenstörm 1989; Liao et al. 2001, 2002; Dickinson 2003), but the wider applicability of this concept remains to be shown. For example, EPS, which contain a substantial amount of more hydrophobic protein (Xu et al. 2011*a,b*; Zhang et al. 2012), has been shown to be responsible for the faster and nonionic aggregation behavior of some marine gels (Ding et al. 2008, 2009; Chen et al. 2011). In Table 2, measured bacteria and phytoplankton EPS protein/carbohydrate ratios, and their corresponding hydrophobic contact area (relative hydrophobicity and amphiphilicity) are summarized. We found a significantly positive correlation between the protein/carbohydrate ratio and hydrophobic contact area ($p < 0.01$). This suggests that the relative hydrophobicity of EPS, which can be responsible for gel formation, particle aggregation and bioflocculation, are regulated, at least in

Table 2. Examples of the composition of characterized bacterial and phytoplankton EPS. "Attached" EPS are that which adheres to cells and are more hydrophobic than "non-attached" EPS, which is free-floating in seawater.

	Species	EPS	Protein/CHO	Hydrophobic contact area (Å)	Ref.
Bacteria	*Pseudomonas fluorescens* Biovar II	Attached	0.45	58.5	Xu et al. (2011*b*)
	Pseudomonas fluorescens Biovar II	Non-attached	0.05	18.3	Xu et al. (2011*b*)
	Sagittula stellata	Attached	0.18	13.9	Xu et al. (2011*b*)
	Sagittula stellata	Non-attached	0.12	11.2	Xu et al. (2011*b*)
Phytoplankton	*Amphora* sp.	Non-attached	Below detection limit	75.2	Zhang et al. (2012)
	Thalassiosira pseudonana	Attached	2.22	86.4	Zhang et al. (2012)
		Non-attached	0.67	79.4	Zhang et al. (2012)
	Dunaliella tertiolecta	Attached	1.02	85.6	Zhang et al. (2012)
		Non-attached	0.75	72.0	Zhang et al. (2012)
	Phaeocystis globasa	Attached	1.55	n.d.	Zhang et al. (2012)
		Non-attached	0.22	n.d.	Zhang et al. (2012)
	Phaeodactylum tricornutum – with Si in media	Attached	0.93	n.d.	Chuang et al. (2014)
	Phaeodactylum tricornutum – with NO Si in media	Attached	0.50	n.d.	Chuang et al. (2014)

part, by relative protein/carbohydrate ratios. Further, these studies found that "attached" or "capsular" EPS (i.e., that which adheres to cells) are usually more hydrophobic than "non-attached" EPS (i.e., dissolved EPS that are free-floating in the water) (Xu et al. 2011*a*,*b*; Zhang et al. 2012), with the former usually having a higher protein/carbohydrate ratio than the latter. The increase in this ratio has been correlated with a decrease in the negative surface charge which in turn favors aggregate formation (Wang et al. 2006).

EPS production

Both prokaryotes and eukaryotes generate EPS abundantly. Marine bacteria generally produce EPS with higher levels of uronic acids (Table 2) compared to those found in EPS produced by marine eukaryotic phytoplankton and non-marine bacteria (Kennedy and Sutherland 1987; Ford et al. 1991; Bhaskar and Bhosle 2005). The production rate and composition of EPS is also dependent on environmental conditions. Nitrogen and phosphorus limitation, light regime and temperature have all been shown to be important (Mague et al. 1980; Myklestad et al. 1989; Myklestad 1995; Staats et al. 2000; Urbani et al. 2005). Additionally, interactions between eukaryotic phytoplankton and associated bacteria alter production of EPS, especially TEP (Grossart 1999; Grossart et al. 2006; Gärdes et al. 2011; Seebah et al. 2014). Gaerdes and co-authors (2011), for example, found that TEP production was reduced in the absence of either live *Marinobacter adhaerens* HP15 (bacteria) or live *Thalassiosira weissflogii* (diatom) compared to treatments where both were present and alive. Some forms of EPS act as high nutrient substrates for microbes (Azam 1998; Verdugo et al. 2004; Azam and Malfatti 2007; Verdugo and Santschi 2010) and differences in its composition and characteristics will lead to variations in its nutritional value. EPS have also been shown to act as a non-specific ligand for iron and other trace elements, thereby potentially enhancing phytoplankton growth (Steigenberger et al. 2010; Strmečki et al. 2010; Hassler et al. 2015; Quigg 2016). The effectiveness of EPS in scavenging trace elements and nanoparticles, may also benefit microbes by protecting them from toxic trace metal ions (e.g., Cu) and/or nanoparticles (Chen et al. 2011; Zhang et al. 2012, 2013; Quigg et al. 2013).

Relatively little is known about how oil and dispersants influence EPS production in phytoplankton. Lab studies indicate the controls and types of EPS may be group and species-specific. For example, diatoms (planktonic and benthic) produce a number of structures using EPS including stalks (unidirectionally deposited, multilayered structures), tubes (structured pseudofilaments), apical pads (small globular structures attaching cells to each other), adhering films, fibrils (Fig. 5), and cell coatings or colony matrices, e.g., in *Chaetoceros socialis* (Drum 1969; Hoagland et al. 1993). While stalks and apical pads help diatoms attach to substrata, tubes and capsules enclose cells. Other roles for EPS in diatoms include motility, habitat stabilization and colony formation (e.g., microphytobenthic communities found along sand flats), and/or anti-desiccation (find more detail in Hoagland et al. 1993; Thornton 2002). Pelagic diatoms release colloidal EPS (Myklestad et al. 1989; Decho 1990; Leppard 1997), some of which form TEP, possibly to physiologically balance energy with carbon and nutrient acquisition (Passow and Laws 2015) and possibly as a lifecycle strategy to enhance sedimentation and survival (Smetacek 1985).

Coccolithophores, a globally distributed calcifying group of phytoplankton, produce EPS and TEP (Engel et al. 2009;

Biermann and Engel 2010). Calcification in coccolithophores requires a layer of acidic polysaccharides encapsulating the cell for production and attachment of coccoliths (de Jong et al. 1979). Coccoliths are embedded in this acidic polysaccharide layer, which is rich in uronic acid that facilitate Ca^{2+} binding (de Jong et al. 1976; van Emburg et al. 1986). The capsule of acidic polysaccharides will stain with Alcian blue in calcifying strains of the coccolithophore *Emiliania huxleyi*. Polysaccharides from this capsule slough off forming free TEP (Engel et al. 2004). Coccolith shedding by *E. huxleyi* typically occurs at the end of bloom events, in particular following nutrient limitation, viral infection, or changes in light availability (Engel et al. 2009). Given how variable EPS production is between taxa, the influence of oil and dispersants is likely dependent upon the composition of the microbial community. While some laboratory-based studies have investigated these responses, much more work is needed to characterize species-specific interactions with these chemicals.

EPS-marine gels

Marine gels consist of three dimensional cross-linked polymers with seawater as the solvent. Marine gel networks are stabilized by physical entanglement of polymers, such as the formation of Ca^{2+} bridges and are sometimes enhanced by hydrophobic interactions (Chin et al. 1998; Orellana et al. 2003; Ding et al. 2008, 2009). The behavior of nanogels, microgels, and TEP (Table 1) may be explained using gel theory (TEP specifically will be discussed in more detail below) (Chin et al. 1998; Verdugo et al 2004; Orellana and Leck 2015). Due to their unique network structure, the gel mass is mostly liquid, but can behave as sticky solids (Orellana and Leck 2015). Characteristics of marine gels include spontaneous assembly within hours to days, volume phase transition at specific pH or temperature points and dispersion following Ca^{2+} chelation (Chin et al. 1998; Orellana et al. 2007; Verdugo 2012; Chen et al. 2015). TEP are composed of acidic polysaccharide-rich macro-gels, that form from dissolved precursors within hours (Passow 2000), change size due to changes in pH (Mari and Robert 2008) or temperature (Piontek et al. 2009; Seebah et al. 2014), and disperse when Ca^{2+} are removed using EDTA (Alldredge et al. 1993). UV radiation cleaves polymers and inhibits marine gel assembly, or even disperse assembled gel networks (Orellana and Verdugo 2003), including TEP (Ortega-Retuerta et al. 2009). Temperature and pH changes also can reduce marine gel formation and induce dispersion (Chen et al. 2015).

About 10–30% of marine dissolved organic carbon self-assembles spontaneously and reversibly in a two-step process (Verdugo and Santschi 2010; Verdugo 2012; Orellana and Leck 2015). Microbial degradation and UV dispersion can result in shorter polymer chains that lead to gel structure instability and low bioavailability. The bioavailability of small molecules (< 600 Da) is high as bacteria may utilize them directly. Dissolved organic matter polymers that cannot be directly incorporated into the cell, can become bioavailable in gel form that act as hotspots for microbial activity. However, polymers that are too large for direct uptake, but fall short to form stable gels, will remain less bioavailable in the DOM pool (Verdugo 2012; Orellana and Leck 2015). Hydrophobic interactions have been shown to significantly impact the gel assembly process (Ding et al. 2008, 2009; Chen et al. 2011, 2015; Orellana et al. 2011). The addition of oil dispersants, which are rich in hydrophobic and hydrophilic moieties (detergents), is expected to disperse gels, especially those where hydrophobic binding plays a major role in driving gel formation (see Fig. 6). Further, the role of oil in stimulating or impeding this process is under investigation.

EPS-transparent exopolymer particles

TEP as "physical gels" are stabilized by physical entanglements and weak hydrophobic interactions and can easily assemble and disperse (Verdugo and Santschi 2010; Verdugo 2012). The interactions need little activation energy (< 50 kJ mol^{-1}) and are reinforced by Ca^{2+} bridges, which cross-link ionized carboxyl groups (negatively charged) in neighboring TEP chains (Chin et al. 1998), and follow along a thermodynamically favorable assembly process (Verdugo 2012). "Chemical gels" formed along the biotic pathway are less common and characterized by covalently cross-linked, irreversibly attached biopolymers (Verdugo 2012; Bar-Zeev et al. 2015). This pathway involves direct sloughing of mucus from cell capsules or colony matrices of phytoplankton or bacteria. This TEP is frequently interlinked by strong covalent bonds (400 kJ mol^{-1}) and assembly is ordinarily irreversible (Verdugo 2012).

TEP occurs as free-floating gel particles that may form abiotically and spontaneously from dissolved precursors, or biologically when colony matrices disintegrate (e.g., from *Chaetoceros socialis* or *Phaeocystis* spp.; Passow and Wassmann 1994). Abiotic formation constitutes the predominant "physical pathway" where TEP forms from dissolved fibrillar polysaccharides released by planktonic organisms (Passow 2000, 2002, 2012; Verdugo and Santschi 2010; Verdugo 2012; Villacorte et al. 2013). TEP and their "non-particulate" precursors exist in a dynamic equilibrium with TEP (Verdugo et al. 2004). Increased release of TEP may occur during or after algae blooms and localized stress conditions, wherein the decoupling of nutrient availability from energy and carbon supplies may lead to an escalation in TEP release (Passow and Laws 2015). *Thalassiosira pseudonana*, for example, releases large quantities of TEP during the end stage of a bloom when nutrients are becoming limited, corresponding to physiological signs of stress such as decreased photosynthetic efficiency and growth rates (Kahl et al. 2008). Similarly, an increase in the total TEP pool was observed in cultures of the diazotrophic cyanobacteria *Trichodesmium* in response to Fe limitation (Berman-Frank et al. 2007). In coccolithophores, coccolith shedding, associated with stress, increases TEP concentrations and coagulation (Engel et al. 2004; Chow et al. 2015). Large accumulations of TEP may also lead to fish kills as it can smoother the gills and suffocate animals (as observed in McInnes and Quigg 2010), or may lead to large foam events after *Phaeocystis* blooms (Riegman et al. 1992), or to huge mucus events in the Adriatic (Stachowitsch et al. 1990).

During aggregation, TEP provides the glue and matrix to hold particles together (Alldredge et al. 1993). Aggregation rates primarily depend on the size, concentration and stickiness of particles (Jackson 2005; Burd and Jackson 2009). The stickiness is defined as the probability that two particles adhere after they have collided (Jackson 1990). TEP stickiness (typically between 0.05 and 0.8) is orders of magnitude higher than the stickiness of other marine organic particles (typically ≤ 0.01) and thus usually dominates aggregation rate (Mari et al. 2014). However, TEP stickiness varies substantially and the relationship between stickiness and chemical properties of TEP are largely unidentified. Older TEP, which is thought to be modified by bacteria, is stickier than fresh TEP produced by autotrophs (Rochelle-Newall et al. 2010). High concentrations of trace metals reduce TEP stickiness (Mari and Robert 2008). In order to estimate the carbon content in TEP, various conversion factors have been empirically determined (Mari 1999; Engel and Passow 2001); conversion factors for natural samples are currently being determined. Given how much the properties of TEP can vary with environmental conditions, more work is needed to understand the modifications imposed by oil and dispersant exposure. This is key in developing a better understanding of the role of TEP in carbon cycling.

EPS-marine snow

In the presence of solid particles, TEP promote the formation of sinking aggregates because of its high stickiness (Logan et al. 1995; Mari et al. 2014), but in the absence of ballasting particles, TEP accumulates in the sea surface microlayer (Wurl et al. 2009). Marine snow (Figs. 1–4) is central for the transport of organic matter to the deep sea. Only particles large enough to sink rapidly reach the deep ocean where they provide food for heterotrophs and lead to the sequestration of carbon, i.e., its removal from the atmosphere for >100 yr. Marine snow frequently accumulates at fronts or in thin layers, especially at sharp density gradients formed by differences in salinity and other hydrographic parameters (references in Turner 2015) and is ubiquitous after the peak phase of many diatoms blooms (Kiørboe et al. 1994; Thornton 2002; Burd and Jackson 2009). Marine snow is formed either by physical coagulation (collision and attachment) of particles into aggregates or by zooplankton activity (Alldredge and Silver 1988; Jackson 1990; Kiørboe 2001; Simons et al. 2002). The formation of MOS via disproportionate mucus production by bacteria in response to oil, has recently been described (Passow 2016).

Towards identifying factors that determine the chemical properties of marine snow, especially of the TEP mucus matrix, we recently examined marine snow chemically (Chen Xu, unpubl.) and spectrally using electrospray ionization mass spectrometry coupled to Fourier transform ion cyclotron resonance mass spectrometry (ESI-FTICR-MS). This process identified between 300 and 400 individual molecules in the pyridine extracted marine snow fraction (which includes microbial cells and extracellular material) collected from the NGOM (Fig. 4A) and from a water accommodated oil fraction (WAF)-amended NGOM (Fig. 4B). While this number is relatively high; it is far less than that observed in the dissolved organic matter extracted from the same NGOM sample (Fig. 4C). The number of peaks assigned to the various types of elemental formulas (CHO, CHON, CHOS, etc.) are represented as pie chart inserts to each plot. These help us to understand the variety of negatively charged (e.g., carboxyl, phosphoric, sulfate, and hydroxyl) and positively charged (e.g., amino) functional groups biopolymers carry. The dominant formulas are those which contain only CHO elements attributable to lipids (high H/C and low O/C) in the region of the van Krevelen diagram (Fig. 4). The next most abundant formulas are associated with proteins: CHON formulas that appear to be mostly concentrated in the protein-like region of the van Krevelen diagram and those containing S atoms probably associated with S-containing proteinaceous substances. It is worth pointing out that the various extraction methods (Xu et al. 2009; Zhang and Santschi 2009) tested for their ability to maintain cell integrity while releasing most attached EPS (Chen Xu, unpubl.) suggest that 1% EDTA is sufficient to remove this material from algae and bacterial cells without cell damage. Nevertheless, these preliminary results indicate that pyridine extracts of marine snow are dominated by molecules mostly associated with biological organisms.

EPS interactions with petrocarbons and hydrocarbons

The DwH spill released "live oil" into the environment, that is, a mixture of oil and natural gas with very high vapor pressures. Oil exposed to normal temperature and pressures loses its gaseous hydrocarbons (natural gas) and is called "dead oil" (Reddy et al. 2012). In situ burning of oil releases pyrogenic hydrocarbons, which differ appreciably from crude oil (Overton et al. 2016). Moreover, the presence of oil may range from thick slicks to thin sheens at the surface to the WAF dispersed within the water body. The water soluble fraction of oil ($< 1\%$), and its derivative products are complex mixtures of alkanes, cyclohexanes, monoaromatic hydrocarbons (e.g., benzene, toluene, xylenes) and PAH's, with only trace amounts of phenols, nitrogen- and sulfur-containing heterocyclic compounds, and heavy metals

(Saeed and Al-Mutairi 1999; Rodrigues et al. 2010). Small oil droplets are also part of WAF. The fate and transport of oil (and WAF) predominately involves dissolution, dispersion, photo-oxidation, biodegradation, sorption to and desorption from particulates (e.g., sediments, minerals, marine snow) (National Research Council (NRC) 2005, Fu et al. 2014; Overton et al. 2016). Dispersion and emulsification of oil may increase the bioavailability of oil products to biodegradation (NRC 2005); marine microbes play and important role in facilitating this activity.

The presence of oil or pyrogenic hydrocarbons and/or petrocarbons released during burning of oil impact aggregation rates and aggregate characteristics. Oil may be incorporated into aggregates during their formation, or collected when aggregates sink through oil plumes, or MOS may form via microbial response to oil (Passow 2016). Arnosti et al. (2016) and others (e.g., Azam 1998; Azam and Malfatti 2007; McGenity et al. 2012; Ziervogel et al. 2012) have described marine snow, including MOS, as "hot spots" for microbial organic matter and/or hydrocarbon degradation pathways in the ocean. Not only are they a location where petrocarbons and hydrocarbons are degraded but also where microbial biomass and EPS concentrations may be elevated (as shown in Fig. 3). Elevated hydrolytic enzyme activities were measured in the MOS during the DwH spill (Ziervogel et al. 2012; Arnosti et al. 2016) with a specific functionality that was different from those persisting in the surrounding seawater (Ziervogel et al. 2012; Arnosti et al. 2016). Patterns of polysaccharide-hydrolyzing enzyme activities were quite distinct between the individual members of microbial degradation networks, and differences in enzyme activities infer very distinct communities of microbes in the different microenvironments. Additionally, microbial communities in and around MOS changed over time (Arnosti et al. 2016), with communities first degrading the more bioavailable petroleum, then the secondary degradation products of the petroleum and microbial exudation products (Arnosti et al. 2016).

There appears to be direct (oxidizing, hydrolyzing, and assimilating) and indirect (emulsifying with EPS) participation by microbes in oil degradation pathways of MOS, as well as a primary and secondary set of degraders. Bacteria with genes for hydrocarbon degradation utilize or metabolize oil (Hazen et al. 2010; Prince et al. 2010; Valentine et al. 2010; Kessler et al. 2011; Lu et al. 2012; Redmond and Valentine 2012; Ziervogel et al. 2012). For example, members of the genus *Cycloclasticus* were found to directly degrade hydrocarbons during the DwH spill, while members of the genus *Halomonas* played an indirect role on aggregates (Arnosti et al. 2016). The activity of the primary degraders sets off a complex cascade of secondary degraders, including but not limited to those of aggregate-associated bacteria and free-living bacteria not incorporated into the aggregates. Free TEP and marine snow provide physical structure, allowing the development of such complex microbial communities needed to degrade oil efficiently, similar to biofilms. The release of EPS by microbes as a response to petrocarbons also physically protects them, promotes attachment, and/or may serve to emulsify and solubilize oil products, thus increasing their bioavailability (Head et al. 2006; McGenity et al. 2012).

The production of EPS biosurfactants may be one important example of a mutualistic interaction between phytoplankton and bacteria in the presence of oil. Surfactants (surface active "soap-like" molecules) with hydrophilic and hydrophobic moieties (see Fig. 6) increase the bioavailability of certain oil components (McGenity et al. 2012). In this respect, these biosurfactant properties are similar to the dispersants used for oil spill remediation. Chemical surfactants are designed to reduce the interfacial tension between oil and water, thereby enhancing dispersion and potentially biodegradation processes (Lewis et al. 2010). Biosurfactants such as EPS or extracellular polysaccharides can emulsify petrocarbons (Head et al. 2006). EPS produced by *Halomonas* sp., for example, have amphiphilic properties, thereby interacting easily with hydrophobic substrates like hydrocarbons, leading to the solubilization and biodegradation of oil components (Gutierrez et al. 2013). The exopolymers with entrained oil droplets form networks that act as an energy and carbon source to other members of the microbial community.

The abundant, large (mm to cm), mucus-rich marine snow which appeared near the surface in the weeks after the DwH spill was produced by the microbial community in response to the released petrocarbons and hydrocarbons (Passow et al. 2012; Gutierrez et al. 2013; Ziervogel and Arnosti 2013). While some phytoplankton species are negatively impacted (e.g., Prouse et al. 1976; González et al. 2009; Hook and Osborn 2012; Ozhan et al. 2014a,b; Garr et al. 2015) others appear to thrive in the presence of oil (Prince et al. 2010; Almeda et al. 2014; Ozhan and Bargu 2014; Ozhan et al. 2014a,b), although it is unclear if phytoplankton benefit directly from oil, or via a symbiotic relationship with prokaryotes. Oil compounds may enter the food chain through bacteria (Graham et al. 2010; Chanton et al. 2012). Alternatively, we propose they may enter through marine snow or phytoplankton, both of which provides food for zooplankton. Food-web interactions will influence degradation rates of oil in a multitude of ways, such as bacterivorous protists that graze on the oil-consuming bacterial community potentially decreasing degradation rates (Beaudoin et al. 2016).

Experiments have revealed microbial MOS contains fossil carbon with the same ^{13}C signature as the oil (Passow 2016); this microbial MOS has a very different appearance from the aggregates formed due to coagulation of individual particles, like phytoplankton (see also Fig. 2 which provides an example). It consisted predominately of amorphous mucus, which was extremely sticky. It was suggested that in vivo production of microbial mucus results in patches of "floating biofilm" which transforms into MOS once the material begins to sink below the air-oil-seawater interface. These too are aggregates with oil (MOS), but not the microbial MOS.

Diatom aggregates forming after the DwH spill incorporated oil, also formed sinking MOS, which was recovered in sediment traps at >1400 m (Passow 2016; Yan et al. 2016). Recent experimental evidence suggests the presence of oil increased the stability and cohesion of MOS, potentially leading to a faster and more efficient transport to sediments (Fu et al. 2014; Daly et al. 2016). Previous studies have found that in sediments phytoplankton is often closely associated with PAHs (Kowalewska and Konat 1997; Kowalewska 1999; Lubecki and Kowalewska 2010; Parsons 2014). The mechanisms leading to such an association were, however, largely unknown. Scavenging and inclusion of oil in phytoplankton aggregates is one potential mechanism (Alldredge and Gotschalk 1988, 1989). Alternatively, oil sorption to phytoplankton cells may also occur. Differences in the amount of oil incorporated into phytoplankton aggregates depends on the species (observed in lab studies), community composition (observed in field studies), cell packaging as well as the quantity and compositions of EPS (e.g., hydrophobic or hydrophilic) and types of oil (Passow 2016).

In Figs. 2–4, we show marine snow formed during mesocosm experiments performed with NGOM seawater "seeded" with microbial populations collected from Galveston Bay, Texas. Apart from controls, there were two treatments of (1) seawater amended water accommodated fraction (WAF) (Fig. 2B), and (2) seawater amended chemically enhanced WAF (CEWAF) prepared at two different concentrations (Fig. 2C,D). Results showed that despite higher EPS production, oil and/or dispersants preferentially partitioned into the colloidal and suspended particulate fractions rather than into the rapidly forming and sinking MOS (Chen Xu, unpubl.). When the seawater was ameliorated with WAF and CEWAF, we observed marine snow that was visibly and chemically different in composition, quality and quantity to the controls (Figs. 2, 4, respectively). We found that WAF and CEWAF had been incorporated into this material (particularly Fig. 2C), which is evident when micro-aggregates were viewed microscopically (Fig. 3C,D). As part of ongoing studies, this material was carefully characterized. A representative finding of the molecular characterization of the marine snow is given in Fig. 6. Initial analysis reveals that the EPS produced in response to CEWAF were more amphiphilic (higher protein content) and thus enhanced the dispersion of CEWAF and sedimentation of this material. The sedimented MOS contained about 30% of oil compounds, as determined through radiocarbon mass balance. On a cautionary note, these results would not necessarily be expected to apply to different conditions (such as near-shore vs offshore plankton community, nutrient levels, temperature, etc.).

Effects of dispersant on EPS-oil interactions

While the dispersant Corexit was primarily applied in response to the DwH oil spill; a mixture of SPC1000 and Mare-Clean200 were also used (The chaos of clean up (TCCP) 2011). Dispersant was added directly to the oil carpet at the sea surface as well as into the leak at various depths (Kujawinski et al. 2011). Dispersants are mixtures, containing as many as 57 different chemical ingredients that are partially soluble in both oil and water. Corexit consists of nonionic ($\sim$ 48%) and anionic ($\sim$ 35%) surfactants with enough solvent or petroleum distillate ($\sim$ 17%) to make a homogeneous dispersant mixture (Singer et al. 1991). When sprayed on an oil slick, dispersants enhance the dispersion of oil, e.g., its penetration onto and into the water. For example, Ramachandran et al. (2004) applied Corexit 9500 in a 1 : 50 ratio and found that it enhanced oil dispersion into the water by 68%. The increase was likely due to the presence of oil droplets in emulsion and the increased dissolution of petrocarbons and/or hydrocarbons from the surfaces of the numerous droplets (i.e., surface area effects). Increased dispersion is meant to lead to increased biodegradation of oil in water (Lewis et al. 2010), although this is currently controversial.

Corexit retarded, reduced, and/or completely inhibited the formation of MOS on a range of time scales, but response patterns vary (Passow 2016). The presence of Corexit may inhibit aggregation by dispersing micro-gels. Further, Corexit has been found to alter the microbial community responding to WAF by instead selecting for microorganisms that utilize Corexit as a substrate rather than oil (Kleindienst et al. 2015a,b). Other studies have shown the growth and viability of hydrocarbon-degrading microorganisms are reduced in the presence of Corexit (Hamdan and Fulmer 2011). In contrast, Baelum et al. (2012) observed that Corexit had no negative effects on microbial growth. These differences may be due to the respective microorganisms used in each study: Baelum et al. (2012) used natural deep water communities, while Hamdan and Fulmer (2011) used hydrocarbon degrading isolates. However, a host of other factors including oil type, weathering and composition, environmental conditions (temperature, pressure), and differences in EPS production and formation may also have played roles.

Surfactant behavior of EPS and dispersants

EPS may be characterized in terms of surfactant behavior in the presence of oil and dispersant using quantitative techniques. Effectiveness and efficiency of EPS surfactant behavior or that of chemical dispersants may be characterized through interfacial tension (IFT) and critical micelle concentration (CMC) rheology measurements. The *effectiveness* of EPS surfactant (or dispersing agent) is thought to be greater when a lower IFT initiates the CMC. When the CMC is reached at lower EPS concentrations, the *efficiency* of EPS as a surfactant is greater. To demonstrate these concepts, we show EPS behaving as a biosurfactant in Fig. 6, that is, amphiphilic molecules (Fig. 6A) which, through intermolecular forces arrange their hydrophilic (anionic or non-ionic) polar head group in the water and their hydrophobic (nonpolar) tail, which shun water, crossing the interfacial boundary to interact with the air or another hydrophobic substance, such as oil (Phase 1; Fig. 6). Amphiphilic EPS surfactant will form micelles (Fig. 6A,C), thereby dispersing hydrophobic molecules in the water (Fig. 6B).

The IFT is greatest in Phase 1 (Fig. 6), at the interface of the oil and seawater where there is little or no EPS. Amphiphilic molecules as part of EPS secreted by microbes diffuse to the interface (Phase 2; Fig. 6). The presence of these molecules on the surface disrupts the cohesive energy and lowers the IFT (Phase 2; Fig. 6). The purpose of surfactant or chemical dispersant application is to lower the oil/water IFT to promote entrainment of oil droplets into the water column and dispersion of the hydrophobic oil components (NRC 2005). When the surfactant concentration increases to the point that the oil/water liquid interface is saturated, the CMC is reached (between stages 2 and 3; Fig. 6). The concentration of dispersant at which the surfactant molecules form a uniform monolayer at the oil/water interface is the CMC (Phase 3; Fig. 6).

Micelles are generated from the EPS/surfactant molecules and the IFT can be decreased to the point that oil droplets are entrained in the micelles which range in size from $\sim$ 3 nm to 50 nm (Phase 3; Fig. 6). It is within stage 3 that self-assembly occurs, wherein the EPS or dispersant surfactants interact with organic matter with the inclusion of water to reversibly form gels or aggregates (Fig. 6D), such as macrogels ($\sim$ 3–6 um), nanogels ($\sim$ 100–200 nm), TEP, or marine snow (Verdugo 2012). Micelles form with a roughly spherical shape (Fig. 6C). However, depending on the physico-chemical properties of the EPS/surfactant molecules, including the tail length and packing in the micelles, the micellar morphologies may be more cylindrical, laminar, cone-shaped, worm-like, and more (Jusufi et al. 2008), which probably accounts for the variance in morphology of TEP and marine snow from different microbial sources. The anionic moieties of the EPS head groups anneal through bridging with divalent cations prevalent in seawater, dominantly Ca^{2+} and Mg^{2+} ions, to form crystalized aggregate networks that increase the aggregate mass enough to sediment and may provide a mechanism to deter biodegradation (Chin et al. 1998) in addition to polymerization, and condensation of the size due to increasing pressure with depth (Chin et al. 1998).

Effect of reactive oxygen species on EPS production and oil interactions

Reactive oxygen species (ROS), have been found to lead to the production of EPS in natural conditions (Liu et al. 2007; Chen et al. 2009; Gong et al. 2012), and thus may potentially play a role in EPS-oil interactions. ROS includes singlet oxygen ($^1O_2$*), superoxide radical ($\cdot O_2^-$), hydroxide radical ($\cdot$OH) and hydrogen peroxide (H_2O_2), and can be through metabolic and photochemical reactions (Kieber et al. 2003). ROS are capable of oxidizing a wide variety of compounds with relatively low selectivity, and can damage macromolecules including DNA, proteins and lipids. To relieve the oxidative stress of ROS, microbes excrete protective EPS to physically block or chemically quench the hazardous ROS (Liu et al. 2007; Chen et al. 2009; Gong et al. 2012). While ROS generated from high energy regions of visible light (UVB) assist in degrading macromolecules, ROS from light with somewhat lower energy (UVA) can crosslink macromolecules, e.g., proteins from marine bacteria (Sun et al., unpubl.) to form sinking particles. In this context, crosslinking refers to chemically binding to different regions of the macromolecule in a random fashion.

An imbalance between ROS and antioxidant production can lead to protein and lipid polymerization. This has been clearly demonstrated in the health sciences literature (e.g., plaques in the human brain and blood vessels, as reviewed in Valko et al. 2007). The polymerization may arise from cross-linking of radicals, which can be generated through ROS via triplet excited states of the biomolecules. A recent study has shown the proteins from marine bacteria may form aggregate via ROS during irradiation (Sun et al., unpubl.). Radical species that promote polymerization reactions can also be related to the presence of metals, such as Fe. Several studies on dissolved organic matter have shown that Fe takes part in the photo-flocculation process (Gao and Zepp 1998; Helms et al. 2013; Sun et al. 2014). Iron not only accelerates ROS production, such as the Fenton reaction, but can also bind to organic ligands as complexes (Chen et al. 2014; Sun and Mopper 2016). Recent studies by Waggoner et al. (2015) and Chen et al. (2014) have demonstrated, through molecular level analyses, that black-carbon like macromolecules can be formed from terrestrial dissolved organic matter in the presence of sunlight and Fe. Light effects on phytoplankton growth, EPS and ROS production occur simultaneously, and thus, would need to be distinguished to evaluate their effects on stickiness, aggregation, degradation and ultimately the fate of oil.

Additionally, ROS might accumulate during and after an oil spill under sunlight, and result in an increased oxidative stress to the microbes. A recent report showed that substantial amounts of $\cdot$OH (1.2×10^{-16} to 2.4×10^{-16} M) was produced from the DwH spill and was at least one magnitude higher than concentrations of clean seawater (Ray and Tarr 2014). ROS may also play an important role in starting a biological cascade that induces the production of TEP. Diatoms undergoing oxidative stress produced large amounts of TEP following an induction of caspase (a family of enzymes implicated in triggering programmed cell death) activity (Kahl et al. 2008). High caspase activity also preceded a large release of TEP in Fe-limited *Trichodesmium* cultures (Berman-Frank et al. 2007).

Effect of riverine input on EPS production and oil interactions

A freshwater diversion event (as a result of flood waters upstream) shortly after the DwH oil spill introduced significant quantities of not only freshwater but also nutrients, sediments and clay minerals into the NGOM (Hu et al. 2011; Gong et al. 2014; Walsh et al. 2015), all of which potentially affected productivity, EPS production and aggregation. Terrestrial sediment and clay mineral as well as resuspension of benthic materials in shallow near shore waters combined with oil may have led to the formation of oil-mineral aggregates (OMAs), or to the additional ballasting of marine snow (Muschenheim and Lee 2002; Khelifa et al. 2005; Niu et al. 2011; Gong et al. 2014; Passow 2016). OMAs have lower surface areas than oil droplets and hence undergo lower rates of biodegradation due to reduced oil availability. Nutrients, introduced from the Mississippi and Atchafalaya Rivers, may have led to enhanced productivity and potentially to increased EPS formation (Hu et al. 2011; Passow 2016). Alleviation of nutrient stress lasted a long time. In August 2010, 3 weeks after the oil well was capped, an area > 11,000 km^2 exhibited elevated chlorophyll concentrations, which were attributed directly and indirectly to the oil spill (Hu et al. 2011). It has been estimated via a modeling study that 12% of the biotic particle exported resulted from increased loadings of Mississippi River nutrients (Walsh et al. 2015).

Knowledge gaps

The microbial community and their EPS production can be viewed as the first responders to pollutants (oil, metals, nanomaterials) in the marine environment. EPS provides attachment (i.e., coagulation) to particle surfaces (biofilms, cell adhesion, marine snow flocs, MOSSFA), with the degree of attachment controlled directly or indirectly by the protein/carbohydrate ratio through reactions related to the surfactant behavior of EPS and ROS-mediated cross-linking reactions. Such interactions can be thought of as part of simultaneous processes occurring in a tightly balanced system that responds through biological and chemical interactions to any change in external or internal conditions (Varela et al. 1974; Maturana and Varela 1980; Santschi et al. 2003). Thus, EPS production by the microbial community and its composition is dynamic. This makes this system challenging to study because most approaches cannot easily distinguish between the biotic and chemical factors promoting the release of EPS. We suggest the following knowledge gaps that will need to be filled through interdisciplinary studies to better understand these complex systems:

1. *Operational definitions of EPS* (Table 1): It is still is not well understood how different operationally defined subgroups of EPS are related to each other (see also Bar-Zeev et al. 2015). This ambiguous classification makes the task of identifying their relationships and interactions extremely complex and challenging.
2. *The best way to measure the surfactant nature, or amphiphilicity, of EPS* (see Tables 1, 2 for some initial details): It is not clear what measures best express the amphiphilic and surface-active nature of EPS: hydrophobic contact area, protein/carbohydrate ratio, aromaticity, surface charge, stainability with Alcian Blue or Coomassie blue.
3. *The best way to measure TEP*: Natural TEP consists of a varying mix of substances. For determination of TEP, historically the primary reference compound is Gum Xanthan, chosen because it looks and behaves more like TEP than other (e.g., Alginic Acid) tested compounds (Passow and Alldredge 1995). However, Gum Xanthan, which is derived from a terrestrial bacterium, does not occur in the ocean, while Alginic Acidis a common marine acidic polysaccharide (Guo et al. 2002; Hung et al. 2003; Santschi et al. 2003). The amount of TEP expressed as Gum Xanthan equivalents is up to a factor of 5 smaller than that expressed as Alginic Acid equivalents, depending on the relative ratios of carrageenan (sulfate groups), Alginic Acid (carboxyl groups), or other compounds, including phosphonates, as well as sample pretreatment (Hung et al. 2003). Consequently, the conversions from both, Gum Xanthan or Alginic Acid equivalents to TEP-C are hampered, because depending on the chemical composition of TEP the carbon content of TEP per Alcian Blue molecule varies and thus the conversion between TEP measured as Gum Xanthan or Alginic Acid equivalents and TEP-carbon is variable (Hung et al. 2003). As with EPS, the approach requires refinement.
4. *The relationship between stickiness and chemical properties of TEP*: TEP stickiness varies substantially and the relationship between stickiness and chemical properties of TEP are largely unidentified. We suggest that the protein/carbohydrate ratios can serve as a proxy for "stickiness," but it will still need to be shown how much relative hydrophobicity (through hydrophobic interactions) or ROS-mediated cross-linking (through chemical bond formation) contribute to the "stickiness" or aggregation potential leading to marine snow and MOSSFA in the presence of oil. We do know that Ca^{2+}-bridging is an important component for the formation of TEP—but we are not sure what and how this may impact stickiness.
5. *The environmental controls of microbial EPS production.* There remains a significant lack of understanding of why different microbes produce different types and amounts of EPS under different environmental conditions. Temperature, light, depth, mixing, pH, redox, are all external factors that can influence the formation of EPS and TEP by the marine biota.
6. *The relationships between EPS and the bacteria that degrade oil*: A complex network of microbes are using the

different components of oil and metabolites of the oil degraders will develop and facilitate biodegradation of the oil. Thus, while the presence of bacteria that can degrade the oil would indicate that they are also the originators of the EPS, such evidence is rather indirect.

7. *The relationship between phytoplankton-derived EPS and bacteria*: While there is evidence that bacteria attached to phytoplankton may be the major contributors to the EPS released in response to oil and Corexit pollution, the evidence is more indirect than direct, as different species can produce EPS without leaving a specific marker for its origin. Further, phytoplankton can produce TEP directly in the absence of bacteria; this exudation to TEP precursors is a physiological response to an imbalance in energy and carbon acquisition (Passow and Laws 2015). The details of these relationships need to be investigated. Understanding the complex relationship between phytoplankton and bacteria will help us to understand if they behave synergistically or antagonistically with regards to EPS release.
8. *The effects of light on phytoplankton growth, ROS, and EPS production and ultimately, on oil transformations*: Visible light, including UVB and UVA, produce different radical oxygen species in aquatic systems. In order relieve the oxidative stress of ROS, microbes can excrete protective EPS to physically block or chemically quench hazardous ROS. Contrary to UVB, which has a degrading effect on macromolecules, lower energy ROS generated from UVA appear to have the potential to cross-link proteins (e.g., in EPS), which has the potential to lead to EPS aggregation and marine snow formation. Since light effects phytoplankton growth, EPS and ROS production occur simultaneously, and their combined effects on the aggregation, degradation, and emulsification potential of EPS and subsequent fate and transport of oil needs to be evaluated.
9. *The relationship between dispersant and the microbial production of EPS and the microbial community composition*: Dispersants have been used to ameliorate the effects and transport of oil. It is however still not clear (1) how the presence of dispersants impacts the ability of bacteria and phytoplankton to generate EPS, (2) how the chemical dispersion of oil eliminates the stimulus that triggers the release of EPS, (3) how (and if) dispersants physically lead to the dispersion of EPS, and (4) how the presence of dispersants leads to a shift in the microbial composition towards strains less or more likely to generate mucus or degrade oil.
10. *The importance of riverine-derived terrestrial detrital matter and nutrients*: It is not clear to what extent the presence of riverine-derived terrestrial detrital material facilitates the formation and sinking of marine snow aggregates; the importance of seasonal timing, the magnitude or duration of the water discharge, and the composition of the discharge all need to be considered. We know for example, that the composition of the discharge varies significantly between the Mississippi and Atchafalaya Rivers despite both having similar source waters. This is thought to be because riverine waters from the former discharge directly into the NGOM while for the latter, the waters are processed in the estuary before heading to the NGOM. The differences in upstream biogeochemical processing have consequences on both the biology and chemistry once mixing with gulf waters occurs (see Quigg et al. 2011 and references therein) including but not limited to nutrient and light limitation of primary productivity. Further, it is not clear what the *role of nutrient stimulation after a spill* e.g., via the release of Mississippi River waters (but see Hu et al. 2011) has on the formation of EPS (marine snow).

Conclusions

In this paper, we focused on the importance of MOS, a biological process, as a transport pathway for oil to the seafloor. Previously, OMAs, tar balls, in addition to other non-biological processes have been considered in the sedimentation of oil. Floating oil residues (e.g., tar particles) were observed during the Ixtoc oil spill near Mexico in the GOM in 1979 (Patton et al. 1981; D'Souza et al. 2016). Ongoing studies (e.g., Daly et al. 2016) are finding indicators in deep sediments suggesting a MOSSFA event may have co-occurred during that spill as well. It is becoming increasingly clear that EPS and marine snow play an important role in determining the fate and transport of oil during and after a spill. This review highlights that there is relatively little known about EPS production in relation to oil and dispersants, especially how its amphiphilicity and surface activity are controlled at the molecular and chemical levels. When dispersants, which are artificial surfactants are used, the response of the microbial community has been even less studied (e.g., Gutierrez et al. 2013; Ziervogel and Arnosti 2013). There is much research remaining in order to develop better response strategies during oil spill events in future.

References

Alldredge, A. L., and C. C. Gotschalk. 1988. *In situ* behavior of marine snow. Limnol. Oceanogr. **33**: 339–351. doi: 10.4319/lo.1988.33.3.0339

Alldredge, A. L., and M. Silver. 1988. Characteristics, dynamics and significance of marine snow. Prog. Oceanogr. **20**: 41–82. doi:10.1016/0079-6611(88)90053-5

Alldredge, A. L., and C. C. Gotschalk. 1989. Direct observations of the mass flocculation of diatom blooms: Characteristics, settling velocities and formation of diatom aggregates. Deep-Sea Res. A **36**: 159–171. doi:10.1016/0198-0149(89)90131-3

Alldredge, A. L., U. Passow, and B. Logan. 1993 The abundance and significance of a class of large, transparent organic particles in the ocean. Deep-Sea Res. I **40**: 1131–1140. doi:10.1016/0967-0637(93)90129-Q

Almeda, R., T. L. Connelly, and E. J. Buskey. 2014. Novel insight into the role of heterotrophic dinoflagellates in the fate of crude oil in the sea. Sci. Rep. **4**: 75–60. doi: 10.1038/srep07560

Armstrong, R. A., M. L. Peterson, C. Lee, and S. G. Wakeham. 2009. Settling velocity spectra and the ballast ratio hypothesis. Deep-Sea Res. II **56**: 1470–1478. doi: 10.1016/j.dsr2.2008.11.032

Arnosti, C., K. Ziervogel, T. Yang, and A. Teske. 2016. Oil-derived marine aggregates–hot spots of polysaccharide degradation by specialized bacterial communities. Deep-Sea Res. II **129**: 179–186. doi:10.1016/j.dsr2.2014.12.008

Asper, V. L., W. G. Deuser, G. H. Knauer, and S. E. Lorenz. 1992. Rapid coupling of sinking particle fluxes between surface and deep ocean waters. Nature **357**: 670–672. doi: 10.1038/357670a0

Azam, F. 1998. Microbial control of oceanic carbon flux: The plot thickens. Science **280**: 694–696. doi:10.1126/science.280.5364.694

Azam, F., and F. Malfatti. 2007. Microbial structuring of marine ecosystems. Nat. Rev. Microbiol. **5**: 782–791. doi: 10.1038/nrmicro1747

Baelum, J., S. Borglin, R. Chakraborty, J. L. Fortney, R. Lamendella, O. U. Mason, and S. A. Malfatti. 2012. Deep-sea bacteria enriched by oil and dispersant from the Deepwater Horizon spill. Environ. Microbiol. **14**: 2405–2416. doi:10.1111/j.1462-2920.2012.02780.x

Bar-Zeev, E., U. Passow, S. Romero-Vargas Castrillón, and M. Elimelech. 2015. Transparent exopolymer particles (TEP): From aquatic environments and engineered systems to membrane biofouling. Environ. Sci. Technol. **49**: 691–707. doi:10.1021/es5041738

Beaudoin, D. J., C. A. Carmichael, R. K. Nelson, C. M. Reddy, A. Teske, and V. P. Edgcomb. 2016. Impact of protists on a hydrocarbon-degrading bacterial community from deep-sea Gulf of Mexico sediments: A mesocosm study. Deep-Sea Res. II **129**: 350–359. doi:10.1016/j.dsr2.2014.01.007

Berman-Frank, I., G. Rosenberg, O. Levitan, L. Haramaty, and X. Mari. 2007. Coupling between autocatalytic cell death and transparent exopolymeric particle production in the marine cyanobacterium *Trichodesmium*. Environ. Microbiol. **9**: 1415–1422. doi:10.1111/j.1462-2920.2007.01257.x

Bhaskar, P. V., and N. B. Bhosle. 2005. Microbial extracellular polymeric substances in marine biogeochemical processes. Curr. Sci. **88**: 45–53.

Biermann, A., and A. Engel. 2010. Effect of CO_2 on the properties and sinking velocity of aggregates of the coccolithophore *Emiliania huxleyi*. Biogeosciences **7**: 1017–1029. doi: 10.5194/bg-7-1017-2010

Burd, A. B., and G. A. Jackson. 2009. Particle aggregation. Ann. Rev. Mar. Sci. **1**: 65–90. doi:10.1146/annurev.marine. 010908.163904

Chanton, J. P., J. Cherrier, R. M. Wilson, J. Sarkodee-Adoo, S. Bosman, A. Mickle, and W. M. Graham. 2012. Radiocarbon evidence that carbon from the Deepwater Horizon spill entered the planktonic food web of the Gulf of Mexico. Environ. Res. Lett. **7**: 045303. doi:10.1088/1748-9326/7/4/045303

Chanton, J., and others. 2015. Using natural abundance radiocarbon to trace the flux of petrocarbon to the sea floor following the Deepwater Horizon oil spill. Environ. Sci. Technol. **49**: 847–854. doi:10.1021/es5046524

Chen, C.-S., and others. 2011. Effects of engineered nanoparticles on the assembly of exopolymeric substances from phytoplankton. PLoS One **6**: e21865. doi:10.1371/journal.pone.0021865

Chen, C.-S., J. M. Anaya, E. Y-T. Chen, E. Farr, and W.-C. Chin. 2015. Ocean warming–acidification synergism undermines dissolved organic matter assembly. PLoS One **10**: e0118300. doi:10.1371/journal.pone.0118300

Chen, H., H. A. Abdulla, R. L. Sanders, S. C. Myneni, K. Mopper, and P. G. Hatcher. 2014. Production of black carbon-like and aliphatic molecules from terrestrial dissolved organic matter in the presence of sunlight and iron. Environ. Sci. Technol. Lett. **1**: 399–404. doi:10.1021/ez5002598

Chen, L. Z., G. H. Wang, S. Hong, A. Liu, C. Li, and Y. D. Liu. 2009. UV-B-induced oxidative damage and protective role of exopolysaccharides in desert cyanobacterium *Microcoleus vaginatus*. J. Integr. Plant Biol. **51**: 194–200. doi:10.1111/j.1744-7909.2008.00784.x

Chin, W.-C., M. C. Orellana, and P. Verdugo. 1998. Spontaneous assembly of marine dissolved organic matter into polymer gels. Nature **391**: 568–572. doi:10.1038/35345

Chiovitti, A., M. J. Higgins, R. E. Harper, R. Wetherbee, and A. Bacic. 2003. The complex polysaccharides of the raphid diatom *Pinnularia viridis* (Bacillariophyceae). J. Phycol. **39**: 543–554. doi:10.1046/j.1529-8817.2003.02162.x

Cho, B. C., and F. Azam. 1988. Major role of bacteria in biogeochemical fluxes in the oceans interior. Nature **332**: 441–443. doi:10.1038/332441a0

Chow, J. S., C. Lee, and A. Engel. 2015. The influence of extracellular polysaccharides, growth rate, and free coccoliths on the coagulation efficiency of *Emiliania huxleyi*. Mar. Chem. **175**: 5–17. doi:10.1016/j.marchem.2015.04.010

Chuang, C. Y., P. H. Santschi, Y. Jiang, Y. F. Ho, A. Quigg, L. D. Guo, M. Ayranov, and D. Schumann. 2014. Important role of biomolecules from diatoms in the scavenging of particle-reactive radionuclides of thorium, protactinium, lead, polonium and beryllium in the ocean: A case study with *Phaeodactylum tricornutum*. Limnol. Oceanogr. **59**: 1256–1266. doi:10.4319/lo.2014.59.4.1256

Cisternas-Novoa, C., C. Lee, and A. Engel. 2015. Transparent exopolymer particles (TEP) and coomassie stainable particles (CSP): Differences between their origin and vertical distributions in the ocean. Mar. Chem. **175**: 56–71. doi: 10.1016/j.marchem.2015.03.009

Daly, K. L., U. Passow, J. Chanton, and D. Hollander. 2016. Assessing the impacts of oil-associated marine snow formation and sedimentation during and after the Deepwater Horizon Oil spill. Anthropocene **13**:18–33. doi: 10.1016/j.ancene.2016.01.006

de Jong, E. W., L. Bosch, and P. Westbroek. 1976. Isolation and characterization of a Ca^{2+}-binding polysaccharide associated with coccoliths of *Emiliania huxleyi* (Lohmann) Kamptner. Eur. J. Biochem. **70**: 611–621. doi:10.1111/j.1432-1033.1976.tb11052.x

de Jong, E., L. van Rens, P. Westbroek, and L. Bosch. 1979. Biocalcification by the marine alga *Emiliania huxleyi* (Lohmann) Kamptner. Eur. J. Biochem. **99**: 559–567. doi: 10.1111/j.1432-1033.1979.tb13288.x

Decho, A. W. 1990. Microbial exopolymer secretions in ocean environments: Their role(s) in food web and marine processes. Oceanogr. Mar. Biol. Ann. Rev. **28**: 73–153.

Decho, A. W., and G. J. Herndl. 1995. Microbial activities and the transformation of organic matter within mucilaginous material. Sci. Total Environ. **165**: 33–42. doi: 10.1016/0048-9697(95)04541-8

Dickinson, E. 2003. Hydrocolloids at interfaces and the influence on the properties of dispersed systems. Food Hydrocoll. **17**: 25–39. doi:10.1016/S0268-005X(01)00120-5

Diercks, A., and V. L. Asper. 1997. *In situ* settling speeds of marine snow aggregates below the mixed layer: Black Sea and Gulf of Mexico. Deep-Sea Res. I **44**: 385–398. doi: 10.1016/S0967-0637(96)00104-5

Ding, Y.-X., W.-C. Chin, A. Rodriguez, C.-C. Hung, P. H. Santschi, and P. Verdugo. 2008. Amphiphilic exopolymers from *Sagitulla stellata* induce self-assembly of DOM polymers and the formation of marine microgels. Mar. Chem. **112**: 11–19. doi:10.1016/j.marchem.2008.05.003

Ding, Y.-X., C.-C. Hung, P. H. Santschi, P. Verdugo, and W.-C. Chin. 2009. Spontaneous assembly of exopolymers from phytoplankton. Terr. Atmos. Ocean. Sci. **20**: 741–747. doi:10.3319/TAO.2008.08.26.01(Oc)

Drum, R. 1969. Electron microscope observations of diatoms. Osterr. Bot. J. **116**: 321–330. doi:10.1007/BF01379632

D'Souza, N. A., and others. 2016. Elevated surface chlorophyll associated with natural oil seeps in the Gulf of Mexico. Nat. Geosci. **9**: 215–218. doi:10.1038/ngeo2631

Dubinsky, E. A., and others. 2013. Succession of hydrocarbon-degrading bacteria in the after-math of the Deepwater Horizon oil spill in the Gulf of Mexico. Environ. Sci. Technol. **47**: 10860–10867. doi:10.1021/es401676y

Engel, A., and U. Passow. 2001. Carbon and nitrogen content of transparent exopolymer particles (TEP) in relation to their Alcian Blue adsorption. Mar. Ecol. Prog. Ser. **219**: 1–10. doi:10.3354/meps219001

Engel, A., B. Delille, S. Jacquet, U. Riebesell, E. Rochelle-Newall, A. Terbrüggen, and I. Zondervan. 2004. Transparent exopolymer particles and dissolved organic carbon production by *Emiliania huxleyi* exposed to different CO_2 concentrations: A mesocosm experiment. Aquat. Microb. Ecol. **34**: 93–104. doi:10.3354/ame034093

Engel, A., J. Szlosek, L. Abramson, Z. Liu, and C. Lee. 2009. Investigating the effect of ballasting by $CaCO_3$ in *Emiliania huxleyi*: I. Formation, settling velocities and physical properties of aggregates. Deep-Sea Res. II **56**: 1396–1407. doi:10.1016/j.dsr2.2008.11.027

Ford, T., E. Sacco, J. Black, T. Kelley, R. Goodacre, R. C. Berkeley, and R. Mitchell. 1991. Characterization of exopolymers of aquatic bacteria by pyrolysis-mass spectrometry. Appl. Environ. Microbiol. **57**: 1595–1601. doi:10.1128/AEM.00316-08

Fu, J., Y. Gong, X. Zhao, S. E. O'Reilly, and D. Zhao. 2014. Effects of oil and dispersant on formation of marine oil snow and transport of oil hydrocarbons. Environ. Sci. Technol. **48**: 14392–14399. doi:10.1021/es5042157

Gao, H., and R. G. Zepp. 1998. Factors influencing photoreactions of dissolved organic matter in a coastal river of the southeastern United States. Environ. Sci. Technol. **32**: 2940–2946. doi:10.1021/es9803660

Gärdes, A., M. H. Iversen, H.-P. Grossart, U. Passow, and M. Ullrich. 2011. Diatom associated bacteria are required for aggregation of *Thalassiosira weissflogii*. ISME J. **5**: 436–445. doi:10.1038/ismej.2010.145

Gong, A. S., C. A. Lanzl, D. M. Cwiertny, and S. L. Walker. 2012. Lack of influence of extracellular polymeric substances (EPS) level on hydroxyl radical mediated disinfection of *Escherichia coli*. Environ. Sci. Technol. **46**: 241–249. doi: 10.1021/es202541r

Gong, Y., X. Zhao, Z. Cai, S. E. O'Reilly, X. Hao, and D. Zhao. 2014. A review of oil, dispersed oil and sediment interactions in the aquatic environment: Influence on the fate, transport and remediation of oil spills. Mar. Poll. Bull. **79**: 16–33. doi:10.1016/j.marpolbul.2013.12.024

González, J., F. G. Figueiras, M. Aranguren-Gassis, B. G. Crespo, E. Fernández, X. A. G. Morán, and M. Nieto-Cid. 2009. Effect of a simulated oil spill on natural assemblages of marine phytoplankton enclosed in microcosms. Estuar. Coast. Shelf Sci. **83**: 265–276. doi:10.1016/j.ecss.2009.04.001

Graham, W. M., R. H. Condon, R. H. Carmichael, I. D'Ambra, H. K. Patterson, L. J. Linn, and F. J. Hermandez, Jr. 2010. Oil carbon entered the coastal planctonic food web during the Deepwater Horizon oil spill. Environ. Res. Lett. **5**: 045301. doi:10.1088/1748-9326/5/4/045301

Grossart, H. P. 1999. Interactions between marine bacteria and axenic diatoms (*Cylindrotheca fusiformis*, *Nitzschia laevis* and *Thalassiosira weissflogii*) incubated under various conditions in the lab. Aquat. Microb. Ecol. **19**: 1–11. doi: 10.3354/ame019001

Grossart, H. P., G. Czub, and M. Simon. 2006. Algae-bacteria interactions and their effects on aggregation and organic matter flux in the sea. Environ. Microbiol. **8**: 1074–1084. doi:10.1111/j.1462-2920.2006.00999.x

Guo, L., C.-C. Hung, P. H. Santschi, and I. D. Walsh. 2002. ^{234}Th scavenging and its relationship to acid polysaccharide abundance in the Gulf of Mexico. Mar. Chem. **78**: 103–119. doi:10.1016/S0304-4203(02)00012-9

Gutierrez, T., D. Berry, T. Yang, S. Mishamandani, L. McKay, A. Teske, and M. D. Aitken. 2013. Role of bacterial exopolysaccharides (EPS) in the fate of the oil released during the Deepwater Horizon oil spill. PLoS One **8**: e67717. doi: 10.1371/journal.pone.0067717

Hamdan, L. J., and P. A. Fulmer. 2011. Effects of Corexit® EC9500A on bacteria from a beach oiled by the Deepwater Horizon spill. Aquat. Microb. Ecol. **63**: 101–109. doi: 10.3354/ame01482

Hansell, D. A., and C. A. Carlson 2001. Marine dissolved organic matter and the carbon cycle. J. Oceanogr. **14**: 41–49. doi:10.5670/oceanog.2001.05

Hassler, C. S., L. Norman, C. A. M. Nichols, L. A. Clementson, C. Robinson, V. Schoemann, R. J. Watson, and M. A. Doblin. 2015. Iron associated with exopolymeric substances is highly bioavailable to oceanic phytoplankton. Mar. Chem. **173**: 136–147. doi:10.1016/j.marchem.2014.10.002

Hazen, T. C., and others. 2010. Deep-sea oil plume enriches indigenous oil-degrading bacteria. Science **330**: 204–208. doi:10.1126/science.1195979

Head, I. M., D. M. Jones, and W. F. M. Roling. 2006. Marine microorganisms make a meal of oil. Nat. Rev. Microbiol. **4**: 173–182. doi:10.1038/nrmicro1348

Helms, J. R., J. Mao, K. Schmidt-Rohr, H. Abdulla, and K. Mopper. 2013. Photochemical flocculation of terrestrial dissolved organic matter and iron. Geochim. Cosmochim. Acta **121**: 398–413. doi:10.1016/j.gca.2013.07.025

Hoagland, K., J. Rosowski, M. Gretz, and S. Roemaer. 1993. Diatom extracellular polymeric substances: Function, fine structure, chemistry, and physiology. J. Phycol. **29**: 537–566. doi:10.1111/j.0022-3646.1993.00537.x

Hook, S. E., and H. L. Osborn. 2012. Comparison of toxicity and transcriptomic profiles in a diatom exposed to oil, dispersants, dispersed oil. Aquat. Toxicol. **124–125**: 139–151. doi:10.1016/j.aquatox.2012.08.005

Hu, C. M., R. H. Weisberg, Y. G. Liu, L. Y. Zheng, K. L. Daly, D. C. English, J. Zhao, and G. A. Vargo. 2011. Did the northeastern Gulf of Mexico become greener after the Deepwater Horizon oil spill? Geophys. Res. Lett. **38**: L09601. doi:10.1029/2011GL047184

Hung, C.-C., L. Guo, P. H. Santschi, N. J. Alvarado-Quiroz, and J. Haye. 2003. Distributions of carbohydrate species in the Gulf of Mexico. Mar. Chem. **81**: 119–135. doi: 10.1016/S0304-4203(03)00012-4

Jackson, G. A. 1990. A model of the formation of marine algal flocs by physical coagulation processes. Deep-Sea Res. A **37**: 1197–1211. doi:10.1016/0198-0149(90)90038-W

Jackson, G. A. 2005. Coagulation theory and models of oceanic plankton aggregation, p. 271–292. *In* I. Droppo, G. Leppard, S. Liss and T. Milligan [eds.], Flocculation in natural and engineered environments. CRC Press.

Joye, S. B., and others. 2016. The Gulf of Mexico ecosystem, six years after the Macondo oil well blowout. Deep Sea Res. **129**: 4–19. doi:10.1016/j.dsr2.2016.04.018

Jusufi, A., A. P. Hynninen, and A. Z. Panagiotopoulos. 2008. Implicit solvent models for micellization of ionic surfactants. J. Phys. Chem. B **112**: 13783−13792. doi:10.1021/jp8043225

Kahl, L., A. Vardi, and O. Schofield. 2008. Effects of phytoplankton physiology on export flux. Mar. Ecol. Prog. Ser. **354**: 3–19. doi:10.3354/meps07333

Kappell, A. D., Y. Wei, R. J. Newton, J. D. Van Nostrand, J. Zhou, S. L. McLellan, and K. R. Hristova 2014. The polycyclic aromatic hydrocarbon degradation potential of Gulf of Mexico native coastal microbial communities after the Deepwater Horizon oil spill. Front. Microbiol., **5**:205. doi:10.3389/fmicb.2014.00205

Kennedy, A. F, and I. W. Sutherland. 1987. Analysis of bacterial exopolysaccharides. Biotechnol. Appl. Biochem. **9**: 12–19. doi:10.1111/j.1470-8744.1987.tb00458.x

Kessler, J. D., D. L. Valentine, M. C. Redmond, M. Du, E. W. Chan, S. D. Mendes, and S. A. Yvon-Lewis. 2011. A persistent oxygen anomaly reveals the fate of spilled methane in the deep Gulf of Mexico. Science **331**: 312–315. doi: 10.1126/science.1199697

Khelifa, A., P. Stoffyn-Egli, P. S. Hill, and K. Lee. 2005. Effects of salinity and clay type on oil–mineral aggregation. Mar. Environ. Res. **59**: 235–254. doi:10.1016/j.marenvres.2004.05.003

Kieber, D. J., B. M. Peake, and N. M. Scully. 2003. Reactive oxygen species in aquatic ecosystems. In UV effects in aquatic organisms and ecosystems. edited by Hebling E. W. and Zagarese H. New York: Royal Soc. Chem. pp. 219–250.

Kiørboe, T. 2001. Formation and fate of marine snow: Small-scale processes with large-scale implications. Sci. Mar. **65**: 57–71. doi:10.3989/scimar.2001.65s257

Kiørboe, T., C. Lundsgaard, M. Olesen, and J. L. S. Hansen. 1994. Aggregation and sedimentation processes during a spring phytoplankton bloom: A field experiment to test coagulation theory. J. Mar. Res. **52**: 297–323. doi:10.1357/0022240943077145

Kleindienst, S., and others. 2015a. Chemical dispersants can suppress the activity of natural oil-degrading microorganisms. Proc. Natl. Acad. Sci. USA **112**: 14900–14905. doi: 10.1073/pnas.1507380112

Kleindienst, S., J. H. Paul, and S. B. Joye. 2015b. Using dispersants after oil spills: Impacts on the composition and activity of microbial communities. Nat. Rev. Microbiol. **13**: 388–396. doi:10.1038/nrmicro3452

Kowalewska, G. 1999. Phytoplankton - the main factor responsible for transport of polynuclear aromatic hydrocarbons from water to sediments in the Southern Baltic ecosystem. ICES J. Mar. Sci. **56**: 219–222. doi:10.1006/jmsc.1999.0607

Kowalewska, G., and J. Konat. 1997. The role of phytoplankton in the transport and distribution of polynuclear aromatic hydrocarbons (PAHs) in the southern Baltic environment. Oceanologia **39**: 267–277.

Kujawinski, E. B., M. C. Kido Soule, D. L. Valentine, A. K. Boysen, K. Longnecker, and M. C. Redmond. 2011. Fate of dispersants associated with the Deepwater Horizon oil spill. Environ. Sci. Technol. **45**: 1298–1306. doi:10.1021/es103838p

Leppard, G. 1997. Colloidal organic fibrils of acidic polysaccharides in surface waters: Electron-optical characteristics, activities, and chemical estimates of abundance. Colloids Surf. **120**: 1–15. doi:10.1016/S0927-7757(96)03676-X

Lewis, A., B. K. Trudel, R. C. Belore, and J. V. Mullin. 2010. Large scale dispersant leaching and effectiveness experiments with oils on calm water. Mar. Pollut. Bull. **60**: 244–254. doi:10.1016/j.marpolbul.2009.09.019

Liao, B. Q., D. G. Allen, I. G. Droppo, G. G. Leppard, and S. N. Liss. 2001. Surface properties of sludge and their role in bioflocculation and settleability. Water Res. **35**: 339–350. doi:10.1016/S0043-1354(00)00277-3

Liao, B. Q., D. G. Allen, G. G. Leppard, I. G. Droppo, and S. N. Liss. 2002. Interparticle interactions affecting the stability of sludge flocs. J. Colloid Interface Sci. **249**: 372–380. doi:10.1006/jcis.2002.8305

Liu, Y., J. Li, X. Qiu, and C. Burda. 2007. Bactericidal activity of nitrogen-doped metal oxide nanocatalysts and the influence of bacterial extracellular polymeric substances (EPS). J. Photochem. Photobiol. A Chem. **190**: 94–100. doi:10.1016/j.jphotochem.2007.03.017

Logan, B. E., U. Passow, A. L. Alldredge, H.-P. Grossart, and M. Simon. 1995. Rapid formation and sedimentation of large aggregates is predictable from coagulation rates (half-lives) of transparent exopolymer particles (TEP). Deep Sea Res. II **42**: 203–214. doi:10.1016/0967-0645(95)00012-F

Long, R. A., and F. Azam. 1996. Abundant protein-containing particles in the sea. Aquat. Microb. Ecol. **10**: 213–221. doi:10.3354/ame010213

Lu, Z., Y. Deng, J. D. Van Nostrand, Z. He, J. Voordeckers, A. Zhou, and L. M. Tom. 2012. Microbial gene functions enriched in the Deepwater Horizon deep-sea oil plume. ISME J. **6**: 451–460. doi:10.1038/ismej.2011.91

Lubecki, L., and G. Kowalewska. 2010. Distribution and fate of polycyclic aromatic hydrocarbons (PAHs) in recent sediments from the Gulf of Gdańsk (SE Baltic). Oceanologia **52**: 669–703. doi:10.5697/oc.52-4.669

MacNaughton, S. J., J. R. Stephen, A. D. Venosa, G. A. Davis, Y.-J. Chang, and D. C. White. 1999. Microbial population changes during bioremediation of an experimental oil spill. Appl. Environ. Microbiol. **65**: 3566–3574.

Mague, T. H., E. Friberg, D. J. Hughes, and I. Morris. 1980. Extracellular release of carbon by marine phytoplankton; a physiological approach. Limnol. Oceanogr. **25**: 262–279. doi:10.4319/lo.1980.25.2.0262

Mari, X. 1999. Carbon content and C:N ratio of transparent exopolymeric particles (TEP) produced by bubbling exudates of diatoms. Mar. Ecol. Prog. Ser. **183**: 59–71. doi:10.3354/meps183059

Mari, X., and M. Robert. 2008. Metal induced variations of TEP sticking properties in the southwestern lagoon of New Caledonia. Mar. Chem. **110**: 98–108. doi:10.1016/j.marchem.2008.02.012

Mari, X., and others. 2014. Effects of soot deposition on particle dynamics and microbial processes in marine surface waters. Global Biogeochem. Cycles **28**: 662–678. doi:10.1002/2014GB004878

Mason, O. U., and others. 2012. Metagenome, metatranscriptome and single-cell sequencing reveal microbial response to Deepwater Horizon oil spill. ISME J. **6**: 1715–1727. doi:10.1038/ismej.2012.59

Maturana, H. R., and F. G. Varela. 1980. Autopoiesis and cognition. The realization of the living, p. 143. D. Reidel Publishing Company.

McGenity, T. J., B. D. Folwell, B. A. McKew, and G. O. Sanni. 2012. Marine crude-oil biodegradation: A central role for interspecies interactions. Aquat. Biosyst. **8**: 10. doi:10.1186/2046-9063-8-10

McInnes, A., and A. Quigg. 2010 Near-annual fish kills in small embayments: Casual versus causal factors. J. Coast. Res. **26**: 957–966. doi:10.2112/JCOASTRES-D-10-00006.1

Mopper, K., J. Zhou, K. Sri Ramana, U. Passow, H. G. Dam, and D. T. Drapeau. 1995. The role of surface-active carbohydrates in the flocculation of a diatom bloom in a mesocosm. Deep-Sea Res. II **42**: 47–73. doi:10.1016/0967-0645(95)00004-A

Muschenheim, D. K., and K. Lee. 2002. Removal of oil from the sea surface through particulate interactions: Review and prospectus. Spill Sci. Technol. Bull. **8**: 9–18. doi:10.1016/S1353-2561(02)00129-9

Myklestad, S. M. 1995. Release of extracellular products by phytoplankton with special emphasis on polysaccharides. Sci. Total Environ. **165**: 155–164. doi:10.1016/0048-9697(95)04549-G

Myklestad, S., O. Holm-Hansen, K. M. Vårum, and B. E. Volcani. 1989. Rate of release of extracellular amino acids and carbohydrates from the marine diatom *Chaetoceros affinis*. J. Plankton Res. **11**: 763–773. doi:10.1093/plankt/11.4.763

Nagasaki, K., Y. Tomaru, N. Katanozaka, Y. Shirai, K. Nishida, S. Itakura, and M. Yamaguchi. 2004. Isolation and characterization of a novel single-stranded RNA virus infecting the bloom-forming diatom *Rhizosolenia setigera*.

Appl. Environ. Microbiol. **70**: 704–11. doi:10.1128/AEM.70.2.704-711.2004

National Research Council (NRC). 2005. Oil spill dispersants: Efficacy and effects. The National Academies Press. Available from www.nap.edu/openbook.php?record_id=11283.

Niu, H., Z. Li, K. Lee, P. Kepkay, and J. V. Mullin. 2011. Modelling the transport of oil– mineral-aggregates (OMAs) in the marine environment and assessment of their potential risks. Environ. Model. Assess. **16**: 61–75. doi:10.1007/s10666-010-9228-0

Orellana, M. V., and P. Verdugo. 2003. Ultraviolet radiation inhibits the exchange between dissolved and particulate organic matter. Limnol. Oceanogr. **48**: 1618–1623. doi:10.4319/lo.2003.48.4.1618

Orellana, M. V., E. J. Lessard, E. Dycus, W. C. Chin, M. S. Foy, and P. Verdugo. 2003. Tracing the source and fate of biopolymers in seawater: Application of an immunological technique. Mar. Chem. **83**: 89–99. doi:10.1016/S0304-4203(03)00098-7

Orellana, M. V., T. W. Petersen, A. H. Diercks, S. Donohoe, P. Verdugo, and G. ven den Engh. 2007. Marine microgels: Optical and proteomic fingerprints. Mar. Chem. **105**: 229–239. doi:10.1016/j.marchem.2007.02.002 doi:10.1016/j.marchem.2007.02.002

Orellana, M. V., P. A. Matrai, C. Leck, C. D. Rauschenberg, A. M. Lee, and E. Coz. 2011. Marine microgels as a source of cloud condensation nuclei in the high arctic. Proc. Natl. Acad. Sci. USA **108**: 13612–13617. doi:10.1073/pnas.1102457108

Orellana, M. V., and C. Leck. 2015. Marine microgels, p. 451–480. *In* D. A. Hansell and C. A. Carlson [eds.], Biogeochemistry of marine dissolved organic matter, 2nd ed., Academic Press, LTD, Elsevier Science, 24–28 Oval Road, London, NW1 7DX, England, 451–480. doi:10.1016/B978-0-12-405940-5.00009-1

Ortega-Retuerta, E., U. Passow, C. M. Duarte, and I. Reche. 2009. Impact of UVB radiation on transparent exopolymer particles. Biogeosciences **6**: 3071–3080. doi:10.5194/bg-6-3071-2009

Overton, E. B., T. L. Wade, J. R. Radović, B. M. Meyer, M. S. Miles, and S. R. Later. 2016. Chemical composition of Macondo and other crude oils and compositional alterations during oil spills. Oceanogr. **29**: 50–63. http://dx.doi.org/10.5670/oceanog.2016.62

Ozhan, K., and S. Bargu. 2014. Distinct responses of Gulf of Mexico phytoplankton communities to crude oil and the dispersant Corexit® EC9500A under different nutrient regimes. Ecotoxicology **23**: 370–384. doi:10.1007/s10646-014-1195-9

Ozhan, K., M. L. Parsons, Bargu. 2014a. How were phytoplankton affected by the Deepwater Horizon oil spill? BioScience **64**: 829–836. doi:10.1186/2046-9063-8-10

Ozhan, K., S. M. Miles, H. Gao, and S. Bargu. 2014b. Relative phytoplankton growth responses to physically and chemically dispersed South Louisiana sweet crude oil. Environ. Monit. Assess. **186**: 3941–3956. doi:10.1186/2046-9063-8-10

Parsons, M., R. Turner, and E. Overton. 2014. Sediment-preserved diatom assemblages can distinguish a petroleum activity signal separately from the nutrient signal of the Mississippi River in coastal Louisiana. Mar. Poll. Bull. **85**: 164–171. doi:10.1016/j.marpolbul.2014.05.057

Passow, U. 2000. Formation of transparent exopolymer particles (TEP) from dissolved precursor material. Mar. Ecol. Prog. Ser. **192**: 1–11. doi:10.3354/meps192001

Passow, U. 2002. Transparent exopolymer particles (TEP) in aquatic environments. Prog. Oceanogr. **55**: 287–333. doi:10.1016/S0079-6611(02)00138-6

Passow, U. 2012. The abiotic formation of TEP under different ocean acidification scenarios. Mar. Chem. **128-129**: 72–80. doi:10.1016/j.marchem.2011.10.004

Passow, U. 2016. Formation of rapidly-sinking, oil-associated marine snow. Deep Sea Res. II. **129**: 232–240. doi:10.1016/j.dsr2.2014.10.001

Passow, U., and P. Wassmann. 1994. On the trophic fate of *Phaeocystis pouchetii* (Hariot): IV. The formation of marine snow by *P. pouchetii*. Mar. Ecol. Prog. Ser. **104**: 151−163. doi:10.3354/meps104153

Passow, U., A. L. Alldredge, and B. E. Logan. 1994. The role of particulate carbohydrate exudates in the flocculation of diatom blooms. Deep-Sea Res. **41**: 335–357. doi:10.1016/0967-0637(94)90007-8

Passow, U., and A. L. Alldredge. 1995. A dye-binding assay for the spectrophotometric measurement of transparent exopolymer particles (TEP). Limnol. Oceanogr. **40**: 1326–1335. doi:10.4319/lo.1995.40.7.1326

Passow, U., K. Ziervogel, V. Asper, and A. Diercks. 2012. Marine snow formation in the aftermath of the Deepwater Horizon oil spill in the Gulf of Mexico. Environ. Res. Lett. **7**: 035301. doi:10.1088/1748-9326/7/3/035301

Passow, U., and E. A. Laws. 2015. Ocean acidification as one of multiple stressors: Response of *Thalassiosira weissflogii* (diatom). Mar. Ecol. Prog. Ser. **541**: 75–90. doi:10.3354/meps11541

Passow, U., and R. D. Hetland. 2016. What happened to all of the oil? Oceanogr. **29**: 88–95. doi:10.5670/oceanog.2016.73

Passow, U., and K. Ziervogel. In press. Marine snow sedimented oil released during the Deepwater Horizon spill. Oceanogr. **29**: 118–125. doi:10.5670/oceanog.2016.76

Patton, J., M. Rigler, P. Boehm, and D. Fiest. 1981. Ixtoc 1 oil spill: Flaking of surface mousse in the Gulf of Mexico. Nature **290**: 235–238. doi:10.1038/290235a0

Peterson, C. H., S. D. Rice, J. W. Short, D. Esler, J. L. Bodkin, B. E. Ballachey, and D. B. Irons. 2003. Long-term ecosystem response to Exxon Valdez oil spill. Science **302**: 2082–2086. doi:10.1126/science.1084282

Pilskaln, C. H., C. Lehmann, and M. W. Silver. 1998. Spatial and temporal dynamics in marine aggregate abundance, sinking rate, and flux: Monterey Bay, Central California. Deep-Sea Res. II **45**: 1803–1837. doi:10.1016/S0967-0645(98)80018-0

Piontek, J., N. Händel, G. Langer, J. Wohlers, U. Riebesell, and A. Engel. 2009. Effects of rising temperature on the formation and microbial degradation of marine diatom aggregates. Aquat. Microb. Ecol. **54**: 305–318. doi: 10.3354/ame01273

Prince, R. C., A. Gramain, and T. J. McGenity. 2010. Prokaryotic hydrocarbon degraders, p. 1669–1692. *In* K. N. Timmis [ed.], Handbook of hydrocarbon and lipid microbiology. http://www.springer.com/978-3-540-77584-3

Prouse, N. J., D. C. Gordon, Jr., and P. D. Keizer. 1976. Effects of low concentrations of oil accommodated in sea water on the growth of unialgal marine phytoplankton cultures. J. Fish. Board Can. **33**: 810–818. doi:10.1139/f76-098

Quigg, A. 2016. Micronutrients, p. 211–231. *In* M. A. Borowitzka, J. Beardall and J. A. Raven [eds.], The physiology of microalgae. Developments in applied phycology series. Springer.

Quigg, A., J. B. Sylvan, A. B. Gustafson, T. R. Fisher, S. Tozzi, and J. W. Ammerman. 2011. Going west: Nutrient limitation of primary production in the Northern Gulf of Mexico and the importance of the Atchafalaya River. Aquat. Geochem. **17**: 519–544. doi:10.1007/s10498-011-9134-3

Quigg, A., and others. 2013. Direct and indirect toxic effects of engineered nanoparticles on algae: Role of natural organic matter. ACS Sustain. Chem. Eng. **1**: 686–702. doi: 10.1021/sc400103x

Quiroz, N. A., C.-C. Hung, and P. H. Santschi. 2006. Binding of Thorium(IV) to carboxylate, phosphate and sulfate functional groups from marine exopolymeric substances (EPS). Mar. Chem. **100**: 337–353. doi:10.1016/j.marchem.2005.10.023

Ramachandran, S. D., P. V. Hodson, C. W. Khan, and K. Lee. 2004. Oil dispersant increases PAH uptake by fish exposed to crude oil. Ecotoxicol. Environ. Safety **59**: 300–308. doi: 10.1016/j.ecoenv.2003.08.018

Ray, P. Z., and M. A. Tarr. 2014. Petroleum films exposed to sunlight produce hydroxyl radical. Chemosphere **103**: 220–227. doi:10.1016/j.chemosphere.2013.12.005

Reddy, C. M., and others. 2012. Composition and fate of gas and oil released to the water column during the Deepwater Horizon oil spill. Proc. Natl. Acad. Sci. USA **109**: 20229–20234. doi:10.1073/pnas.1101242108

Redmond, M. C., and D. L. Valentine. 2012. Natural gas and temperature structured a microbial community response to the Deepwater Horizon oil spill. Proc. Natl. Acad. Sci. USA **109**: 20292–20297. doi:10.1073/pnas.1108756108

Riegman, R., A. M. Noordeloos, and G. C. Cadee. 1992. *Phaeocystis* blooms and eutrophication of the continental coastal zones of the North Sea. Mar. Biol. **112**: 479–484. doi:10.1007/BF00356293

Rivers, A. R., S. Sharma, S. G. Tringe, J. Martin, S. B. Joye, and M. A. Moran. 2013. Transcriptional response of bathypelagic marine bacterioplankton to the Deepwater Horizon oil spill. ISME J. **7**: 2315–2329. doi:10.1038/ismej.2013.129

Rochelle-Newall, E. J., X. Mari, and O. Pringault. 2010. Sticking properties of transparent exopolymeric particles (TEP) during aging and biodegradation. J. Plankton Res. **32**: 1433–1442. doi:10.1093/plankt/fbq060

Rodrigues, R. V., K. C. Miranda-Filho, E. P. Gusmão, C. B. Moreira, L. A. Romano, and L. A. Sampaio. 2010. Deleterious effects of water-soluble fraction of petroleum, diesel and gasoline on marine pejerrey *Odontesthes argentinensis* larvae. Sci. Total Environ. **408**: 2054–2059. doi:10.1016/j.scitotenv.2010.01.063

Ron, E. Z., and E. Rosenberg. 2002. Biosurfactants and oil bioremediation. Curr. Opin. Biotechnol. **13**: 249–252. doi: 10.1016/S0958-1669(02)00316-6

Saeed, T., and M. Al-Mutairi. 1999. Chemical composition of the water-soluble fraction of the leaded gasolines in seawater. Environ. Int. **25**: 117–129. doi:10.1016/S0160-4120(98)00093-2

Santschi, P. H., E. Balnois, K. J. Wilkinson, J. Zhang, J. Buffle, and L. Guo. 1998. Fibrillar polysaccharides in marine macromolecular organic matter as imaged by atomic force microscopy and transmission electron microscopy. Limnol. Oceanogr. **43**: 868–908. doi:10.4319/lo.1998.43.5.0896

Santschi, P. H., L. Guo, I. D. Walsh, M. S. Quigley, and M. Baskaran. 1999. Boundary exchange and scavenging of radionuclides in continental margin waters of the Middle Atlantic Bight. Implications for organic carbon fluxes. Cont. Shelf Res. **19**: 609–636. doi:10.1016/S0278-4343(98)00103-4

Santschi, P. H., C.-C. Hung, L. Guo, J. Pinckney, G. Schultz, N. Alvarado-Quiroz, and I. Walsh. 2003. Control of acid polysaccharide production, and ^{234}Th and POC export fluxes by marine organisms. Geophys. Res. Lett. **30**: 1044. doi:10.1029/2002GL016046

Seebah, S., C. Fairfield, M. S. Ullrich, and U. Passow. 2014. Aggregation and sedimentation of *Thalassiosira weissflogii* (diatom) in a warmer and more acidified future ocean. PloS One **9**: e112379. doi:10.1371/journal.pone.0112379

Silver, M. 2015. Marine snow: A brief historical sketch. Limnol. Oceanogr. Bull. **24**: 5–10. doi:10.1002/lob.10005

Silver, M., A. Shanks, and J. Trent. 1978. Marine snow: Microplankton habitat and source of small-scale patchiness in pelagic populations. Science **201**: 371–373. doi: 10.1126/science.201.4353.371

Simon, M., H.-P. Grossart, B. Schweitzer, and H. Ploug. 2002. Microbial ecology of organic aggregates in aquatic systems. Aquat. Microb. Ecol. **28**: 175–211. doi:10.3354/ame028175

Singer, M. M., D. L. Smalheer, R. S. Tjeerdema, and M. Martin. 1991. Effects of spiked exposure to an oil dispersant on the early life stages of four marine species.

Environ. Toxicol. Chem. **10**: 1367–1374. doi:10.1897/1552-8618(1991)10[1367:EOSETA]2.0.CO;2

Smetacek, V. 1985. Role of sinking in diatom life history cycles: Ecological, evolutionary and geological significance. Mar. Biol. **84**: 239–251. doi:10.1186/2046-9063-8-10

Smith, D. C., M. Simon, A. L. Alldredge, and F. Azam. 1992. Intense hydrolytic enzyme activity on marine aggregates and implications for rapid particle dissolution. Nature **359**: 139–142. doi:10.1038/359139a0

Staats, N., L. J. Stal, and L. R. Mur. 2000. Exopolysaccharide production by the epipelic diatom *Cylindrotheca closterium*: Effects of nutrient conditions. J. Exp. Mar. Biol. Ecol. **249**: 13–27. doi:10.1016/S0022-0981(00)00166-0

Stachowitsch, M., N. Fanuko, and M. Richter. 1990. Mucus aggregates in the Adriatic Sea: An overview of stages and occurrences. Mar. Ecol. **11**: 327–350. doi:10.1111/j.1439-0485.1990.tb00387.x

Steigenberger, S., P. J. Statham, C. Völker, and U. Passow. 2010. The role of polysaccharides and diatom exudates in the redox cycling of Fe and the photoproduction of hydrogen peroxide in coastal waters. Biogeosciences **7**: 109–119. doi:10.5194/bg-7-109-2010

Stenstörm, T. A. 1989. Bacterial hydrophobicity, an overall parameter for the measurement of adhesion potential to soil particles. Appl. Environ. Microbiol. **55**: 142–147. doi:10.1186/2046-9063-8-10

Strmečki, S., M. Plavšić, S. Steigenberger, and U. Passow. 2010. Characterization of phytoplankton exudates and carbohydrates in relation to their complexation of copper, cadmium and iron. Mar. Ecol. Prog. Ser. **408**: 33–46. doi:10.1186/2046-9063-8-10

Sun, L., H. Chen, H. A. Abdulla, and K. Mopper. 2014. Estimating hydroxyl radical photochemical formation rates in natural waters during long-term laboratory irradiation experiments. Environ. Sci. Process. Impacts **16**: 757–763. doi:10.1039/C3EM00587A

Sun, L., and K. Mopper. 2016. Studies on hydroxyl radical formation and correlated photoflocculation process using degraded wood leachate as a CDOM Source. Front. Mar. Sci. **2**. 10.3389/fmars.2015.00117

The chaos of clean up (TCCP). 2011. Analysis of Potential Health and Environmental impacts of chemicals in dispersant materials. (Rep Toxipedia Consulting Services and Earthjustice).

Thornton, D. C. O. 2002. Diatom aggregation in the sea: Mechanisms and ecological implications. Eur. J. Phycol. **37**: 149–161. doi:10.1017/S0967026202003657

Turner, J. 2015. Zooplankton fecal pellets, marine snow, phytodetritus and the ocean's biological pump. Prog. Oceanogr. **130**: 205–248. doi:10.1016/j.pocean.2014.08.005

Urbani, R., E. Magaletti, P. Sist, and A. M. Cicero. 2005. Extracellular carbohydrates released by the marine diatoms *Cylindrotheca closterium, Thalassiosira pseudonana* and *Skeletonema costatum*: Effect of P-depletion and growth status. Sci. Total Environ. **353**: 300–306. doi:10.1016/j.scitotenv.2005.09.026

Valentine, D. L., and others. 2010. Propane respiration jump-starts microbial response to a deep oil spill. Science **330**: 208–211. doi:10.1126/science.1196830

Valentine, D. L., G. B. Fisher, S. C. Bagby, R. K. Nelson, C. M. Reddy, S. P. Sylva, and M. A. Woo. 2014. Fallout plume of submerged oil from Deepwater Horizon. Proc. Natl. Acad. Sci. USA **111**: 15906–15911. doi:10.1073/pnas.1414873111

Valko, M., D. Leibfritz, J. Moncol, M. T. D. Cronin, M. Mazur, and J. Telser. 2007. Free radicals and antioxidants in normal physiological functions and human disease. Int. J. Biochem. Cell Biol. **39**: 44–84. doi:10.1016/j.biocel.2006.07.001

van Emburg, P. R., E. W. de Jong, and W. T. Daems. 1986. Immunochemical localization of a polysaccharide from biomineral structures (coccoliths) of *Emiliania huxleyi*. J. Struct. Biol. **94**: 246–259. doi:10.1016/0889-1605(86)90071-6

Varela, F. G., H. R. Maturana, and R. Uribe. 1974. Autopoiesis: The organization of living systems, its characterization and a model. BioSystems **5**: 187–196. doi:10.1016/0303-2647(74)90031-8

Verdugo, P. 2007. Dynamics of marine biopolymer networks. Polym. Bull. **58**: 139–143. doi:10.1007/s00289-006-0615-2

Verdugo, P. 2012. Marine microgels. Ann. Rev. Mar. Sci. **4**: 375–400. doi:10.1146/annurev-marine-120709-142759

Verdugo, P., A. L. Alldredge, F. Azam, D. L. Kirchman, U. Passow, and P. H. Santschi. 2004. The oceanic gel phase: A bridge in the DOM–POM continuum. Mar. Chem. **92**: 67–85. doi:10.1016/j.marchem.2004.06.017

Verdugo, P., M. V. Orellana, W.-C. Chin, T. W. Petersen, G. van den Eng, R. Benner, and J. I. Hedges. 2008. Marine biopolymer self-assembly: Implications for carbon cycling in the ocean. Faraday Discuss. **139**: 393–398. doi:10.1039/b800149a

Verdugo, P., and P. H. Santschi. 2010. Polymer dynamics of dissolved organic carbon networks and gel formation in seawater. Deep-Sea Res. II **57**: 1486–1493. doi:10.1016/j.dsr2.2010.03.002

Villacorte, L. O., M. D. Kennedy, G. Amy, and J. C. Schippers. 2009. The fate of transparent exopolymer particles (TEP) in integrated membrane systems: Removal through pre-treatment processes and deposition on reverse osmosis membranes. Water Res. **43**: 5039–5052. doi:10.1016/j.watres.2009.08.030

Villacorte, L. O., Y. Ekowati, H. Winters, G. L. Amy, J. C. Schippers, and M. D. Kennedy. 2013. Characterization of transparent exopolymer particles (TEP) produced during algal bloom: A membrane treatment perspective. Desalin. Water Treat. **51**: 1021–1033. doi:10.1080/19443994. 2012.699359

Waggoner, D. C., H. Chen, A. S. Willoughby, and P. G. Hatcher. 2015. Formation of black carbon-like and alicyclic aliphatic compounds by hydroxyl radical initiated degradation of lignin. Org. Geochem. **82**: 69–76. doi:10.1016/j.orggeochem.2015.02.007

Walsh, J. J., and others. 2015. A simulation analysis of the plankton fate of the Deepwater Horizon oil spills. Cont. Shelf Res. **107**: 50–68. doi:10.1016/j.csr.2015.07.002

Wang, Z., L. Liu, J. Yao, and W. Cai. 2006. Effects of extracellular polymeric substances on aerobic granulation in sequencing batch reactors. Chemosphere **63**: 1728–1735. doi:10.1016/j.chemosphere.2005.09.018

White, H. K., and others. 2012. Impact of the Deepwater horizon oil spill on a deep-water coral community in the Gulf of Mexico. Proc. Natl. Acad. Sci. USA **109**: 20303–20308. doi:10.1073/pnas.1118029109

Wurl, O., L. Miller, R. Röttgers, and S. Vagle. 2009. The distribution and fate of surface-active substances in the sea-surface microlayer and water column. Mar. Chem. **115**: 1–9. doi:10.1016/j.marchem.2009.04.007

Xu, C., P. H. Santschi, K. A. Schwehr, and C.-C. Hung. 2009. Optimized isolation and purification procedure for obtaining strongly actinide binding exopolymeric substances (EPS) from two bacteria (*Sagittula stellata* and *Pseudomonas fluorescens* Biovar II). Bioresour. Technol. **100**: 6010–6021. doi:10.1016/j.biortech.2009.06.008

Xu, C., and others. 2011a. Controls of ^{234}Th removal from the oligotrophic ocean by polyuronic acids and modification by microbial activity. Mar. Chem. **123**: 111–126. doi:10.1016/j.marchem.2010.10.005

Xu, C., S. Zhang, C.-Y. Chuang, E. J. Miller, K. A. Schwehr, and P. H. Santschi. 2011b. Chemical composition and relative hydrophobicity of microbial exopolymeric substances (EPS) isolated by anion exchange chromatography and their actinide-binding affinities. Mar. Chem. **126**: 27–36. doi:10.1016/j.marchem.2011.03.004

Yan, B., and others. 2016. Sustained deposition of contaminants from the Deepwater Horizon spill. Proc. Natl. Acad. Sci. USA **113**: E3332–E3340. doi:10.1073/pnas.1513156113

Yang, T., L. M. Nigro, T. Gutierrez, L. D'Ambrosio, S. B. Joye, R. Highsmith, and A. Teske. 2016. Pulsed blooms and persistent oil-degrading bacterial populations in the water column during and after the Deepwater Horizon blowout. Deep-Sea Res. II **129**: 282–291. doi:10.1016/j.dsr2.2014.01.014

Zhang, S., C. Xu, and P. H. Santschi. 2008. Chemical composition and 234Th(IV) binding of extracellular polymeric substances (EPS) produced by the marine diatom *Amphora* sp. Mar. Chem. **112**: 81–92. doi:10.1016/j.marchem.2008.05.009

Zhang, S., and P. H. Santschi. 2009. Application of cross-flow ultrafiltration for isolating exopolymeric substances (EPS) from a marine diatom (*Amphora sp.*). Limnol. Oceanogr.: Methods **7**: 419–429. doi:10.4319/lom.2009.7.419

Zhang, S., Y. L. Jiang, C.-S. Chen, J. Spurgin, K. A. Schwehr, A. Quigg, W.-C. Chin, and P. H. Santschi. 2012. Aggregation and dissolution of quantum dots in marine environments: Importance of extracellular polymeric substances. Environ. Sci. Technol. **46**: 8764−8772. doi:10.1021/es301000m

Zhang, S., Y. Jiang, C.-S. Chen, D. Creeley, K. A. Schwehr, A. Quigg, W.-C. Chin, and P. H. Santschi. 2013. Ameliorating effects by extracellular polymeric substances excreted by *Thalassiosira pseudonana* on algal toxicity of CdSe Quantum Dots. Aquat. Toxicol. **126**: 214–223. doi:10.1016/j.aquatox.2012.11.012

Zhou, J., K. Mopper, and U. Passow. 1998. The role of surface-active carbohydrates in the formation of transparent exopolymer particles by bubble adsorption of seawater. Limnol. Oceanogr. **43**: 1860–1871. doi:10.1186/2046-9063-8-10

Ziervogel, K., L. McKay, B. Rhodes, C. L. Osburn, J. Dickson-Brown, C. Arnosti, and A. Teske. 2012. Microbial activities and dissolved organic matter dynamics in oil-contaminated surface seawater from the Deepwater Horizon oil spill site. PLoS One **7**: e34816. doi:10.1371/journal.pone.0034816

Ziervogel, K., and C. Arnosti. 2013. Enhanced protein and carbohydrate hydrolyses in plume-associated deep waters initially sampled during the early stages of the Deepwater Horizon oil spill. Deep-Sea Res. II **129**: 368–373. doi:10.1186/2046-9063-8-10

Acknowledgments

Importantly, we thank the reviewers, senior editor (Claudia Benitez-Nelson), and the editor-in-chief (Patricia Soranno), whose comments greatly improved the manuscript. This research was made possible by a grant from The Gulf of Mexico Research Initiative to support consortium research entitled ADDOMEx (Aggregation and Degradation of Dispersants and Oil by Microbial Exopolymers) Consortium. We thank Kendra Kopp for help with preparing Fig. 2 and Rachel Windham for making Fig. 6. Jessica Hillhouse formatted the references in the bibliography.

12

Climate-induced warming of lakes can be either amplified or suppressed by trends in water clarity

Kevin C. Rose,[1] Luke A. Winslow,[2] Jordan S. Read,[2] Gretchen J. A. Hansen*[3,a]

[1]Department of Biological Sciences, Rensselaer Polytechnic Institute, Troy, New York; [2]U.S. Geological Survey, Office of Water Information, Middleton, Wisconsin; [3]Bureau of Science Services, Wisconsin Department of Natural Resources, Madison, Wisconsin

Scientific Significance Statement

Although climate change is rapidly warming many surface waters globally, lakes have been found to be warming at different rates. Lakes respond differently to air temperatures due in part to often large differences in water clarity and lake depth. However, because water clarity in lakes in many regions is also changing, it is unclear how combined changes in air temperature and water clarity influence temperatures in lakes of different depths. We used process-based modeling and empirical observations to show that water clarity trends of about 1% per year in either direction can amplify or suppresses warming at rates comparable to climate-induced warming and that trends in water clarity may be as important as rising air temperatures in determining how waterbodies respond to climate change.

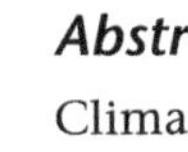

Abstract

Climate change is rapidly warming aquatic ecosystems including lakes and reservoirs. However, variability in lake characteristics can modulate how lakes respond to climate. Water clarity is especially important both because it influences the depth range over which heat is absorbed, and because it is changing in many lakes.

*Correspondence: kev.c.rose@gmail.com

Author Contribution Statement: Together, all authors conceived and designed the experiments. LAW and JSR conceived and designed the modeling framework, and LAW, JSR, and GJAH aggregated driver data and lake metadata for simulations. LAW ran the model simulations and KCR analyzed the data and drafted the manuscript. All authors contributed to the manuscript.

The hydrodynamic model used, General Lake Model (GLM), is available at: http://aed.see.uwa.edu.au/research/models/GLM/ and GLM source code is available at: https://github.com/AquaticEcoDynamics/GLM.

Observational data for Crystal and Sparkling Lakes is available through the North Temperate Lakes Long Term Ecological Research site (https://lter.limnology.wisc.edu/). Specifically, we used the NTL LTER water clarity dataset (number 5730), referred to as the "North Temperate Lakes LTER: Light Extinction - Trout Lake Area 1981 - current" (available online at: https://lter.limnology.wisc.edu/data/filter/5730) and the temperature dataset (number 5731), referred to as "Physical Limnology of the North Temperate Lakes Primary Study Lakes" (available online at https://lter.limnology.wisc.edu/data/filter/5721). Light extinction coefficients (K_d, units: m^{-1}) were converted to estimates of Secchi depth (units: m) using the relationship $K_d = 1.7$/Secchi depth.

[a]Present address: Fish and Wildlife Division, Minnesota Department of Natural Resources, St. Paul, Minnesota, USA

Here, we show that simulated long-term water clarity trends influence how both surface and bottom water temperatures of lakes and reservoirs respond to climate change. Clarity changes can either amplify or suppress climate-induced warming, depending on lake depth and the direction of clarity change. Using a process-based model to simulate 1894 north temperate lakes from 1979 to 2012, we show that a scenario of decreasing clarity at a conservative yet widely observed rate of 0.92% yr^{-1} warmed surface waters and cooled bottom waters at rates comparable in magnitude to climate-induced warming. For lakes deeper than 6.5 m, decreasing clarity was sufficient to fully offset the effects of climate-induced warming on median whole-lake mean temperatures. Conversely, a scenario increasing clarity at the same rate cooled surface waters and warmed bottom waters relative to baseline warming rates. Furthermore, in 43% of lakes, increasing clarity more than doubled baseline bottom temperature warming rates. Long-term empirical observations of water temperature in lakes with and without clarity trends support these simulation results. Together, these results demonstrate that water clarity trends may be as important as rising air temperatures in determining how waterbodies respond to climate change.

Climate change is warming lakes (Schneider and Hook 2010; Butcher et al. 2015; O'Reilly et al. 2015), with important implications for their ecology and the goods and services they provide to society (Woodward et al. 2010; Carpenter et al. 2011; Shimoda et al. 2011). In lakes, warming is driven primarily by rising air temperatures (Schmid et al. 2014; Toffolon et al. 2014; Butcher et al. 2015; Layden et al. 2015; O'Reilly et al. 2015). However, warming rates can vary widely among lakes (O'Reilly et al. 2015). This is in part because variability in lake characteristics can modulate how lakes respond to air temperature changes (Bayer et al. 2013; Schmid et al. 2014; Kraemer et al. 2015a, 2015b). In particular, water clarity can regulate how water bodies respond to atmospheric warming by controlling how solar radiation is absorbed in the water column (Persson and Jones 2008; Rinke et al. 2010; Read and Rose 2013). For example, single lake studies have shown that large clarity declines contribute to warmer surface waters but volumetrically cooler lakes, even during periods of rising air temperatures (Tanentzap et al. 2008; Rinke et al. 2010).

Water clarity is changing in many lakes and reservoirs, with both increasing and decreasing trends frequently reported (McCullough et al. 2013; Olmanson et al. 2013; Lottig et al. 2014; Rose et al. in press). While large and abrupt step-changes in water clarity are possible in certain systems (Scheffer et al. 1993), long-term clarity trends with annual rates of change of about 1% are common (Lottig et al. 2014). Water clarity affects lake water temperature by influencing the vertical partitioning of heat and outward heat fluxes. For example, lakes with low water clarity trap heat closer to the surface and have larger outward heat fluxes (Read and Rose 2013). However, the temperature profiles of diverse lakes may respond differently to identical changes in clarity. Lake depth may be an important characteristic that regulates how clarity influences lake temperatures. Specifically, depth may constrain the maximum vertical difference in temperature that changes in water clarity can produce, with deeper lakes being more sensitive to clarity changes (Gorham 1964).

Until now, the role of clarity in determining lake responses to climate change has not been tested, in part due to the paucity of geographically distributed long-term water temperature and clarity records. Process-based hydrodynamic models may provide a solution to overcome this data limitation. These models are commonly applied tools to understand and predict climate change impacts on lakes when observational data are sparse (Fang and Stefan 2009; Bayer et al. 2013; Fink et al. 2014; Read et al. 2014; Schmid et al. 2014; Butcher et al. 2015). Process-based simulation models can accurately predict water temperatures for spatially distributed lakes for several reasons. First, monitoring data exist for key components of lake heat budgets (e.g., solar radiation). Second, hydrodynamic model complexity is relatively constrained (compared with biological or biogeochemical models). This approach is especially powerful when validated with long-term observational data. Additionally, new modeling approaches enable the simulation of thousands of lakes simultaneously (Read et al. 2014).

Here, we employ a scenario approach to ask if conservative long-term trends in water clarity can meaningfully influence the thermal response of a diverse and broadly representative population of lakes to climate change. We then complement scenarios with observational data demonstrating the model's ability to simulate realistic water

Table 1. Values used for GLM model parameters for simulation of the 1894 north temperate lakes from 1979 to 2012.

Parameter	Value
Minimum layer thickness (m)	0.2
Maximum layer thickness (m)	Varied; 0.3–1.5
Bulk aerodynamic transport coefficients	0.0013
Convective overturn mixing efficiency	0.125
Wind stirring efficiency	0.23
Shear production efficiency	0.2
Kelvin–Helmholtz turbulent billows mixing efficiency	0.3
Hypolimnetic turbulence mixing efficiency	0.5

temperature trends in two neighboring lakes, one with a long-term clarity trend and one without. We predicted that increasing clarity would amplify whole-lake and bottom temperature warming, but suppress surface temperature warming. Conversely, we predicted that decreasing clarity would suppress whole-lake average and bottom temperature warming, but amplify surface temperature warming. Finally, we predicted that lake depth would regulate how sensitive lake temperatures are to changing water clarity because depth influences the maximum vertical difference in heat partitioning.

Methods

We used the process-based model General Lake Model (GLM; Hipsey et al. 2014) version 2.0.2 to model the thermal characteristics of 1894 north temperate lakes in Wisconsin, U.S.A. at a daily time-step over the period 1979–2012. Important model parameters are described in Table 1. Consistent with existing studies (e.g., O'Reilly et al. 2015), we focus on summertime (July, August, and September) temperatures. Thus, data represent summertime averages and not annual temperatures or trends.

We use GLM to explore trends in surface, bottom, and whole lake mean temperatures. GLM utilizes a vertically layered Lagrangian structure to simulate water temperatures. We defined surface and bottom temperatures as the modeled temperature of the shallowest and deepest model layer respectively. Whole-lake mean temperature refers to the volumetrically averaged temperature and was calculated in each lake using a set of 20 temperature measurements in each water column and accounting for the volume in each layer.

Heat flux model components were formatted as time-series inputs to GLM for wind speed, air temperature, relative humidity, precipitation, and downwelling longwave and shortwave radiation. Data for these model components was accessed from the North American Land Data Assimilation System (NLDAS; ldas.gsfc.nasa.gov/nldas). A statewide air temperature trend was estimated over the same period using data from the National Oceanic and Atmospheric Administration (NOAA) National Centers for Environmental Information (NCEI; https://www.ncdc.noaa.gov/climate-monitoring/).

Lake-specific model inputs included hypsographic curves (or estimates thereof), wind sheltering coefficients, and water clarity. Hypsographic curves were estimated either from bathymetric maps or a cone parameterized by lake surface area and maximum depth (Read et al. 2014). Wind sheltering coefficients were estimated from lake surface area and surrounding vegetation height (Van Den Hoek et al. 2015). Water clarity starting points for each lake (values for year 1979 and all years for the baseline simulation; see below) were derived from in situ Secchi disk depth measurements and Landsat satellite-derived Secchi disk depth estimates. Secchi disk depths were calculated as the mean of all observations for each lake. Lakes ranged in maximum depth from 0.9 m to 106.7 m (median: 7.3 m), in area from 1 ha to 53,394 ha (median: 29 ha), and in initial Secchi disk depth from 0.1 m to 13.4 m (median: 2.5 m).

Lake simulations were not calibrated to individual lakes. Model accuracy was evaluated by root mean squared error (RMSE) for epilimnion, hypolimnion, and whole-lake mean temperature profiles. The RMSE between model simulated and observed temperatures was 2.72°C for all depths ($n = 224{,}812$ measurements in 1137 lakes), 1.84°C for epilimnetic temperatures ($n = 13{,}708$ measurements in 643 lakes), 3.26°C for hypolimnetic temperatures ($n = 13{,}708$ measurements in 643 lakes), and 2.07°C for whole-lake mean temperatures ($n = 23{,}005$ measurements in 747 lakes). For the July–August–September period, RMSE values were 0.98°C for epilimnetic temperature ($n = 188$ measurements in 54 lakes), 2.00°C for the hypolimnetic temperature ($n = 188$ measurements in 54 lakes), and 1.87°C for whole-lake mean temperatures ($n = 188$ measurements in 54 lakes). For further details on the model performance, see Read et al. (2014).We ran three 34-yr scenario simulations for all lakes to assess the potential impacts of long-term changes in clarity on lake temperatures across a landscape of thousands of lakes. In the first baseline scenario, water clarity was held constant, and any trends in lake temperatures are due to changes in atmospheric forcing. We also created scenarios of increasing and decreasing clarity, where Secchi depth (in m) was increased or decreased by 0.92% yr^{-1}. Secchi disk depth was then converted to the diffuse attenuation coefficient (k_d, units: m^{-1}) using the equation $k_d = 1.7$/Secchi depth. The annual rate of clarity change in the scenarios was chosen to represent a conservative, realistic long-term trend in water clarity. This rate of change was calculated from survey of 3251 lakes spread across eight states in the central United States (Lottig et al. 2014). Specifically, the 0.92% yr^{-1} represents the posterior estimate for the mean slope observed across the lake survey, analyzed using a hierarchical model of $\log_e$-transformed Secchi depth as a linear function of year and allowing intercepts, slopes, and model error variances to differ among lakes. While the period of water clarity measurements differed among lakes, 99% of measurements were from the 1970s and afterwards. In our scenario simulations, final (year 2012) Secchi disk depths ranged from 0.2 m to 17.3 m (median: 3.2 m) for the increasing clarity scenario and ranged from 0.1 m to 9.4 m (median: 1.8 m) for the decreasing clarity scenario.

We used the Theil-Sen slope estimator (Sen 1968) to quantify annual temporal changes in lake thermal characteristics for the three scenarios. This non-parametric method calculates slopes from all permutations of paired observations, and is robust to outliers and non-normality (Wilcox 2011). We report the Theil-Sen slopes for the population of lakes as the difference between the increasing or decreasing scenario value and the baseline scenario value in order to control for long-

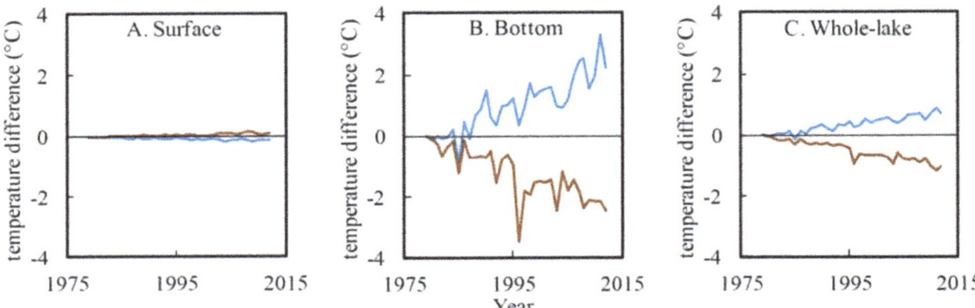

Fig. 1. Trends in surface (**A**), bottom (**B**), and whole-lake (**C**) mean temperature for simulations in increasing (blue), and decreasing (brown) clarity.

Table 2. Estimated median trends (units: °C decade^{-1}) for 1894 lakes simulations of increasing or decreasing water clarity at 0.92% yr^{-1} and no trend in water clarity.

Scenario	Surface	Bottom	Whole lake average
Increasing water clarity	0.222 (0.8)	0.303 (1.0)	0.263 (0.9)
Decreasing water clarity	0.316 (1.1)	0.097 (0.3)	0.155 (0.5)
No change in water clarity (baseline)	0.267 (0.9)	0.216 (0.7)	0.237 (0.8)

Values in parentheses indicate the temperature change (units: °C) over the 34 yr simulation period, 1979–2012.

term trends in water temperature that were not due to changing water clarity (i.e., those due to climate warming). We also report median Theil-Sen slopes for the baseline scenario to assess how modeled lake temperature characteristics changed over the period 1979–2012. A locally weighted polynomial regression (LOWESS; a component of the R "stats" package) was applied to each plot of Theil-Sen slopes vs. depth to visualize how slopes varied for each treatment across the population of lakes. Calculations were all run in R, version 3.2.2 (R Core Team 2015).

In addition to our scenarios modeling thousands of lakes, empirical observations of water temperature and clarity from two adjacent lakes in the NTL LTER site were used as a case study to assess if GLM accurately predicts observed temperature trends in lakes with and without clarity trends. The NTL LTER sampling program has collected data on seven lakes in Northern Wisconsin regularly since 1981 (https://lter.limnology.wisc.edu/). Two of these lakes, Crystal (latitude: 46.001, longitude: −89.614) and Sparkling (latitude: 46.005, longitude: −89.699) are located 6.6 km from each other and have similar maximum depth (20.4 m and 20.0 m, respectively), area (37 ha and 64 ha, respectively), and Secchi depth (long term mean of 10.3 m and 8.2 m, respectively). Crystal Lake has experienced a long-term decline in water clarity while Sparkling has not. At the same time, both lakes have been exposed to a regionally warming climate. We used these lakes as a natural experimental (Crystal) and control (Sparkling) to assess the impacts of water clarity loss and atmospheric forcing on lake temperature characteristics. We assessed changes in observed mean summer (July, August, September) near-surface, near-bottom, and whole-lake mean temperatures in both lakes over the period 1981–2011 [2012 was excluded because a whole-lake temperature manipulation experiment began in Crystal Lake starting in 2012 (Lawson et al. 2015)]. Near-surface was defined as measurements $<$ 0.5 m deep and near-bottom was defined as measurements $>$ 16 m deep in both lakes. We also calibrated GLM to these lakes and used site-specific lake water clarity data to test if GLM could accurately predict water temperatures and temperature trends in these lakes. Site-specific calibrations are described in the Supporting Information.

Results

Lake thermal responses to clarity changes were heterogeneous across systems as well as within the water column of individual systems (Fig. 1). Whole-lake mean summer temperatures increased by a median of 0.237°C decade^{-1} under the baseline scenario, meaning that lakes warmed by a median of about 0.8°C over the 34 yr period absent of changes in water clarity (Table 2). Surface and bottom temperatures warmed at median rates of 0.267 and 0.216°C decade^{-1}, respectively, under the baseline scenario (Fig. 1; Table 1). Over this period, observed regional summer air temperatures warmed by 0.356°C decade^{-1} from 1979 to 2012.

Bottom and whole-lake temperatures were more sensitive to clarity changes than surface water temperatures (Fig. 2). Decreasing clarity by 0.92% annually accelerated the median surface temperature warming by 18% but suppressed bottom temperature warming by 55% relative to the baseline warming rate. Inclusive of baseline warming, over the 34-yr study period this translates to a median increase in surface

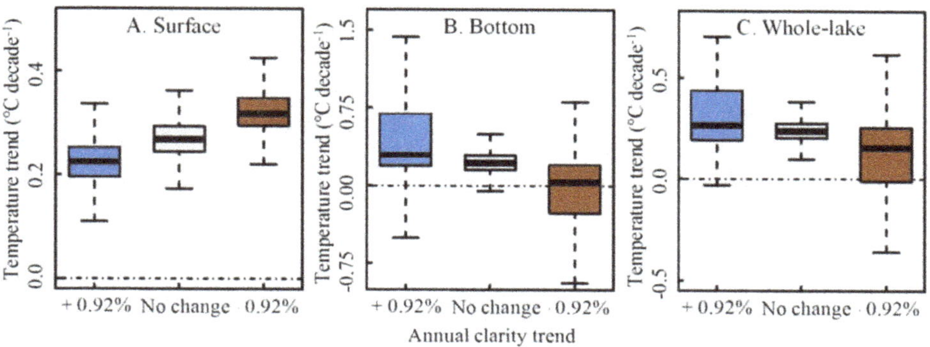

Fig. 2. Median differences in temperature through time between increasing (blue) or decreasing clarity (brown) simulations and the baseline simulation (gray) for surface (**A**), bottom (**B**), and whole-lake mean temperature (**C**).

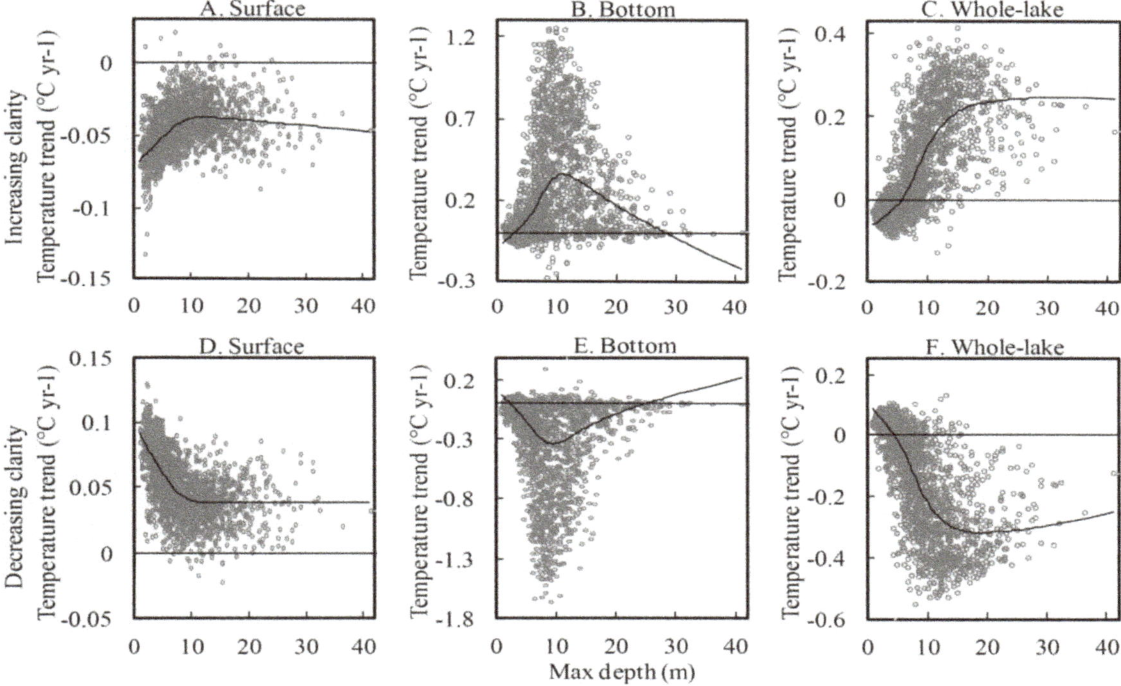

Fig. 3. Maximum depth strongly influenced surface (left, **A** and **D**), bottom (center, **B** and **E**) and whole-lake (right, **C** and **F**) temperature trends (units: °C decade^{-1}) in response to increasing (top row, **A–C**) or decreasing (bottom row, **D–F**) water clarity. Black lines are locally weighted polynomial regressions (LOWESS). The deepest two lakes (106.7 m and 71.9 m deep) were removed from plotting to better highlight the distribution of lakes.

temperature of 1.1°C, and a much smaller median increase in bottom temperature of 0.3°C in lakes with decreasing clarity (Table 2). In 48% of lakes, declining water clarity more than fully offset the effects of climate warming, and lake bottom waters actually cooled over the study period in spite of climate warming. Increasing clarity had opposite effects. Increasing clarity by 0.92% annually suppressed the median surface warming rate by 17%, and accelerated the median bottom temperature warming rate by 41% relative to the baseline. Inclusive of baseline warming, over the 34-yr study period, median surface temperatures increased by 0.8°C and bottom temperatures increased by 1.0°C in lakes with increasing clarity (Table 2). In 43% of lakes, increasing clarity more than doubled baseline bottom temperature warming rates.

Because of the opposite effects of water clarity changes on surface and bottom waters, clarity changes often influenced mean whole-lake temperatures less so than they did bottom waters alone. Furthermore, whole-lake temperatures frequently moved in the opposite direction compared with surface water temperatures. In 26% of lakes, decreasing clarity more than fully suppressed whole-lake warming, and whole-lake temperatures actually cooled as lakes became less clear despite warming in the baseline scenario. Across all lakes, increasing clarity

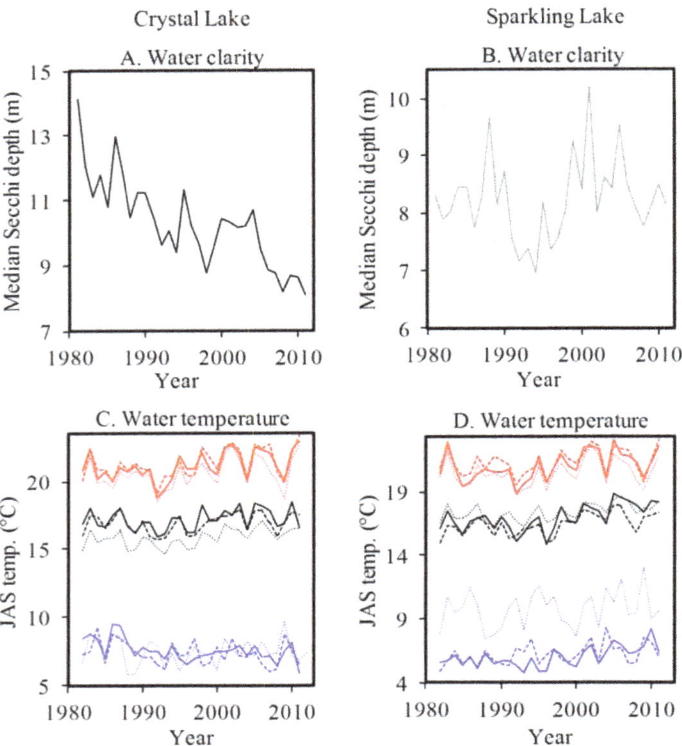

Fig. 4. Crystal Lake (left column) and Sparkling Lake (right column) water clarity (**A, B**) and temperatures (**C, D**), for observed temperatures (solid lines), modeled temperatures using the generic model parameterization (dotted lines) and lake-specific model parameterization (**C, D**; dashed lines) for surface (red), whole-lake average (black), and bottom (blue) temperatures. Temperatures are for the July–August–September time period over the period 1981–2011. Sparkling Lake's clarity did not change and its surface and bottom waters warmed, concurrent with warming air temperatures. Crystal Lake exhibited declining clarity; its surface waters warmed while its bottom waters cooled.

accelerated the median whole-lake warming rate by 11%, while decreasing clarity suppressed the median warming rate by 35% (relative to the baseline). Inclusive of baseline warming, whole-lake temperature increased by a median of 0.9°C in response to increasing clarity, nearly double the increase (0.5°C) seen in lakes undergoing decreasing clarity (Table 2).

Water bodies responded to clarity changes quite differently from one another, highlighting the importance of site-specific features in regulating lake thermal responses to climate. In particular, bottom temperature responses to clarity changes varied substantially among lakes. For example, while decreasing clarity accelerated the surface temperature warming rate in over 99% of lakes, it suppressed baseline bottom temperature warming in 78% of lakes and amplified bottom temperature warming in 22%. Similar to lake bottom temperatures, whole-lake temperature responses to changing clarity were quite variable. Decreasing water clarity suppressed baseline whole-lake warming in 65% of lakes, while it accelerated warming in 35%.

Table 3. Observed and simulated trends in water temperature (units: °C decade^{-1}) in Crystal and Sparkling Lakes over the period 1982–2011.

	Surface	Bottom	Whole lake average
Crystal			
Observed trend	0.58 (1.8)	−0.44 (1.4)	0.24 (0.7)
Modeled trend	0.56 (1.7)	−0.30 (0.9)	0.23 (0.7)
RMSE	0.43	0.76	0.57
Sparkling			
Observed trend	0.55 (1.7)	0.46 (1.4)	0.69 (2.1)
Modeled trend	0.43 (1.3)	0.3 (0.9)	0.54 (1.7)
RMSE	0.57	0.80	0.78

Values in parentheses indicate the temperature change (units: °C) over the 31 yr simulation period, 1981–2011. RMSE refers to root mean squared error (units: °C).

Maximum depth strongly influenced lake thermal sensitivity to clarity changes (Fig. 3). Whole-lake temperatures of deep lakes were particularly sensitive to clarity changes. For example, for lakes > 6.5 m deep, decreasing clarity actually reduced the median whole-lake temperature for this subset of lakes despite substantial warming under the baseline scenario. The median baseline warming rate for this population of lakes ($n = 1061$) was 0.251°C decade^{-1}, which is slightly higher than the total lake population (0.237°C decade^{-1}). In contrast, bottom temperatures in lakes of intermediate depth (3–18 m) were particularly sensitive to clarity changes (Fig. 3). Decreasing clarity in lakes of this depth range was sufficient to cool lake bottom waters despite baseline warming, while increasing clarity more than doubled bottom temperature warming relative to the baseline rate. Shallow lakes (< 3 m) were relatively insensitive to clarity trends (Fig. 3).

Long term in situ measurements of water clarity and temperature at two North Temperate Lakes Long Term Ecological Research sites, Crystal and Sparkling Lakes, support simulation results showing that declining clarity suppresses warming and can even lead to cooling trends despite regional warming (Fig. 4). Over the period 1981–2011, water clarity in Sparkling Lake did not change significantly ($p = 0.157$), while water clarity decreased significantly by 2.2% yr^{-1} in Crystal Lake ($p < 0.001$). Summertime surface and bottom water temperatures in Sparkling Lake increased by 0.421°C decade^{-1} ($p = 0.035$) and 0.234°C decade^{-1} ($p = 0.036$), respectively. Relative to Sparkling Lake, decreasing water clarity accelerated warming in the surface waters of Crystal Lake by a rate of 0.494°C decade^{-1} (17%; $p = 0.042$), while also substantial cooling bottom waters (rate of −0.517°C decade^{-1}, $p = 0.043$). When calibrated to Crystal and Sparkling Lakes, model predictions matched observed data and trends well (Table 3). In both lakes, RMSE was higher for bottom water temperatures than for surface or whole lake average temperature, but lower than the observed change in temperature over the simulation period.

Discussion

Our results show that trends in water clarity can substantially influence the thermal trajectories of lakes and reservoirs, especially in systems more than a few meters deep. Studies of climate change impacts on lakes and reservoirs have primarily focused on large, deep lakes (Schneider and Hook 2010; Shimoda et al. 2011; Fink et al. 2014; Layden et al. 2015; but see O'Reilly et al. 2015). Many lakes in which warming trends have been documented also have experienced long-term changes in water clarity (Shimoda et al. 2011). For example, Lake Tanganyika, located in the African Rift Valley, has experienced a long term increase in water clarity equivalent to 1.1% yr^{-1} over the period 1913–1996 (Verburg et al. 2003) and an increase in surface temperature of 0.129°C $decade^{-1}$ over the period 1912–2013 (Kraemer et al. 2015a, 2015b). Consistent with an effect of increasing clarity, deeper waters (50–80 m) have warmed faster than surface waters (Kraemer et al. 2015a, 2015b). In this case, the surface temperature may have warmed faster and deeper waters more slowly than they would have if water clarity had not increased.

Because changing water clarity influences the vertical partitioning of heat, the magnitude of clarity changes on temperature depended in part on lake depth (Fig. 3). Bottom temperatures of deep lakes were relatively insensitive to clarity changes because they received little solar radiation even when clarity was high, and therefore trends in water clarity produced only small changes in temperature. Meanwhile, shallow lakes (< 3 m) were relatively insensitive to clarity trends because they mixed more frequently than deeper lakes. For example, the median duration of stratification was about 40% shorter in lakes < 3 m deep than it was in lakes > 3 m deep. The combination of these two factors explains why lakes of intermediate depth (3–18 m) were particularly sensitive to clarity changes.

Increasing water clarity can have the same direction and magnitude of effect on whole-lake warming as climate-induced summertime warming of lakes. Given that water clarity is frequently used as a metric of water quality and clarity targets are used as management objectives, efforts to improve water clarity may have the unintended consequence of accelerating whole-lake warming rates. This effect may explain why lakes are warming faster than the atmosphere in some regions (Austin and Colman 2007; O'Reilly et al. 2015). For example, reoligotrophication of the large Central European Lake Constance has resulted in warmer temperatures at 7.5–10 m depth, even after correcting for changes in air temperature forcing (Rinke et al. 2010).

The scenario results are complemented by our two case study lakes, where our model accurately simulated observed changes in water temperatures resulting from the combined influence of climate change and clarity change. When calibrated to Crystal and Sparkling individually, model error rates (RMSE) were lower than the observed change in temperature. In contrast, in other studies simulating water temperatures model RMSE can exceed observed temperature change even when a model is calibrated to an individual lake (Bennington et al. 2010). However, it should be noted that RMSE values can vary with the time scale over which values are calculated (Piccolroaz et al. in press). Additional confidence in the model performance's comes from the fact that the modeled baseline warming rates are consistent with estimates of the warming rates in lakes relative to the atmosphere (70–85%; Schmid et al. 2014). When paired with observational data, simulation models serve as an important tool to assess interactions among lake characteristics and future trends in lake temperature and to characterize trends across large numbers of lakes where observational data are unavailable.

Several factors may contribute to uncertainty in model predictions. A generic calibration was applied to all lakes, but some parameterizations, such as those describing energy and momentum fluxes, may vary based on lake-specific features. This was dealt with by using site-specific data (e.g., wind sheltering; Van Den Hoek et al. 2015) when possible. However, a previous analysis of simulation results using lakes in this study has shown that lake-specific RMSE was significantly correlated with features such as lake depth and water clarity, while lake-specific RMSE was uncorrelated with features such as lake area and surrounding canopy height (Read et al. 2014). Here, lake depth explained variability in lake temperature sensitivity to changing water clarity (Fig. 3). Deep lakes are especially sensitive to parameterizations of water clarity, and less sensitive to parameterizations of other characteristics such as wind-sheltering (Read et al. 2014). The initial parameterization of water clarity suffers from the fact that water clarity estimates were unavailable for every lake in every year. Furthermore, water clarity can vary seasonally (Morris and Hargreaves 1997; Lathrop et al. 1999), which was unaccounted for in the model and may have contributed to the relatively large RMSE in bottom temperature measurements and in deeper lakes. Improvements in the ability to evaluate water clarity at both intra- and inter-annual scales should improve the accuracy of hydrodynamic models with improvements most noticeable in deep lakes.

We simulated a relatively conservative rate of change in water clarity, 0.92% yr^{-1}. This rate represents a mean estimate observed across a survey of thousands of lakes and reservoirs throughout eight states in the north-central United States (Lottig et al. 2014). However, faster rates are certainly plausible. For example, Crystal Lake has experienced a 2.2% yr^{-1} decline in water clarity over the period 1981–2011 (Fig. 4a). Faster clarity changes will have the effect of altering temperatures at faster rates, and permanent water clarity changes will also permanently alter thermal characteristics.

While our scenarios simulated identical rates and direction of change in all 1894 lakes, it is unlikely that water clarity in all lakes in a region will change simultaneously. One recent study found that 29% of Wisconsin lakes have exhibited trends in water clarity, with the vast majority (23% of all lakes) declining in clarity (Rose et al. in press). Agricultural land use intensification and increasing precipitation were attributed as the primary drivers of water clarity loss. In other regions, changes in dissolved organic matter play an important role in water clarity trends (Morris et al. 1995; Schindler et al. 1997). Thus, observed trends in water clarity may increase or decrease depending in large part on future changes in land management practices, acid deposition, and precipitation (Monteith et al. 2007; Carpenter et al. 2015). Studies that include other regions have found that about 10% of lakes have long term trends in water clarity (Olmanson et al. 2013; Lottig et al. 2014). Interestingly, a recent global analysis found that a small minority of lakes (about 10%, but only one trend significant) have exhibited cooling trends (O'Reilly et al. 2015). Changing clarity may be a factor contributing to the observed cooling trends in many of these lakes.

Because water clarity can accelerate or suppress water temperature trends, clarity may also regulate how aquatic taxa experience climate changes. In particular, the bottom waters of deep lakes with declining clarity may provide a refuge from climate-induced warming for thermally sensitive aquatic organisms. However, this thermal-maintenance comes with tradeoffs. Declining clarity may increase the risk of hypoxia and decrease primary productivity supporting higher trophic levels (Verburg et al. 2003) and reduce the efficiency of visually foraging fish (Diehl 1988).

The effects of climate change on lake and reservoirs depend on both climate drivers and in-lake characteristics such as water clarity. Real-world interactions between clarity and climate change on lake temperatures are complicated by the fact that climate change alters not only air temperatures but also the amount, timing, frequency, and type of precipitation (Karl et al. 2009; Bayer et al. 2013). In many regions, extreme precipitation events are increasing (Karl et al. 2009), which can rapidly flush materials from watersheds, reducing water clarity in receiving water bodies. Watershed land cover and land use practices, particularly agriculture, are also an important control on water clarity trends (Olmanson et al. 2013). Therefore, feedbacks between land use/land cover and the effects of increasing air temperatures and changing precipitation patterns may ultimately regulate the direction and magnitude of warming of lakes and reservoirs. Our results highlight the importance of clarity trends in controlling temperatures in northern temperate lakes and the diversity of responses to the interaction of clarity and climate both across and within lakes. Understanding the implications of the range of future lake temperature warming rates among water bodies represents an important future challenge.

References

Austin, J. A., and S. M. Colman. 2007. Lake Superior summer water temperatures are increasing more rapidly than regional temperatures: A positive ice-albedo feedback. Geophys. Res. Lett. **34**: 1–5.

Bayer, T. K., C. W. Burns, and M. Schallenberg. 2013. Application of a numerical model to predict impacts of climate change on water temperatures in two deep, oligotrophic lakes in New Zealand. Hydrobiologia **713**: 53–71.

Bennington, V., G. A. McKinley, N. Kimura, and C. H. Wu. 2010. General circulation of Lake Superior: Mean, variability, and trends from 1979 to 2006. J. Geophys. Res. Ocean **115**: 1–14.

Butcher, J. B., D. Nover, T. E. Johnson, and C. M. Clark. 2015. Sensitivity of lake thermal and mixing dynamics to climate change. Clim. Change **129**: 295–305.

Carpenter, S. R., E. H. Stanley, and M. J. Vander Zanden. 2011. State of the world's freshwater ecosystems: Physical, chemical, and biological changes. Annu. Rev. Environ. Resour. **36**: 75–99.

Carpenter, S. R., E. G. Booth, C. J. Kucharik, and R. C. Lathrop. 2015. Extreme daily loads: Role in annual phosphorus input to a north temperate lake. Aquat. Sci. **77**: 71–79.

Diehl, S. 1988. Foraging efficiency of three freshwater fishes: Effects of structural complexity and light. Oikos **53**: 207–214.

Fang, X., and H. G. Stefan. 2009. Simulations of climate effects on water temperature, dissolved oxygen, and ice and snow covers in lakes of the contiguous U.S. under past and future climate scenarios. Limnol. Oceanogr. **54**: 2359–2370.

Fink, G., M. Schmid, B. Wahl, T. Wolf, and A. Wüest. 2014. Heat flux modifications related to climate-induced warming of large European lakes. Water Resour. Res. **50**: 2072–2085.

Gorham, E. 1964. Morphometric control of annual heat budgets in temperate lakes. Limnol. Oceanogr. **9**: 525–529.

Hipsey, M., L. Bruce, and D. Hamilton. 2014. GLM general lake model: Model overview and user information. Version 2.0. University of Western Australia. Pages 1–42.

Karl, T. R., J. M. Melillo, and T. C. Peterson [eds.]. 2009. Global climate change impacts in the United States. Cambridge Univ. Press.

Kraemer, B. M., O. Anneville, S. Chandra, and others. 2015a. Morphometry and average temperature affect lake stratification responses to climate change. Geophys. Res. Lett. **42**: 4981–4988. doi:10.1002/2015GL064097

Kraemer, B. M., S. Hook, T. Huttula, and others. 2015b. Century-long warming trends in the upper water column of Lake Tanganyika. PLoS One **10**: e0132490. doi:10.1371/journal.pone.0132490

Lathrop, R. C., S. R. Carpenter, and D. M. Robertson. 1999. Summer water clarity responses to phosphorus, Daphnia grazing, and internal mixing in Lake Mendota. Limnol. Oceanogr. **44**: 137–146.

Lawson, Z. J., M. J. Vander Zanden, C. A. Smith, E. Heald, T. R. Hrabik, S. R. Carpenter, and J. Rosenfeld. 2015. Experimental mixing of a north-temperate lake: Testing the thermal limits of a cold-water invasive fish. Can. J. Fish. Aquat. Sci. **72**: 926–937.

Layden, A., C. Merchant, and S. MacCallum. 2015. Global climatology of surface water temperatures of large lakes by remote sensing. Int. J. Climatol. **35**: 4464–4479. doi: 10.1002/joc.4299

Lottig, N. R., T. Wagner, E. Norton Henry, K. Spence Cheruvelil, K. E. Webster, J. A. Downing, and C. A. Stow. 2014. Long-term citizen-collected data reveal geographical patterns and temporal trends in lake water clarity. PLoS One **9**: 155–165.

McCullough, I. M., C. S. Loftin, and S. A. Sader. 2013. Landsat imagery reveals declining clarity of Maine's lakes during 1995–2010. Freshw. Sci. **32**: 741–752.

Monteith, D. T., J. L. Stoddard, C. D. Evans, and others. 2007. Dissolved organic carbon trends resulting from changes in atmospheric deposition chemistry. Nature **450**: 537–541.

Morris, D. P., H. Zagarese, C. E. Williamson, E. G. Balseiro, B. R. Hargreaves, B. Modenutti, R. Moeller, and C. Queimalinos. 1995. The attenuation of solar UV radiation in lakes and the role of dissolved organic carbon. Limnol. Oceanogr. **40**: 1381–1391.

Morris, D. P., and B. R. Hargreaves. 1997. The role of photochemical degradation of dissolved organic carbon in regulating the UV transparency of three lakes on the Pocono Plateau. Limnol. Oceanogr. **42**: 239–249.

Olmanson, L. G., P. L. Brezonik, and M. E. Bauer. 2013. Geospatial and temporal analysis of a 20-year record of Landsat-based water clarity in Minnesota's 10,000 lakes. J. Am. Water Resour. Assoc. **50**: 748–761.

O'Reilly, C. M., S. Sharma, D. K. Gray, and others. 2015. Rapid and highly variable warming of lake surface waters around the globe. Geophys. Res. Lett. 773–781. doi: 10.1002/2015GL066235

Persson, I., and I. D. Jones. 2008. The effect of water colour on lake hydrodynamics: A modelling study. Freshw. Biol. **53**: 2345–2355.

Piccolroaz, S., E. Calamita, B. Majone, A. Gallice, A. Siviglia, and M. Toffolon. In press. Prediction of river water temperature: A comparison between a new family of hybrid models and statistical approaches. Hydrol. Process., in press, doi:10.1002/hyp.10913

R Core Team. 2015. R: A language and environment for statistical computing.

Read, J. S., and K. C. Rose. 2013. Physical responses of small temperate lakes to variation in dissolved organic carbon concentrations. Limnol. Oceanogr. **58**: 921–931.

Read, J. S., L. A. Winslow, G. J. A. Hansen, J. Van Den Hoek, P. C. Hanson, L. C. Bruce, and C. D. Markfort. 2014. Simulating 2,368 temperate lakes reveals weak coherence in stratification phenology. Ecol. Modell. **291**: 142–150.

Rinke, K., P. Yeates, and K. O. Rothhaupt. 2010. A simulation study of the feedback of phytoplankton on thermal structure via light extinction. Freshw. Biol. **55**: 1674–1693. doi:10.1111/j.1365-2427.2010.02401.x

Rose, K. C., S. R., M. W. Diebel, and M. G. Turner. In press. Annual precipitation as a regulator of spatial and temporal drivers of lake water clarity. Ecol. Appl., 1–10.

Scheffer, M., S. H. Hosper, M.-L. Meijer, B. Moss, and E. Jeppesen. 1993. Alternative equilibria in shallow lakes. Trends Ecol. Evol. **8**: 275–279.

Schindler, D. W., P. J. Curtis, S. E. Bayley, B. R. Parker, K. G. Beaty, and M. P. Stainton. 1997. Climate-induced changes in the dissolved organic carbon budgets of boreal lakes. Biogeochemistry **36**: 9–28.

Schmid, M., S. Hunziker, and A. Wüest. 2014. Lake surface temperatures in a changing climate: A global sensitivity analysis. Clim. Change **124**: 301–315.

Schneider, P., and S. J. Hook. 2010. Space observations of inland water bodies show rapid surface warming since 1985. Geophys. Res. Lett. **37**: 1–5.

Sen, P. K. 1968. Estimates of the regression coefficient based on Kendall's Tau. J. Am. Stat. Assoc. **63**: 1379–1389.

Shimoda, Y., M. E. Azim, G. Perhar, M. Ramin, M. A. Kenney, S. Sadraddini, A. Gudimov, and G. B. Arhonditsis. 2011. Our current understanding of lake ecosystem response to climate change: What have we really learned from the north temperate deep lakes? J. Great Lakes Res. **37**: 173–193.

Tanentzap, A. J., N. D. Yan, B. Keller, R. Girard, J. Heneberry, J. M. Gunn, D. P. Hamilton, and P. A. Taylor. 2008. Cooling lakes while the world warms: Effects of forest regrowth and increased dissolved organic matter on the thermal regime of a temperate, urban lake. Limnol. Oceanogr. **53**: 404–410. doi:10.4319/lo.2008.53.1.0404

Toffolon, M., S. Piccolroaz, B. Majone, A. Soja, F. Peeters, M. Schmid, and A. Wüest. 2014. Prediction of surface temperature in lakes with different morphology using air temperature. Limnol. Oceanogr. **59**: 2185–2202.

Van Den Hoek, J., J. S. Read, L. A. Winslow, P. Montesano, and C. D. Markfort. 2015. Examining the utility of satellite-based wind sheltering estimates for lake hydrodynamic modeling. Remote Sens. Environ. **156**: 551–560.

Verburg, P., R. E. Hecky, and H. Kling. 2003. Ecological consequences of a century of warming in Lake Tanganyika. Science **301**: 505–507.

Wilcox, R. 2011. Theil-Sen estimator, p. 484–485. *In* R. Wilcox [ed.], Introduction to robust estimation and hypothesis testing. Academic Press.

Woodward, G., D. M. Perkins, and L. E. Brown. 2010. Climate change and freshwater ecosystems: Impacts across multiple levels of organization. Philos. Trans. R. Soc. Lond. B. Biol. Sci. **365**: 2093–2106.

Acknowledgments

The authors acknowledge support from the Department of Biological Sciences at Rensselaer Polytechnic Institute, the Wisconsin Department of Natural Resources, the U.S. National Science Foundation awards EF-1638704 to KCR and DEB-0822700 to the North Temperate Lakes Long Term Ecological Research site, and the U.S. Geological Survey's Office of Water Information. Funding from the Department of the Interior Northeast Climate Science Center supported this study's lake modeling efforts.

13

Carbon accumulation in Amazonian floodplain lakes: A significant component of Amazon budgets?

Luciana M. Sanders,[1,2] Kathryn H. Taffs,[1] Debra J. Stokes,[3] Christian J. Sanders,[2] Joseph M. Smoak,[4] Alex Enrich-Prast,[5] Paul A. Macklin,[2] Isaac R. Santos,[2] Humberto Marotta[6,7]*

[1]Southern Cross Geoscience, Southern Cross University, Lismore, New South Wales, Australia; [2]National Marine Science Centre, School of Environment, Science and Engineering, Southern Cross University, Coffs Harbour, New South Wales, Australia; [3]Marine Ecology Research Centre, Southern Cross University, Lismore, New South Wales, Australia; [4]Department of Environmental Science, University of South Florida, St. Petersburg, Florida, USA; [5]Department of Environmental Change, Linköping University, Linköping, Sweden; [6]Ecosystems and Global Change Laboratory (LEMG-UFF) / International Laboratory of Global Change (LINCGlobal), Biomass and Water Management Research Center (NAB-UFF), Graduated Program in Geosciences (Environmental Geochemistry), Universidade Federal Fluminense (UFF), Niterói, Rio de Janeiro, Brazil; [7]Sedimentary and Environmental Processes Laboratory (LAPSA-UFF), Department of Geography, Graduated Program in Geography, Universidade Federal Fluminense (UFF), Niterói, Rio de Janeiro, Brazil

Scientific Significance Statement

In the Amazon Basin, floodplains and the lakes within them are major sources of greenhouse gases. However, it is not known whether carbon burial in the floodplain lakes is large enough to offset their greenhouse gas emissions. Here, we show that Amazonian floodplain lakes can partly offset greenhouse gas emissions through significant carbon burial rates (266 ± 57 g C m^{-2} yr^{-1}) that are greater than any other measured lake type. By extrapolating our observations, we show that burial in Amazon floodplain lakes may play an important role in the carbon budget of the Amazon Basin.

Abstract

The Amazon floodplains cover approximately 10% of the Amazon Basin and are composed of predominantly anoxic sediments that may store large amounts of carbon. Our study combines ^{210}Pb derived sedimentation rates from four recently analyzed sediment cores ($n = 4$) with previously published organic carbon (OC) burial estimates ($n = 18$) to provide a broad, first order estimate of carbon accumulation in Amazon floodplain lakes. The OC burial rates were 266 ± 57 g C m^{-2} yr^{-1}. This rate is several folds greater than those reported for lakes in arctic, boreal, temperate, and tropical regions. The large amount and spatial variation of OC burial rates in these floodplain lakes highlights the need for increased sampling efforts to better measure these potentially important components of the Amazon Basin carbon budget.

*Correspondence: l.sanders.13@student.scu.edu.au

Author Contribution Statement: LMS, KT, DS, CJS and HM designed and planned study; AEP and HM performed the sampling in the field; LMS, CJS, PAM, and JMS did the laboratory analysis; PAM made Fig. 1. LMS wrote the paper. All authors reviewed and edited the manuscript.

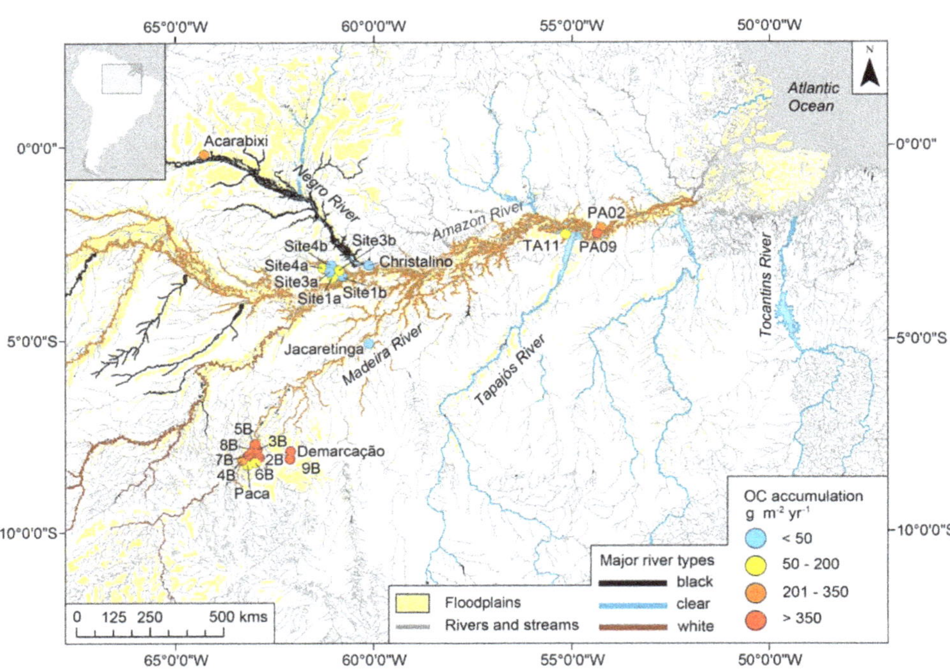

Fig. 1. Amazon Basin including the floodplain areas and the study sites of the original data ($n = 4$) and from a compilation of published work ($n = 18$). The location and description of each study site is listed in Table 1.

The forest of the Amazon Basin assimilates 14% of the world's CO_2 via annual gross photosynthesis (Song-Miao et al. 1990; Zhao and Running 2010). Much of this organic matter remains within the Amazon Basin as plant biomass and soil carbon (Moreira-Turcq et al. 2004; Junk 2013; Zocatelli et al. 2013; Espírito-Santo et al. 2014). Indeed, almost half of the carbon across the Amazon Basin may be stored underground as roots and soils (Zhao and Running 2010; Espírito-Santo et al. 2014; Whitaker et al. 2014). During periods of intense rain and flooding, organic material is exported from catchments via leaching and runoff (Mayorga et al. 2005; Ward et al. 2013). Rivers within the Amazon Basin carry megatons of organic material, a large portion of which is deposited on the floodplains and in floodplain lakes when the seasonal flooding subsides (Hedges et al. 1986; Aalto et al. 2003; Moreira-Turcq et al. 2004; Mortillaro et al. 2012; Moreira-Turcq et al. 2013; Sobrinho et al. 2016). As a result, the floodplains and their lakes may be significant carbon sinks in the Amazon Basin because of slow organic material decomposition in mostly anaerobic sediments (Mertes 1994; Hamilton et al. 2002; Dong et al. 2012; Ferland et al. 2014).

Floodplain lakes are inundated year-round and behave as organic matter producers as well as potential carbon storage systems along the Amazon Basin (Moreira-Turcq et al. 2004; Dong et al. 2012; Getirana and Paiva 2013). Although the importance of floodplains for carbon cycling has been addressed (Zocatelli et al. 2013; Abril et al. 2014; Marotta et al. 2014), the contribution of the Amazon floodplain lakes to the Amazon, and therefore the global carbon cycle, has not been taken into full consideration (Abril et al. 2014). Our objective of this study was to quantify and compile measures of organic carbon (OC) burial rates in a wide range of Amazonian floodplain lakes along gradients of size and location and compare them to published values from other lakes across the globe. The dataset includes floodplain lakes of various sizes, positioned in close proximity to the major rivers of the Amazon Basin (Madeira, Negro, Amazon, Tapajós and Solimoes Rivers; Fig. 1) and in regions within the major water types (white, black, and clearwater; Table 1). All OC burial rate estimates included in this dataset are based on sediment accumulation rates derived from ^{210}Pb dating methods, representing the past century.

Methods

Sediment cores from our 2012 field survey were collected in four floodplain lakes within the Amazon Basin (Table 1). The dataset included in this work include these cores from our own survey ($n = 4$) and carbon accretion rates reported in the literature ($n = 18$), encompassing a broad range of floodplain lakes in the Amazon Basin. The total dataset of 22 cores include sampling locations in proximity to the major rivers of the Basin (e.g., Madeira, Solimões, Amazon, Negro, and Tapajós), which also represent three different river water types (white, black, and clearwater; Table 1).

Table 1. Description of samples used to calculate average organic carbon burial in floodplain lakes of the Amazon Basin. Refer to the Supporting Information Metadata and Figure 1 for location of sites.

Site	Core id	Main river	Water type	Soil OC (%)	OC burial (g m^{-2} yr^{-1})	OC method	Source
Santa Ninha	TA11	Amazon	White water	0.8	91	IRMS	Cordeiro et al. (2008)
Calado	Site 1a	Solimoes	Black water	12.6	70	CHN analyzer	Smith et al. (2002)
Calado	Site 3a	Solimoes	Black water	6.8	49	CHN analyzer	Smith et al. (2002)
Calado	Site 4a	Solimoes	Black water	2.2	25	CHN analyzer	Smith et al. (2002)
Calado	Site 1b	Solimoes	Black water	12.5	55	CHN analyzer	Smith et al. (2002)
Calado	Site 3b	Solimoes	Black water	6.7	56	CHN analyzer	Smith et al. (2002)
Calado	Site 4b	Solimoes	Black water	2.3	11	CHN analyzer	Smith et al. (2002)
Pacoval	PA02	Amazon	White water	1.5	100	IRMS	This work
Lago Verde	PA09	Tapajos	Clear water	3.7	475	IRMS	This work
Acarabixi	Acarabixi	Negro	Black water	25.0	265	IRMS	Cordeiro et al. (2008)
Demarcacao	Demarcacao	Machado	Black water	6.9	500	LOI	This work
Cristalino	Cristalino	Negro	Black water	2.5	28	CHN analyzer	Devol et al. (1988)
Jacaretinga	Jacaretinga	Madeira	White water	2.5	43	CHN analyzer	Devol et al. (1988)
Paca	Paca	Jamari	Clear water	7.8	193	LOI	This work
Paca	2B	Jamari	Clear water	9.4	385	LOI	Bonotto and Vergotti (2015)
Araca	3B	Jamari	Clear water	5.7	1123	LOI	Bonotto and Vergotti (2015)
Brasileira	4B	Jamari	Clear water	5.0	260	LOI	Bonotto and Vergotti (2015)
Tucunare	5B	Jamari	Clear water	3.6	551	LOI	Bonotto and Vergotti (2015)
Nazare	6B	Madeira	Black water	3.5	158	LOI	Bonotto and Vergotti (2015)
Conceicao	7B	Madeira	Black water	7.7	662	LOI	Bonotto and Vergotti (2015)
S. Catarina	8B	Madeira	Black water	6.1	390	LOI	Bonotto and Vergotti (2015)
Demarcacao	9B	Machado	Black water	7.6	365	LOI	Bonotto and Vergotti (2015)

The four sediment cores in our own survey (Table 1) were collected with a 7 cm diameter and 50 cm (length) acrylic tube by means of percussion and rotation to minimize compression. Sediment cores were then sectioned in 2 cm intervals in the field until the bottom of each core ~ 40 cm depth. The sediment cores were subsequently analyzed for radionuclide concentrations in an HPGe gamma detector and sediment dates were calculated using the ^{210}Pb dating method (Appleby and Oldfield 1992).

Freeze dried and ground sediments were packed and sealed in gamma tubes. The ^{210}Pb and ^{226}Ra activities were calculated by using a factor that includes the gamma detector efficiency, as previously determined from certified reference material IAEA-300 (Baltic Sea Sediment) and the gamma-ray intensity. The ^{210}Pb and ^{226}Ra activities were measured using the 46.5 KeV and 351.9 KeV gamma peaks, respectively. Prior to radionuclide measurements, samples were set aside for at least 3 weeks, to allow for ^{222}Rn to ingrow and establish secular equilibrium between ^{226}Ra and its granddaughter ^{214}Pb. The excess ^{210}Pb activities were determined by subtracting the ^{226}Ra concentrations (i.e., supported ^{210}Pb) from the total ^{210}Pb concentrations. The sediment accumulation rates (cm yr^{-1}), taken from the ^{210}Pb constant initial concentration (CIC) dating method, and the dry bulk density (g cm^{-3}) in each interval (cm) were used to determine mass accumulation rates (Appleby and Oldfield 1992). OC was determined either through the dry combustion method at 550°C for 2 h, using a conversion factor of 1.724 (Schumacher 2002) and/or Flash Elemental Analyzer, along with a ratio mass spectrometer (IRMS) (Thermo Fisher Delta) (*see* Table 1). OC accumulation rates were determined from the sediment accretion rates (cm yr^{-1}), dry bulk densities (g cm^{-3}) and OC content, following procedures detailed elsewhere (Sanders et al. 2016).

Our dataset compilation includes work that contained the minimal parameters to determine carbon burial rates in the Amazon floodplain lakes. Carbon burial data was taken from peer reviewed literature that used the ^{210}Pb dating method and that contained organic material or OC content data along with dry bulk densities (Table 1). We found a total of 18 sediment cores from four peer reviewed papers that contained these parameters (Table 1). An analysis of variance at a confidence level of 0.05 was performed to test whether types of lakes were different. Once differences were encountered, a two tailed Tukey-Kramer was performed to separate individual pairs of lake types at 95% level of confidence.

There are currently no accurate estimates on floodplain lake areal extents. However, of the approximately 840,000 km^2 of the Amazon floodplain wetlands, between 56,000 and 73,000 km^2 may be classified as permanently

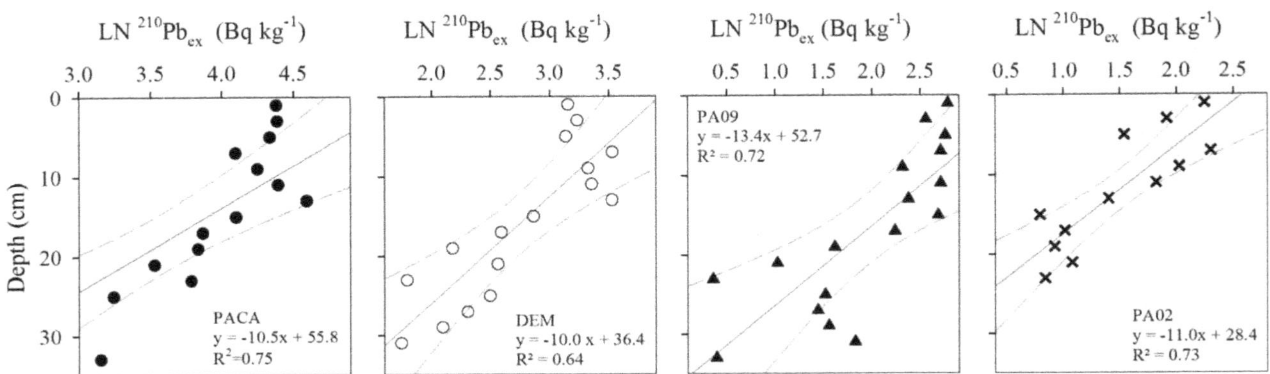

Fig. 2. Excess ^{210}Pb profiles used to determine the sediment accumulation rates through the CIC approach (Appleby and Oldfield 1992). The excess ^{210}Pb was fitted via the least square procedure and the slope of the log-linear curve was used to calculate sediment accumulation rates (*see* Methods for details).

flooded open water systems during low and high waters, respectively (Hess et al. 2015). In order to provide conservative upscaling to the entire Amazon, we assumed a minimum permanently flooded open water area (56,000 km^2) to represent a broad, first order estimate of the Amazon floodplain lake areal extent.

Results and discussion

A downcore decrease in $^{210}Pb_{(ex)}$ activity was found in the four sediment cores from the surface to a depth of 29 cm (Fig. 2). Estimated OC burial rates from these cores ranged from 100 g m^{-2} yr^{-1} to 500 g m^{-2} yr^{-1} (Table 1). We combined our OC burial rates with those reported in other Amazon studies, giving an average rate of 266 (±) 57 (SE) g C m^{-2} yr^{-1} over the past century, with median and geometric means at 175 g C m^{-2} yr^{-1} and 145 g C m^{-2} yr^{-1}, respectively. The OC burial rates values ranged from 11 g C m^{-2} yr^{-1} to 1123 g C m^{-2} yr^{-1}, with a true population mean from 154 to 378 (95% C.I.). These rates are several fold greater and significantly higher (Tukey-Kramer; $p < 0.05$) than those reported for lakes in other climatic regions, including arctic (Sobek et al. 2009), boreal (Ferland et al. 2014), temperate (Dietz et al. 2015) and tropical (Alcocer et al. 2014) lakes (6 g C m^{-2} yr^{-1}, 2 g C m^{-2} yr^{-1}, 33 g C m^{-2} yr^{-1}, and 24 g C m^{-2} yr^{-1}, respectively) (*see* Fig. 3 for details), and in subtropical floodplain lakes (15 g C m^{-2} yr^{-1}) (Dong et al. 2012). Therefore, in spite of large spatial variability, OC burial rates in the Amazon lakes are clearly several fold greater than those reported for other lacustrine systems worldwide (Fig. 3).

The data used for this study includes carbon burial rates in the main regions of the Amazon Basin, including lakes associated with whitewater ($n = 3$), blackwater ($n = 13$ cores), and clearwater ($n = 6$) river floodplains (Table 1). Our results show that lakes of the floodplain of white, black and clearwater rivers accumulated carbon at a rate of 78 ± 14,

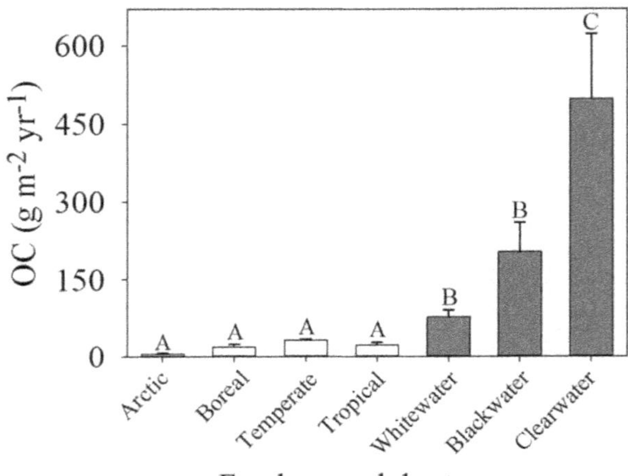

Fig. 3. Organic carbon burial rates from lakes of the floodplain of three main river water types, i.e., whitewater 78 ± 14 ($n = 3$), blackwater 203 ± 57 ($n = 13$ cores), and clearwater 498 ± 124 ($n = 6$) using original and previously published data (dark grey bars) compared to other freshwater lake types; arctic $n = 7$: 5.9 ± 1.3 SE (Sobek et al. 2014), boreal $n = 12$: 19.6 ± 4.8 SE (Ferland et al. 2014), temperate $n = 116$: 33 ± 2.1 SE (Dietz et al. 2015), and tropical $n = 2$: 23.5 ± 5.5 SE (Sobek et al. 2009; Alcocer et al. 2014) (white bars). Different capital letters indicate significant difference ($p < 0.05$) in carbon burial between lake types.

203 ± 57 and 498 ± 124 (SE) g C m^{-2} yr^{-1}, respectively (Fig. 3B). Clearwater floodplain lakes had significantly higher carbon burial rates than black and white water lakes (Tukey-Kramer; $p < 0.05$). Black and white river floodplain lakes were not significantly different ($p > 0.05$). Delineating the regions of different lake types is essential for decreasing uncertainty in calculating carbon burial in Amazon floodplain lakes. For instance, whitewater lakes contain large amounts of dissolved minerals and suspended particulates that originates from the Andes mountain range (Junk et al. 2015). These

waters are rich in clay, high in nutrients and very fertile (Guyot et al. 2007). In contrast, blackwater rivers originate from tropical forest runoff, and are high in humic material, have a low pH, suspended matter and are poor in nutrients (Abril et al. 2014). Clearwater rivers are considered intermediate in terms of fertility due to the low concentrations of dissolved minerals (Junk 2013) as these highly transparent waters drain regions of relatively low soil erosion.

Upscaling these OC burial rates to the Amazon Basin area is very challenging because of large uncertainties in the area of the lakes, their ephemeral nature and the variable sequestration rates reported in the literature. In order to provide conservative upscaling, we assumed a minimum Amazon floodplain lake area of 56,000 km^2. This estimate is related to a conservative value for open water systems during low waters, and does not include small lakes due to image resolution issues (Hess et al. 2015). Using this conservative first order estimate, Amazon floodplain lakes accumulate approximately 15 Tg yr^{-1} of OC. While the permanent lakes included here are likely to have greater OC burial rates than ephemeral floodplains surrounding the lakes, some carbon is also expected to accumulate on the ephemeral floodplains not included in our upscaling exercise.

In summary, this study has established that the floodplain lakes located within the Amazon Basin accumulate significant quantities of carbon compared to other freshwater lakes, and are an important component of the carbon cycle in the Amazon Basin. However, our dataset cannot establish local variability in burial rates and the large variability likely reflects this dynamic system (Constantine et al. 2014). Indeed, OC burial may be driven by lake morphology (Blais and Kalff 1995), but very little information is available on lake morphology of the Amazon floodplain lakes. Our first order estimate implies that Amazon floodplain lakes may accumulate 266 (±) 57 (SE) g C m^{-2} yr^{-1} which is in the same range as the net carbon sink of the Amazon forests (270 g C m^{-2} yr^{-1}) (Ometto et al. 2005), and carbon evasion, CO_2 and CH_4 degassing, from the Amazon rivers and wetlands (120 g C m^{-2} yr^{-1}) (Richey et al. 2002; Abril et al. 2014). The high OC burial rates of these floodplain lakes indicate that even though these systems only represent about 1% of the total Amazon Basin area, they may contribute disproportionately to the overall Amazon, and therefore, global carbon cycle. Because of the small sample size currently available ($n = 22$ lakes; Table 1), additional datasets are required to refine our estimates and decrease uncertainty when upscaling carbon burial rates.

References

Aalto, R., L. Maurice-Bourgoin, T. Dunne, D. R. Montgomery, C. A. Nittrouer, and J. L. Guyot. 2003. Episodic sediment accumulation on Amazonian flood plains influenced by El Niño/Southern Oscillation. Nature **425**: 493–497. doi:10.1038/nature02002

Abril, G., and others. 2014. Amazon River carbon dioxide outgassing fuelled by wetlands. Nature **505**: 395–398. doi: 10.1038/nature12797

Alcocer, J., A. C. Ruiz-Fernández, E. Escobar, L. H. Pérez-Bernal, L. A. Oseguera, and V. Ardiles-Gloria. 2014. Deposition, burial and sequestration of carbon in an oligotrophic, tropical lake. J. Limnol. **73**: 21–33. doi:10.4081/jlimnol.2014.783

Appleby, P. G., and F. Oldfield. 1992. Application of lead-210 to sedimentation studies, p. 731–783. *In* M. Ivanovich and S. Harmon [eds.], Uranium series disequilibrium: Application to earth, marine and environmental science. Oxford Science Publications.

Blais, J. M., and J. Kalff. 1995. The influence of lake morphometry on sediment focusing. Limnol. Oceanogr. **40**: 582–588. doi:10.4319/lo.1995.40.3.0582

Bonotto, D. M., and M. Vergotti. 2015. 210Pb and compositional data of sediments from Rondonian lakes, Madeira River basin, Brazil. Appl. Radiat. Isot. **99**: 5–19.

Constantine, J. A., T. Dunne, J. Ahmed, C. Legleiter, and E. D. Lazarus. 2014. Sediment supply as a driver of river meandering and floodplain evolution in the Amazon Basin. Nat. Geosci. **7**: 899–903. doi:10.1038/ngeo2282

Cordeiro, R. C., B. Turcq, K. Suguio, A. Oliveira da Silva, A. Sifeddine, and C. Volkmer-Ribeiro. 2008. Holocene fires in East Amazonia (Carajás), new evidences, chronology and relation with paleoclimate. Glob. Planet. Change **61**: 49–62.

Devol, A. H., J. E. Richey, W. A. Clark, S. L. King, and L. A. Martinelli. 1988. Methane emissions to the troposphere from the Amazon floodplain. J. Geophys. Res. **93**: 1583–1592.

Dietz, R. D., D. R. Engstrom, and N. J. Anderson. 2015. Patterns and drivers of change in organic carbon burial across a diverse landscape: Insights from 116 Minnesota lakes. Glob. Biogeochem. Cycles **29**: 708–727. doi: 10.1002/2014GB004952

Dong, X., N. J. Anderson, X. Yang, X. Chen, and J. Shen. 2012. Carbon burial by shallow lakes on the Yangtze floodplain and its relevance to regional carbon sequestration. Glob. Change Biol. **18**: 2205–2217. doi:10.1111/j.1365-2486.2012.02697.x

Espírito-Santo, F. D. B., and others. 2014. Size and frequency of natural forest disturbances and the Amazon forest carbon balance. Nat. Commun. **5**: 1–6. doi:10.1038/ncomms4434

Ferland, M. E., Y. T. Prairie, C. Teodoru, and P. A. Del Giorgio. 2014. Linking organic carbon sedimentation, burial efficiency, and long-term accumulation in boreal lakes. J. Geophys. Res. Biogeosci. **119**: 836–847. doi: 10.1002/2013JG002345

Getirana, A. C. V., and R. C. D. Paiva. 2013. Mapping large-scale river flow hydraulics in the Amazon Basin. Water Resour. Res. **49**: 2437–2445. doi:10.1002/wrcr.20212

Guyot, J. L., J. M. Jouanneau, L. Soares, G. R. Boaventura, N. Maillet, and C. Lagane. 2007. Clay mineral composition of river sediments in the Amazon Basin. Catena **71**: 340–356. doi:10.1016/j.catena.2007.02.002

Hamilton, S. K., S. J. Sippel, and J. M. Melack. 2002. Comparison of inundation patterns among major South American floodplains. J. Geophys. Res. Atmos. **107**: 5-1–5-14. doi:10.1029/2000JD000306

Hedges, J. I., W. A. Clark, P. D. Quay, J. E. Richey, A. H. Devol, and U. D. M. Santos. 1986. Compositions and fluxes of particulate organic material in the Amazon River. Limnol. Oceanogr. **31**: 717–738. doi:10.4319/lo.1986.31.4.0717

Hess, L. L., J. M. Melack, A. G. Affonso, C. Barbosa, M. Gastil-Buhl, and E. M. L. M. Novo. 2015. Wetlands of the lowland Amazon Basin: Extent, vegetative cover, and dual-season inundated area as mapped with JERS-1 synthetic aperture radar. Wetlands **35**: 745–756. doi:10.1007/s13157-015-0666-y

Junk, W. J. 2013. Current state of knowledge regarding South America wetlands and their future under global climate change. Aquat. Sci. **75**: 113–131. doi:10.1007/s00027-012-0253-8

Junk, W. J., F. Wittmann, J. Schöngart, and M. T. F. Piedade. 2015. A classification of the major habitats of Amazonian black-water river floodplains and a comparison with their white-water counterparts. Wetlands Ecol. Manage. **23**: 677–693. doi:10.1007/s11273-015-9412-8

Marotta, H., L. Pinho, C. Gudasz, D. Bastviken, L. J. Tranvik, and A. Enrich-Prast. 2014. Greenhouse gas production in low-latitude lake sediments responds strongly to warming. Nat. Clim. Change **4**: 467–470. doi:10.1038/nclimate2222

Mayorga, E., A. K. Aufdenkampe, C. A. Masiello, A. V. Krusche, J. I. Hedges, P. D. Quay, J. E. Richey, and T. A. Brown. 2005. Young organic matter as a source of carbon dioxide outgassing from Amazonian rivers. Nature **436**: 538–541. doi:10.1038/nature03880

Mertes, L. A. K. 1994. Rates of flood-plain sedimentation on the central Amazon River. Geology **22**: 171–174. doi:10.1130/0091-7613(1994)022<0171:ROFPSO>2.3.CO;2

Moreira-Turcq, P., J. M. Jouanneau, B. Turcq, P. Seyler, O. Weber, and J. L. Guyot. 2004. Carbon sedimentation at Lago Grande de Curuai, a floodplain lake in the low Amazon region: Insights into sedimentation rates. Palaeogeogr. Palaeoclimatol. Palaeoecol. **214**: 27–40. doi:10.1016/j.palaeo.2004.06.013

Moreira-Turcq, P., M. P. Bonnet, M. Amorim, M. Bernardes, C. Lagane, L. Maurice, M. Perez, and P. Seyler. 2013. Seasonal variability in concentration, composition, age, and fluxes of particulate organic carbon exchanged between the floodplain and Amazon River. Glob. Biogeochem. Cycles **27**: 119–130. doi:10.1002/gbc.20022

Mortillaro, J. M., F. Rigal, H. Rybarczyk, M. Bernardes, G. Abril, and T. Meziane. 2012. Particulate organic matter distribution along the lower Amazon River: Addressing aquatic ecology concepts using fatty acids. PLoS One **7**: e46141. doi:10.1371/journal.pone.0046141

Ometto, J. P. H. B., A. D. Nobre, H. R. Rocha, P. Artaxo, and L. A. Martinelli. 2005. Amazonia and the modern carbon cycle: Lessons learned. Oecologia **143**: 483–500. doi:10.1007/s00442-005-0034-3

Richey, J. E., J. M. Melack, A. K. Aufdenkampe, V. M. Ballester, and L. L. Hess. 2002. Outgassing from Amazonian rivers and wetlands as a large tropical source of atmospheric CO2. Nature **416**: 617–620. doi:10.1038/416617a

Sanders, C. J., and others. 2016. Examining 239 + 240Pu, 210Pb and historical events to determine carbon, nitrogen and phosphorus burial in mangrove sediments of Moreton Bay, Australia. J. Environ. Radioactiv. **151**: 623–629. doi:10.1016/j.jenvrad.2015.04.018

Schumacher, B. A. 2002. Methods for the determination of total organic carbon (TOC) in soils and sediments, p. 1–25. United States Environmental Protection Agency, Ecological Risk Assessment Support Center Office of Research and Development.

Sobek, S., E. Durisch-Kaiser, R. Zurbrügg, N. Wongfun, M. Wessels, N. Pasche, and B. Wehrli. 2009. Organic carbon burial efficiency in lake sediments controlled by oxygen exposure time and sediment source. Limnol. Oceanogr. **54**: 2243–2254. doi:10.4319/lo.2009.54.6.2243

Sobek, S., N. J. Anderson, S. M. Bernasconi, and T. Del Sontro. 2014. Low organic carbon burial efficiency in arctic lake sediments. J. Geophys. Res. Biogeosci. **119**: 1231–1243. doi:10.1002/2014JG002612

Sobrinho, R. L., and others. 2016. Spatial and seasonal contrasts of sedimentary organic matter in floodplain lakes of the central Amazon basin. Biogeosciences **13**: 467–482. doi:10.5194/bg-13-467-2016

Song-Miao, F., S. C. Wofsy, P. S. Bakwin, D. J. Jacob, and D. R. Fitzjarrald. 1990. Atmosphere-biosphere exchange of CO2 and O3 in the central Amazon forest. J. Geophys. Res. **95**: 16851−16864. doi:10.1029/JD095iD10p16765

Smith, L. K., J. M. Melack, and D. E. Hammond. 2002. Carbon, nitrogen, and phosphorus content and 210Pb-derived burial rates in sediments of an Amazon floodplain lake. Amazoniana **17**: 413–436.

Ward, N. D., and others. 2013. Degradation of terrestrially derived macromolecules in the Amazon River. Nat. Geosci. **6**: 530–533. doi:10.1038/ngeo1817

Whitaker, J., and others. 2014. Microbial community composition explains soil respiration responses to changing carbon inputs along an Andes-to-Amazon elevation gradient. J. Ecol. **102**: 1058–1071. doi:10.1111/1365-2745.12247

Zhao, M., and S. W. Running. 2010. Drought-induced reduction in global terrestrial net primary production from 2000 through 2009. Science **329**: 940–943. doi:10.1126/science.1192666

Zocatelli, R., P. Moreira-Turcq, M. Bernardes, B. Turcq, R. C. Cordeiro, S. Gogo, J. R. Disnar, and M. Boussafir. 2013. Sedimentary evidence of soil organic matter input to the Curuai Amazonian floodplain. Org. Geochem. **63**: 40–47. doi:10.1016/j.orggeochem.2013.08.004

Acknowledgments

LMS is supported by an APA and IPRS scholarships. CJS and IRS were supported by the Australian Research Council (DE160100443, DP150103286, and LE140100083). HM and AEP are supported by FAPERJ, CNPq, and CAPES.

14

A whole-lake experiment confirms a small centric diatom species as an indicator of changing lake thermal structure

Jasmine E. Saros,[1] Robert M. Northington,[1] Dennis S. Anderson,[1] Nicholas John Anderson[2]*

[1]Climate Change Institute & School of Biology and Ecology, University of Maine, Orono, Maine; [2]Department of Geography, Loughborough University, Leicestershire, LE11 3TU, United Kingdom

Scientific Significance Statement

In many lakes across the Northern Hemisphere, paleolimnological records have revealed that the relative abundances of small centric diatoms, such as the cosmopolitan species *Discostella stelligera*, have changed over the past century. Such changes have been thought to be a result of climate change based on inferences from small-scale experiments and contemporary observational studies. However, there have been no whole-ecosystem tests of the mechanisms underlying such changes in diatoms. This study provides evidence from a whole-lake experiment that shows the importance of water column stratification on *D. stelligera*, thus providing a mechanistic link between climate and observed patterns in lake fossil records.

Abstract

In many lakes across the Northern Hemisphere, paleolimnological records have revealed that the relative abundances of the small centric diatom, *Discostella stelligera*, changed over the past century, with these widespread shifts attributed to climate change. Specifically, small-scale experiments and current spatial distribution patterns suggested that this species is more abundant when lake mixing depths are shallower, but a direct test of this hypothesis at the whole-lake scale was lacking. We conducted a whole-lake mixing manipulation in a remote arctic lake that normally has relatively shallow thermal stratification and abundant *D. stelligera* populations during the summer. We employed a "Before-After-Control-Impact" design using an experimental lake and a control lake. Lake thermal structure and diatom populations were monitored in both lakes in summer 2013 without manipulating either lake, and again in summer 2014 when the experimental lake was manipulated to achieve deeper mixing depths. The abundance of *D. stelligera* declined during the manipulated period of deeper mixing in 2014, while it increased during the same time frame in the control lake. The same pattern was not observed for the four other diatom taxa found in both lakes in both years. Our results confirm the use of *D. stelligera* as an indicator of changing lake thermal structure, and suggest that the broader application of this tool to lake sediment records will yield greater insight into longer-term variability in the response of lake ecosystems to climate.

*Correspondence: jasmine.saros@maine.edu

Author contribution statement: JES developed the research question and designed the experiment with input from NJA. JES and RMN conducted the fieldwork and experiment, and RMN and DSA analyzed samples and processed data. JES wrote the manuscript with input from all co-authors.

Introduction

Monitoring and remote sensing approaches have documented climate-driven changes in many lake ecosystems over recent decades, with effects ranging from increasing surface water temperatures (Schneider and Hook 2010) and altered lake thermal structure (Kraemer et al. 2015) to changing delivery of organic and inorganic materials from watersheds (Strock et al. 2016) and shifts in phenology (Berger et al. 2010). Our primary tool, however, for assessing longer-term effects of climate change on lakes and for understanding patterns of natural variability in lake ecosystems remains the paleolimnological record. This underscores the importance of developing reliable proxies and fossil indicators of climate-driven change that are preserved in lake sediments.

Fossil diatom assemblages are often used to reconstruct chemical (Fritz et al. 1991; Anderson and Rippey 1994) and habitat (Stone and Fritz 2004) changes in lakes over centennial to millennial timescales. These inferences rest on relationships established from contemporary ecological studies of current distribution patterns and, to a lesser extent, experimental results. The links between climate-driven lake ecosystem changes and diatom community structure, however, have been more difficult to decipher (Saros and Anderson 2015). This is due in part to the myriad effects of climate on lakes and hence diatom communities, as well as the often limited spatial (e.g., one integrated sample, no vertical resolution) and temporal (e.g., once or twice per year) resolution of diatom distribution assessments. Pairing experimental approaches with these contemporary distribution assessments becomes necessary to identify controlling factors and mechanisms.

The relative abundances of the small centric diatom *Discostella stelligera* (Cleve and Grunow) Houk and Klee have undergone widespread increases in alpine, boreal, temperate, and particularly arctic areas since 1850 (Smol et al. 2005; Rühland et al. 2008; Perren et al. 2009; Saros et al. 2012), with these shifts often used as indicators of warming-induced changes in lakes. It is important to note, however, that *D. stelligera* has also declined over the same time frame in many lakes experiencing warming (Perren et al. 2009; Hobbs et al. 2010; Saros et al. 2012), raising questions about the mechanisms involved. Correlative studies suggest positive effects of longer ice-free seasons on this species (Rühland et al. 2008), but given the highly coherent nature of changing ice off dates across lakes in a region (Magnuson et al. 1990; Arp et al. 2013), it remains unclear how this would lead to variable response of this species across a landscape.

Many studies using surveys and small-scale experiments have now provided evidence for the importance of greater water column stability and/or shallower epilimnion thickness for the success of small centric diatom taxa (Carney et al. 1988; Winder et al. 2009), including *D. stelligera* (Saros et al. 2012, 2014). Because of regional variability in how increasing air temperatures alter surface winds and delivery of materials from the watershed that affect lake water transparency, changes in water column stability and mixing depths may vary across regions experiencing similar warming and lead to different patterns in *D. stelligera* sedimentary profiles across lakes (Perren et al. 2009; Saros et al. 2012; Saros and Anderson 2015). The response of this opportunistic species, however, can also be complex and variable (Saros and Anderson 2015), with interactive effects between light and nutrients on this species often apparent (Saros et al. 2014; Malik and Saros 2016). This suggests the need for a direct, whole-ecosystem scale test of thermal structure effects on the response of this important and widely used indicator species.

We conducted a whole-lake mixing manipulation in a remote arctic lake that normally has relatively shallow thermal stratification and abundant *D. stelligera* populations during the summer. We employed a "Before-After-Control-Impact" (BACI) design (Stewart-Oaten et al. 1986), with an experimental lake and a control lake. Lake thermal structure and diatom populations were monitored in both lakes in summer 2013 without manipulating either lake, and again in summer 2014 when the experimental lake was manipulated to achieve deeper mixing depths. We predicted that shallower mixing depths would favor *D. stelligera* in this region, and that the relative abundance of this species would decline during the manipulated period of deeper mixing.

Methods

Site description

Two arctic lakes near Kangerlussuaq, southwest Greenland (67.01°N, 50.69°W) were selected for this experiment (Fig. 1) because *D. stelligera* is common and abundant in many lakes in this area (Perren et al. 2009), and the remote nature of the area ensured no direct disturbance from human activity. Lakes across this area are generally dilute and oligotrophic.

Lake SS16 was selected as the experimental lake for this study for four reasons. Previous research indicated that *D. stelligera* is the dominant diatom in this lake (Perren et al. 2009), making up more than 40% of diatom assemblages deposited in lake sediments over the past century. The lake is also relatively small (surface area of 0.033 km^2, maximum depth of 13 m), making it a manageable size for a whole-lake thermal manipulation. The lake thermally stratifies during summer with a mixing depth of 4 m. Finally, an unnamed lake in close proximity to Lake SS16 is similar in all three features, providing an ideal "control" system to which to compare the manipulated Lake SS16. We refer to this second, unmanipulated lake as Control Lake (surface area of 0.022 km^2, maximum depth of 9.8 m), and hereafter refer to Lake SS16 as Experimental Lake.

The two lakes are situated at approximately 66.91°N, 50.44°W, in an area only accessible by helicopter. The two lakes are hydrologically connected early in the summer by

Fig. 1. Overview of location and study: (a) location of Control and Experimental (Exp) lakes in southwest Greenland, and close-up of the two lakes (black scale bar = 100 m; image source Earthexplorer.usgs.gov); (b) Experimental Lake; (c) *Discostella stelligera* (yellow scale bar = 5 μm; photo by Jeffery R. Stone); (d) Solar Bee being deployed into Experimental Lake by helicopter; (e) Solar Bee in Experimental Lake. Photos by Benjamin Burpee unless otherwise noted.

very low surface flow from Control Lake to Experimental Lake; in this semi-arid region (total annual precipitation < 200 mm yr^{-1}) there is generally very little surface flow connecting lakes. Both lakes are circumneutral and dilute (conductivity of 70–80 μS cm^{-1}) with low to moderate total phosphorus (<12 μg L^{-1}). Concentrations of dissolved organic carbon (DOC) are about 10 mg L^{-1} in both lakes. The *D. stelligera* populations in both lakes have cell sizes that range from 4 μm to 6 μm; these cells have the features of *D. stelligera* as opposed to *D. pseudostelligera*. Both lakes contain populations of the planktivorous three-spined stickleback (*Gasterosteus aculeatus*), and hence zooplankton communities of both lakes are dominated by rotifers (primarily *Asplanchna* and *Polyarthra* taxa), with no cladocerans and only copepod nauplii observed.

Whole-lake experiment

A BACI design was used to assess whether deeper mixing affects the relative abundances of *D. stelligera*. Data from Experimental Lake and Control Lake during Year 1 (2013) serve as the "Before" data, while data collected from both lakes in Year 2 (2014) during the manipulation of Experimental Lake serve as the "After" data. Control Lake, not manipulated at any point, provides the "Control" data, while Experimental Lake undergoing manipulation provides the "Impact" data.

During both years, lakes were monitored with automated instrumentation for a 4-week period, starting shortly after ice off and spanning many weeks of thermally stratified conditions. The period of observation extended only to late July to avoid the initial stages of thermocline erosion that typically begin in August. Ice off on Experimental Lake was 13 June in 2013 and 15 June in 2014; ice off on Control Lake was 10 June in 2013 and 12 June in 2014. A string of temperature probes (Onset HOBO Pendant) recording at hourly resolution was deployed in each lake at 1-m intervals from the lake surface to 9 m to assess stratification patterns. Rotating sediment traps (Technicap PPS 4/3) were deployed in each lake to determine planktic diatom community structure. Sediment traps were used because they provide a more integrated plankton sample over time compared to water column samples, and they allowed for automated, unattended collection at this remote site. The trap in Experimental Lake was deployed at 8-m depth, while the trap in Control Lake was at 7 m due to the shallower maximum depth of this lake. A new collection bottle in the trap rotated into position every 5 days, providing six 5-day composite samples from 19 June to 19 July in both years for both lakes. Rotation dates of the traps were kept the same in both years. We relied on the short duration of the experiment and low temperatures (4–5°C) at trap depth for sample preservation; more than 2×10^6 μm^3 mL^{-1} trap material of soft-bodied algae also accumulated in each trap, with larger quantities measured in the first two traps of each season. Microscopic examination indicated no signs of degradation across samples.

We returned to the site briefly on 24 June 2013 and 26 June 2014 to collect epi-, meta-, and hypolimnion plankton samples in each lake to verify the presence and assess the vertical distribution of *D. stelligera*.

Water column measurements of light, chemistry and chlorophyll were conducted manually twice each summer, shortly after ice off and at the end of the sampling period. While the timing of the early season sampling is the same in both years, the timing of the late season sampling differs between years owing to logistical constraints (i.e., timing of

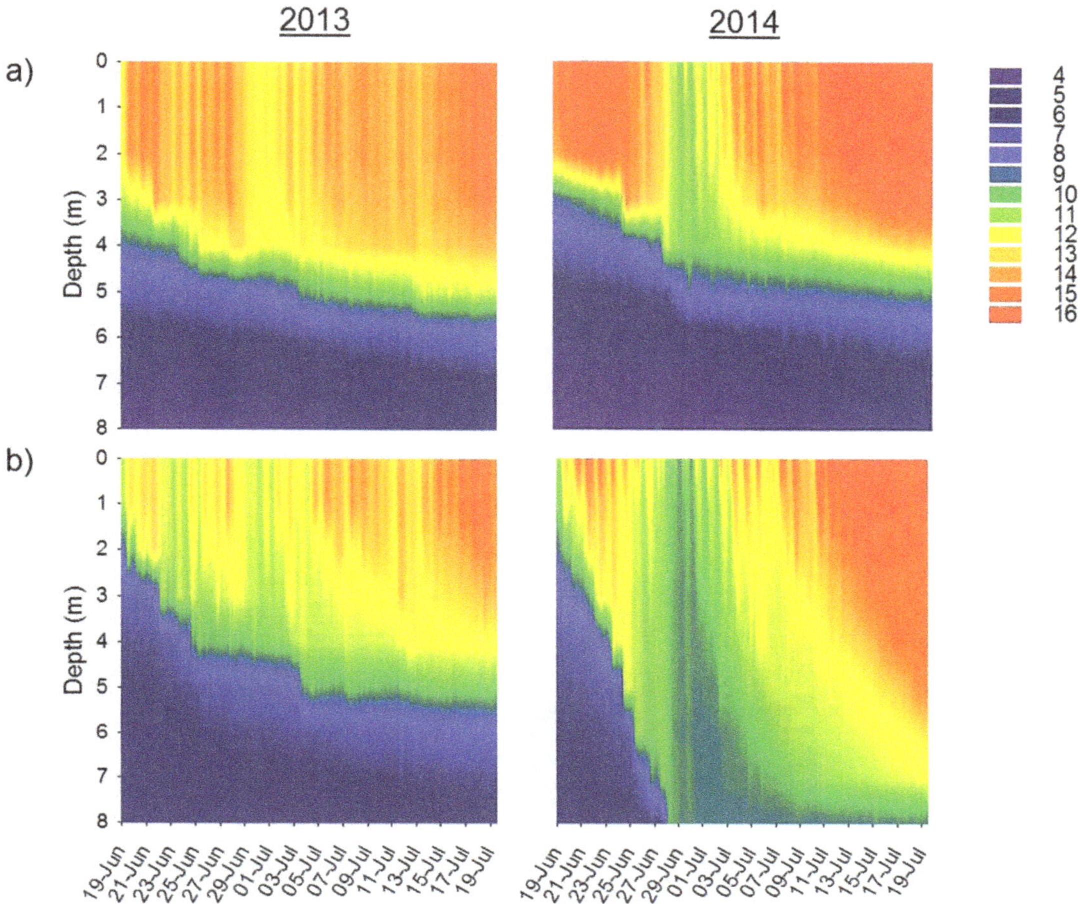

Fig. 2. Temperature profiles from 19 June to 19 July in both years of (a) Control Lake, and (b) Experimental Lake. Neither lake was manipulated in 2013; Experimental Lake was manipulated in 2014. Color scale in units of 1°C.

availability of air transportation; occurred 31 July 2013 and 15 August 2014). Conductivity was measured with a multi-parameter sonde (Hydrolab MS5). Vertical attenuation of photosynthetically active radiation (PAR) was measured with a radiometer (Biospherical Instruments BIC). Water samples for total and dissolved nutrients, DOC, and chlorophyll *a* (Chl *a*) were collected from the epilimnion (at 2 m) using a van Dorn horizontal sampling bottle.

In Year 2 (2014), the thermal structure of Experimental Lake was manipulated using a solar-powered hydraulic lift system (Medora Corporation SolarBee 10000v18), which pumps water up from a selected depth and redistributes it to the surface. This is the same system used in other whole-lake manipulations that assessed the biological effects of changing thermocline depth (Cantin et al. 2011; Jobin and Beisner 2014; Sastri et al. 2014). The SolarBee was deployed by helicopter into the deep area of the lake on 15 June 2014. The intake hose for the system was set to 8 m, the target depth for deepened mixing. Once the system was anchored, adjusted, and properly positioned, it was started (evening of 17 June). The first plankton sample began collecting in the sediment traps at 12:01 a.m on 19 June.

Sample processing

Each sediment trap bottle was homogenized before an 8-mL sub-sample was removed with an electronic pipet. Subsamples were treated with 30% hydrogen peroxide to remove organic material and a known concentration of microspheres was added to each processed sample to calculate absolute abundances of each diatom taxon. A minimum of 300 diatom valves were counted for each sample under oil immersion at a magnification of 630X.

From the early and late season manually collected water samples, nutrients and DOC were measured as described in Burpee et al. (2016). Chl *a* was extracted in acetone and analyzed with a spectrophotometer (Varian Cary50). We used an average of early and late season light attenuation coefficients for each lake to calculate average light intensities in the epilimnion (as in Tilzer and Goldman (1978)) for each 5-day period.

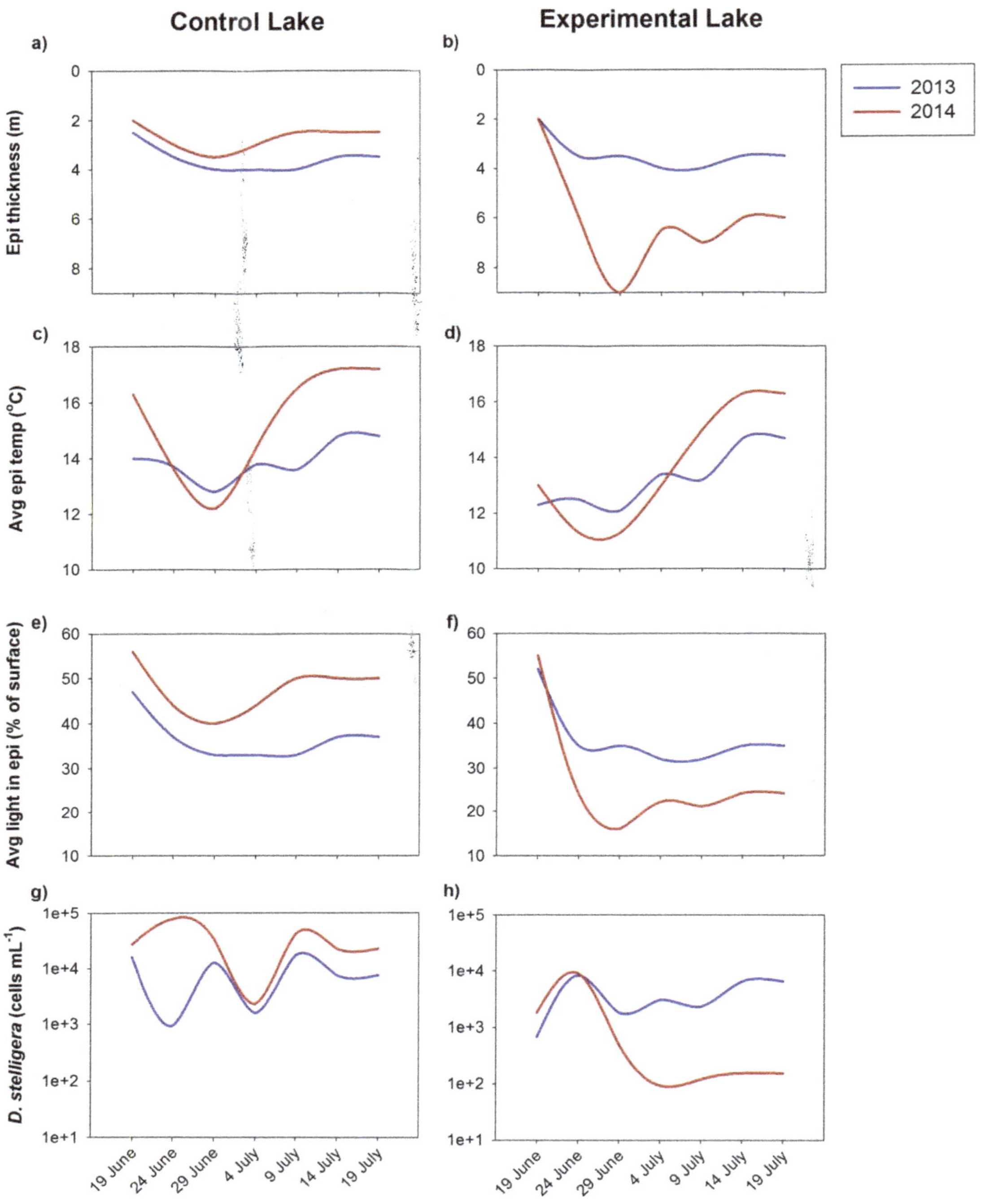

Fig. 3. Response of Control Lake and Experimental Lake in both years, with neither lake manipulated in 2013 and only Experimental Lake manipulated in 2014: (a, b) epilimnion thickness; (c, d) average epilimnetic temperature; (e, f) calculated average light intensity experienced by a cell circulating in the mixed layer; (g, h) cell densities of *Discostella stelligera* in sediment traps (note log scale of *y*-axis). Data are shown as spline curves; values in panels a through d are based on averages from each 5-day sediment trap interval, matching the sampling resolution of the primary response variable (*Discostella stelligera* cell densities).

Data analysis

The response of *D. stelligera* to the thermal manipulation was assessed with a two-factor analysis of variance (ANOVA), with treatment (i.e., lake) and year as the two factors. A significant ($p < 0.05$) interaction between these two factors would indicate an effect of the 2014 manipulation in Experimental Lake on *D. stelligera* population size. Cell densities were log-transformed to meet the assumptions of normality.

Table 1. Water column data collected early (mid-June) and late (late July 2013 or mid-August 2014) summer each year in both lakes. Neither lake was manipulated in 2013; only Experimental Lake (Exp Lake) was manipulated in 2014, with the "late" samples being the post-manipulation data (denoted by shading). Chemical parameters are measured from samples collected at a depth of 2 m.

	2013				2014			
	Control lake		Exp lake		Control lake		Exp lake	
	Early	Late	Early	Late	Early	Late	Early	Late
Conductivity (μS cm^{-1})	65	67	69	71	64	70	71	73
TP (μg L^{-1})	8	11	8	7	4	4	4	4
TN (μg L^{-1})	580	610	480	443	585	602	475	546
DIN (μg L^{-1})	6	5	3	6	19	<2	4	<2
DOC (mg L^{-1})	11	12	11	12	10	10	11	11
Chlorophyll (μg L^{-1})	5.2	3.5	4.5	2.9	4.6	5.6	3.9	5.6
1% PAR (m)	6.02	6.85	5.83	6.44	6.61	7.71	6.75	6.73

The responses of four other diatom taxa common to both lakes in both years were also tested with this approach. In all ANOVA analyses (regardless of taxon), the number of degrees of freedom was 12 (16 total data points - 1 (treatment) −1 (year) - 1(trt × year) - 1 (intercept)). With sediment traps collecting over the same time intervals and dates each year, sample period should not be an issue with this design but we tested its effect by adding it as a term in the ANOVA for *D. stelligera* and found that it was not significant (p=0.75).

Results

From 19 June to 19 July 2013, epilimnion thickness varied between 2 to 4 m in both Experimental Lake and Control Lake (Figs. 2, 3). In 2014, epilimnion thickness in Control Lake again varied from 2 to 4 m, being similar to but often slightly shallower than in 2013. This shallower mixing in Control Lake was likely due to the higher air temperatures in 2014 compared to 2013 (Danish Meterological Institute, http://www.dmi.dk/groenland/arkiver/vejrarkiv/), resulting in higher July surface water temperatures in 2014 in both lakes (Figs. 2, 3). In the manipulated Experimental Lake in 2014, epilimnion thickness began similarly to that in 2013 (between 2-3 m depth from 19 to 23 June both years), but then deepened to vary between 4 and 8 m between 24 and 28 June 2014. On 29 June, stratification broke down and the lake mixed until 5 July; stratification redeveloped and between 6 and 19 July, varied largely between 6 m and 7 m, with occasional days at 5 m.

Water column metrics collected in early and late summer each year revealed no difference in lake water chemistry as a consequence of the thermal manipulation in Experimental Lake (Table 1). The 1%PAR depth deepened over the summer in all cases except the manipulated Experimental Lake in 2014, in which it remained the same from early to late summer (Table 1). The water column plankton samples collected 6–8 d into the monitored period revealed maximum *D. stelligera* cell densities in the metalimnia of both lakes on 24 June 2013, and epilimnetic maxima in both lakes on 26 June 2014 (Supporting Information Fig. S1).

Calculated average light intensities in the mixed layer ranged from 30% to 50% of surface irradiance in both lakes over summer 2013 (Fig. 3). In 2014, they ranged from 40% to 55% of surface irradiance in Control Lake. Once deeper mixing began (25 June) in Experimental Lake in 2014, average light intensities in the mixed layer ranged from 16% to 24% of surface irradiance.

Cell densities of *D. stelligera* in Experimental Lake declined during the period of manipulated, deeper mixing ($p = 0.0025$ for treatment x year interaction; Fig. 3). During the first ten sampling days of each year, cell densities of this species in Experimental Lake sediment traps were comparable. From 29 June to 19 July when consistently deeper mixing was achieved, cell densities ranged from 4 to 40 times lower (absolute densities lower by 1,300 cells mL^{-1} to 6,300 cells mL^{-1} trap material) in 2014 compared to 2013. In contrast, in Control Lake, cell densities in the traps were higher in 2014 compared to 2013 ($p = 0.01$).

Four other diatom taxa were common to both lakes in both years: *Fragilaria tenera v. nanana* (Lange-Bertalot) Lange-Bertalot and S. Ulrich, *Stauroforma exiguiformis* (Lange-Bertalot) R.J. Flower, V.J. Jones and Round, *Fragilaria capucina v. gracilis* (Oestrup) Hustedt, and *Tabellaria flocculosa* (Roth) Kütz. strain III sensu Koppen (Fig. 4). The treatment × year interaction was not significant for *F. tenera v. nanana* ($p = 0.17$), *S. exiguiformis* ($p = 0.23$), or *T. flocculosa* ($p = 0.75$), but was significant for *F. capucina v. gracilis* ($p = 0.01$). This interaction resulted from the positive effect of deeper mixing in 2014 on this species in Experimental Lake. We note, however, that this species was abundant both years in the stratified Control Lake, leaving the relationship, if any, between mixing and this species unclear.

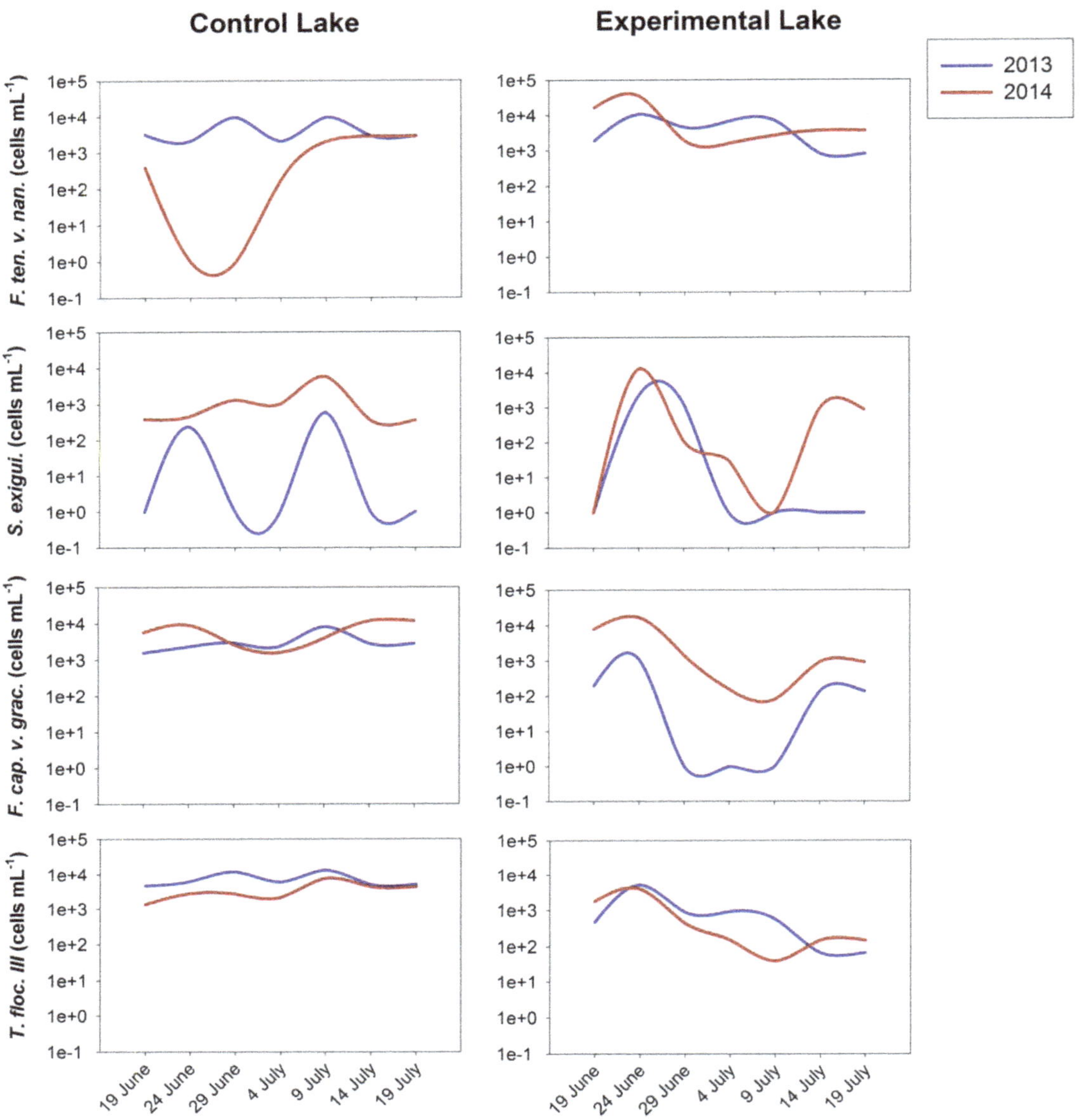

Fig. 4. Sediment trap cell densities of other diatom taxa common to both lakes in both years, with neither lake manipulated in 2013 and only Experimental Lake manipulated in 2014. From top of panel to bottom, taxa include *Fragilaria tenera v. nanana, Stauroforma exiguiformis, Fragilaria capucina v. gracilis,* and *Tabellaria flocculosa* strain III. Note log scale of y-axis.

Discussion

The results of this whole-lake experiment are consistent with those of surveys and small-scale experiments (Winder et al. 2009; Saros et al. 2012, 2014) and support the importance of lake thermal structure for the success of *D. stelligera* populations. In these lakes as well as others (Winder et al. 2009; Saros et al. 2012), multiple lines of evidence continue to support the use of increasing relative abundances of *D. stelligera* in lake sediment records as an indicator of shallower mixing depths. In our study, ice off occurred at essentially the same time both years, yet deeper mixing in Experimental Lake in 2014 resulted in lower *D. stelligera* densities. Furthermore, average epilimnetic temperatures in mid-July were higher in both lakes in 2014 than in 2013, yet *D. stelligera* declined in Experimental Lake, suggesting that direct temperature effects on this species are not the primary control on its success. With early and late season sampling indicating no apparent changes in water chemistry in this case, our results suggest that altered average light exposure as a result of deeper mixing may be one of the main mechanisms by which changing thermal structure affected *D. stelligera* densities in this lake.

Multiple factors, including water transparency, water temperatures, and wind strength, affect mixing depths and the strength of thermal stratification in lake ecosystems (Fee et al. 1996; Kraemer et al. 2015). Widespread changes in lake thermal structure have been occurring over recent decades (Kraemer et al. 2015; Williamson et al. 2015) and are driven to varying degrees by these multiple factors, resulting in different patterns (e.g., deeper thermocline depth, increasing strength of stratification) across systems. These differences in lake thermal response are consistent with the variable patterns observed in *D. stelligera* sedimentary profiles across lakes (reviewed by Saros and Anderson 2015). For example, since the late 1800s, *D. stelligera* increased in relative abundance in the sediments of a small alpine lake experiencing warming, advancing treeline, and likely increasing DOC concentrations, factors that would promote shallower stratification (Saros et al. 2012). In contrast, this species decreased in relative abundance over the same time frame in the sediments of a large lake on Isle Royale in Lake Superior experiencing increasing wind speeds, a factor that would promote deeper mixing depths (Saros et al. 2012).

Our results support the use of *D. stelligera* as an indicator of changing mixing depths in some lake ecosystems. Of the five diatom taxa found in both lakes in both years, *D. stelligera* was the only species showing a consistent, positive response to shallower thermal stratification. Changes in the abundances of other taxa, such as *F. tenera v. nanana* and *T. flocculosa*, were not linked to stratification patterns in these lakes, suggesting that these species were affected by other factors. We do reiterate, however, the cautions advised in Saros and Anderson (2015). *Discostella stelligera* responds to the interactive effects of light and nutrients (Saros et al. 2012, 2014; Malik and Saros 2016), which likely influence how and why it responds to lake thermal structure. As a result, while this species is more abundant during thermally stratified conditions in many lakes (Winder et al. 2009; Saros et al. 2012), there are some lakes in which this species is more abundant during periods of turnover (Köster and Pienitz 2006; Boeff et al. 2016). When possible, the site-specific seasonal distribution patterns of *D. stelligera* should be assessed before applying this tool to a lake sediment core (as in Boeff et al. 2016).

Changes over recent decades in the factors affecting lake stratification are well documented, as are the implications of these changes for plankton production and diversity in lakes (Berger et al. 2010; Cantin et al. 2011). What remains less clear, however, is how recent changes in lake thermal structure compare to longer-term variability. We are already learning more about this longer-term variability through the use of this indicator species in lake sediment records spanning back hundreds (Smol et al. 2005; Rühland et al. 2008; Saros et al. 2012) to thousands (Wang et al. 2008; Stone et al. 2016) of years. The results of this whole-lake experiment provide an important step in confirming the use of this inference tool when coupled with contemporary pelagic sampling of seasonal distributions of this species. The broader application of this tool will yield greater insight into longer-term variability in the response of lake ecosystems to climate.

References

Anderson, N. J., and B. Rippey. 1994. Monitoring lake recovery from point-source eutrophication: The use of diatom-inferred epilimnetic total phosphorus and sediment chemistry. Freshw. Biol. **32**: 625–639. doi:10.1111/j.1365-2427.1994.tb01153.x

Arp, C. D., B. M. Jones, and G. Grosse. 2013. Recent lake ice-out phenology within and among lake districts of Alaska, U.S.A. Limnol. Oceanogr. **58**: 2013–2028. doi:10.4319/lo.2013.58.6.2013

Berger, S. A., S. Diehl, H. Stibor, G. Trommer, and M. Ruhenstroth. 2010. Water temperature and stratification depth independently shift cardinal events during plankton spring succession. Glob. Change Biol. **16**: 1954–1965. doi:10.1111/j.1365-2486.2009.02134

Boeff, K. A., K. E. Strock, and J. E. Saros. 2016. Evaluating planktonic diatom response to climate change across three lakes with differing morphometry. J. Paleolimnol. **56**: 33–47. doi:10.1007/s10933-016-9889-z

Burpee, B. T., J. E. Saros, R. M. Northington, and K. S. Simon. 2016. Microbial nutrient limitation in arctic lakes in a permafrost landscape of southwest Greenland. Biogeosciences **13**: 365–374. doi:10.5194/bg-13-365-2016

Cantin, A., B. E. Beisner, J. M. Gunn, Y. T. Prairie, and J. G. Winter. 2011. Effects of thermocline deepening on lake plankton communities. Can. J. Fish. Aquat. Sci. **68**: 260–276. doi:10.1139/F10-138

Carney, H. J., P. J. Richerson, C. R. Goldman, and R. C. Richards. 1988. Phytoplankton demographic processes and experiments on interspecific competition. Ecology **69**: 664–678. doi:10.2307/1941015

Fee, E.J., R.E. Hecky, S.E.M. Kasian, and D.R. Cruikshank. 1996. Effects of lake size, water clarity, and climatic variability on mixing depths in Canadian Shield lakes. Limnol. Oceanogr. **41**:912–920. doi:10.4319/lo.1996.41.5.0912

Fritz, S.C., S. Juggins, B. W. Battarbee, and D. R. Engstrom. 1991. Reconstruction of past changes in salinity and climate using a diatom-based transfer function. Nature **352**: 706–708. doi:10.1038/352706a0

Hobbs, W. O., and others. 2010. Quantifying recent ecological changes in remote lakes of North America and Greenland using sediment diatom assemblages. PLoS One **5**: e10026. doi:10.1371/journal.pone.0010026

Jobin, V. O., and B. E. Beisner. 2014. Deep chlorophyll maxima, spatial overlap and diversity in phytoplankton exposed to experimentally altered thermal stratification. J. Plankton Res. **36**: 933–942. doi:10.1093/plankt/fbu036

Köster, D., and R. Pienitz. 2006. Seasonal diatom variability and paleolimnological inferences - a case study. J. Paleolimnol. **35**: 395–416. doi:10.1007/s10933-005-1334-7

Kraemer, B. M., and others. 2015. Morphometry and average temperature affect lake stratification responses to climate change. Geophys. Res. Lett. **42**: 4981–4988. doi:10.1002/2015GL064097

Magnuson, J. J., B. J. Benson, and T. K. Kratz. 1990. Temporal coherence in the limnology of a suite of lakes in Wisconsin, USA. Freshw. Biol. **23**: 145–159. doi: 10.1111/j.1365-2427.1990.tb00259.x

Malik, H., and J. E. Saros. 2016. Effects of temperature, light and nutrients on five *Cyclotella sensu lato* taxa assessed with *in situ* experiments in arctic lakes. J. Plankton Res. **38**: 431–442. doi:10.1093/plankt/fbw002

Perren, B. B., M. S. V. Douglas, and N. J. Anderson. 2009. Diatoms reveal complex spatial and temporal patterns of recent limnological change in West Greenland. J. Paleolimnol. **42**: 233–247. doi:10.1007/s10933-008-9273-8

Rühland, K., A. M. Paterson, and J. P. Smol. 2008. Hemispheric-scale patterns of climate-related shifts in planktonic diatoms from North American and European lakes. Glob. Change Biol. **14**: 2740–2754. doi:10.1111/j.1365-2486.2008.01670.x

Saros, J. E., and N. J. Anderson. 2015. The ecology of the planktonic diatom *Cyclotella* and its implications for global environmental change studies. Biol. Rev. **90**: 522–541. doi:10.1111/brv.12120

Saros, J. E., J. R. Stone, G. T. Pederson, K. E. H. Slemmons, T. Spanbauer, A. Schliep, D. Cahl, C. E. Williamson, and D. R. Engstrom. 2012. Climate-induced changes in lake ecosystem structure inferred from coupled neo- and paleoecological approaches. Ecology **93**: 2155–2164. doi: 10.1890/11-2218.1

Saros, J. E., K. E. Strock, J. Mccue, E. Hogan, and N. J. Anderson. 2014. Response of *Cyclotella* species to nutrients and incubation depth in Arctic lakes. J. Plankton Res. **36**: 450–460. doi:10.1093/plankt/fbt126

Sastri, A. R., J. Gauthier, P. Juneau, and B. E. Beisner. 2014. Biomass and productivity responses of zooplankton communities to experimental thermocline deepening. Limnol. Oceanogr. **59**: 1–16. doi:10.4319/lo.2014.59.1.0001

Schneider, P., and S.J. Hook. 2010. Space observations of inland water bodies show rapid surface warming since 1985. Geophys. Res. Lett. **37**: L22405. doi:10.1029/2010GL045059

Smol, J. P., and others. 2005. Climate-driven regime shifts in the biological communities of arctic lakes. Proc. Natl Acad. Sci. U.S.A. **102**: 4397–4402. doi:10.1073/pnas.0500245102

Stewart-Oaten, A., W. W. Murdoch, and K. R. Parker. 1986. Environmental impact assessment: "Pseudoreplication" in time? Ecology **67**: 929–940. doi:10.2307/1939815

Stone, J. R., and S. C. Fritz. 2004. Three-dimensional modeling of lacustrine diatom habitat areas: Improving paleolimnological interpretation of planktic:benthic ratios. Limnol Oceanogr. **49**: 1540–1548. doi:10.4319/lo.2004.49.5.1540

Stone, J. R., J. E. Saros, and G. T. Pederson. 2016. Coherent late Holocene climate-driven shifts in the structure of three Rocky Mountain lakes. The Holocene. **26**: 1103–1111. doi:10.1177/0959683616632886

Strock, K.E., J.E. Saros, S.J. Nelson, S.D. Birkel, J.S. Kahl, and W.H. McDowell. 2016. Extreme weather years drive episodic changes in lake chemistry: Implications for recovery from sulfate deposition and long-term trend sin dissolved organic carbon. Biogeochemistry. **127**: 353–365. doi: 10.1007/s10533-016-0185-9

Tilzer, M. M., and C. R. Goldman. 1978. Importance of mixing, thermal stratification, and light adaptation for phytoplankton productivity in Lake Tahoe (California-Nevada). Ecology **59**: 810–821. doi:10.2307/1938785

Wang, L., and others. 2008. Diatom-based inference of variations in the strength of Asian winter monsoon winds between 17,500 and 6000 calendar years B.P. J. Geophys. Res. **113**(D21): 1–9. doi:10.1029/2008JD010145

Williamson, C.E., E.P. Overholt, R.M. Pilla, T.H. Leach, J.A. Brentrup, L.B. Knoll, E.M. Mette, and R.E. Moeller. 2015. Ecological consequences of long-term browning in lakes. Sci. Rep. **5**: 18666. doi:10.1038/srep18666

Winder, M., J. E. Reuter, and G. S. Schladow. 2009. Lake warming favours small-sized planktonic diatom species. Proc. Roy. Soc. Lond. B. **276**: 427–435. doi:10.1098/rspb.2008.1200

Acknowledgments

We are sincerely grateful to Benjamin Burpee, Rachel Fowler, Hamish Greig, Emily Rice, Kristin Strock, and Kathryn Warner for extraordinary field assistance, and to our helicopter pilots Gregor Beer and Petrus Nobreus. CH2M Hill Polar Services provided logistical support for this project; in particular, Geoff Miller provided extensive assistance with the assembly and transport of the Solar Bee. We thank Corey Simnioniw of the Medora Corporation for modifying the design of the Solar Bee to meet our requirements. Benjamin Burpee, Johanna Cairns, Jeffery Stone and Nora Theodore provided photos and figure elements. Brian McGill provided assistance with statistical analyses. This research was funded by the Arctic System Science program of the US National Science Foundation (grant #1203434 to JES).

15

Can small zooplankton mix lakes?

S. Simoncelli,[1]* *S. J. Thackeray*,[2] *D. J. Wain*[1]
[1]Department of Architecture and Civil Engineering, University of Bath, Claverton Down, Bath, United Kingdom; [2]Centre for Ecology & Hydrology, Lancaster Environment Centre, Bailrigg, Lancaster, United Kingdom

Scientific Significance Statement

Biomixing is the mixing of waters by living organisms in oceans and lakes. Research of the past several decades has provided important insights about the role of biomixing in oceans, showing that vertical migrators, such as crustacean zooplankton, may be able to enhance ocean mixing. However, there is little evidence for the role of biomixing in lakes, including the organisms that might contribute to it, and its potential effects on lake processes. If biomixing occurs in lakes, it has the potential to weaken vertical temperature stratification and enhance fluxes of nutrients and dissolved substances. We argue that there is a need for studies, particularly field studies, on the potential of vertical migrators to generate biomixing in lakes.

Abstract

The idea that living organisms may contribute to turbulence and mixing in lakes and oceans (biomixing) dates to the 1960s, but has attracted increasing attention in recent years. Recent modeling and experimental studies suggest that marine organisms can enhance turbulence as much as winds and tides in oceans, with an impact on mixing. However, other studies show opposite and contradictory results, precluding definitive conclusions regarding the potential importance of biomixing. For lakes, only models and lab studies are available. These generally indicate that small zooplankton or passive bodies generate turbulence but different levels of mixing depending on their abundance. Nevertheless, biogenic mixing is a complex problem, which needs to be explored in the field, to overcome limitations arising from numerical models and lab studies, and without altering the behavior of the animals under study.

Mixing is defined as the combined action of dispersion of dissolved or suspended substances (chemicals or sediment) and enhancement of diffusion of fluid properties, such as heat or salinity (Thorpe 2005). Mixing in lakes plays an important role because it can affect biological and chemical processes (Fischer et al. 1979). External forces acting on lakes can deliver energy into the water column and can drive different local mixing mechanisms depending on the part of the lake under investigation (*see* Fig. 1). The surface layer is the most dynamic and energetic environment; here wind events (A in Fig. 1) usually provide most of the kinetic energy, creating shear, and inducing mixing. During storms, intense mixing can also be generated close to the surface via formation and breaking of surface waves (B) or seiche activity (C). Other processes, such as nocturnal convection (D), when the lake surface cools at night, may alter the potential energy of the water column and affect the lake stratification (Jonas et al. 2003). In the littoral zone, mixing can be enhanced when physical processes (E), such as seiches or wind-generated internal waves, interact with lake physical boundaries and generate boundary mixing with a possible impact on nutrient fluxes (MacIntyre et al. 1999).

The lake interior, below the surface and away from the bottom and shores, responds differently to external forces because of the vertical temperature stratification. The lake interior is

*Correspondence: s.simoncelli@bath.ac.uk

Author Contribution Statement: DJW developed the research ques-tion and designed the research in conjunction with SJT. SS synthesized the literature on the topic and wrote the paper with input from all co-authors.

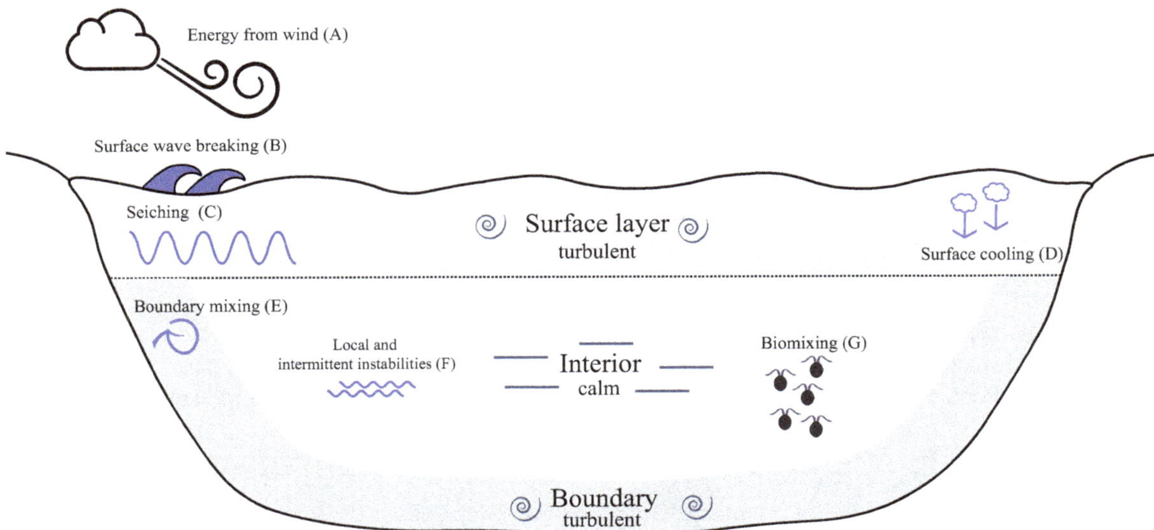

Fig. 1. Sketch illustrating main mixing processes operating in three different lake regions. In the surface layer, energy from wind (**A**) leads to mixing by breaking surface waves (**B**) or via seiching motions (**C**) or convective mixing can act at night when the surface is cooling (**D**). Boundaries are subjected to mixing events (**E**) for example via interactions of internal waves. The lake interior is the calmest region with local and intermittent mixing events (**F**). Vertical migrators (**G**) may provide energy for enhancing the mixing in this layer. Eddies indicate layers with mixing, while straight lines in the interior indicate that energy production is extremely weak and sporadic.

the most quiescent part of a lake where mixing events (F in Fig. 1) are intermittent and localized processes (Wüest and Lorke 2003; Bouffard and Boegman 2012). For this reason, understanding which mechanisms drive interior mixing is of crucial importance for lake ecosystem functioning. Recent research suggests that swimming organisms may operate as a previously neglected mixing mechanism in the interior (Fig. 1, G): by creating hydrodynamic disturbances, such as jets or turbulent eddies, organisms may deliver potential energy to the water column, with a significant contribution toward interior mixing. Recent investigations show that the contribution of horizontal migrators, such as fish, is usually negligible (Gregg and Horne 2009; Pujiana et al. 2015) and attention should instead be focused on vertically migrating zooplankton.

There is currently insufficient understanding of the role of vertically migrating zooplankton as agents of biomixing: these organisms can swim against the stable density stratification, with potential effects on water column mixing and ecological processes. For example, biomixing from vertical migrators may be able to replenish nutrients in surface-depleted waters and stimulate primary production by phytoplankton. If nutrients are brought to the surface, they can also be redistributed via other surface mixing events (such as wind-driven transport or river inflows) to other regions. Oxygen distribution may be altered as well: biomixing enhancement of oxygen fluxes between the surface and metalimnion could reduce deep-water oxygen depletion, with impacts on habitat quality and biogeochemical cycling. Vertical migrators, once they reach the epilimnion, may still enhance turbulence and mixing in unstratified surface waters. Zooplankton-generated turbulent motions can alter ecological interactions by advecting passive bodies such as algae, and increasing encounter rates between zooplankton grazers and their phytoplankton resources (Harris et al. 2000). Given these under-studied possibilities, it is important to study the ecological significance of biomixing in lakes.

Quantifying biomixing is a complex problem because results depend on several factors such as the organisms under investigation, their swimming mode, their concentration and their interactions with the environment. Direct comparisons between current models in the literature and field measurements is not always possible, because probes are not able to sample what happens near the organism's body while swimming.

In the following, we provide a theoretical framework to understand the fundamental physics of biomixing along with some results from in situ ocean observations. We then discuss current studies in lakes and suggest that there is insufficient evidence about the role of biomixing in freshwater bodies. Field observations are needed to overcome some limitations of current studies, and to verify the potential role of biomixing in lakes.

Measuring biomixing

Mixing in lakes can be generally described through a turbulent kinetic energy (TKE) balance, which in the simplest case reads (Osborn 1980; Ivey and Imberger 1991):

$$m = b + \varepsilon \quad (1)$$

where m is the production of TKE, b is the buoyancy flux accounting for the vertical mixing and ε the TKE dissipation rate. External forces, such as wind at the lake surface or eddies generated by swimming organisms in the interior, can provide TKE and contribute to the production term (m) in Eq. 1. Part of the source energy is inevitably dissipated as heat (ε) by viscous processes acting at the molecular level. However, some energy may be converted into potential energy (b) and affect the position of fluid particles. Changes in the potential energy of the water column can partially destroy the stable vertical stratification and lead to mixing (Fig. 2). Dissipation rates ε can be measured in situ through specific devices, such as shear probes or temperature microstructure profilers, but ε does not provide direct information about mixing. When an increase of ε is observed, it means that energy (m) is transferred in the fluid but mixing may not occur, if no input energy is transferred into the component b.

Energy dissipation rates (ε) can however be linked to vertical mixing (b) via the vertical eddy diffusion coefficient K_V (Osborn 1980):

$$K_V = b / N^2 = \Gamma \cdot \varepsilon / N^2 \qquad (2)$$

where $N = [-(g/\rho)\partial\rho/\partial z]^{1/2}$ is the buoyancy frequency describing the vertical stratification which depends on the gravitational acceleration g, water depth z, density ρ and its gradient $\partial\rho/\partial z$. The estimation of the eddy diffusion coefficient K_V is relevant for the quantification of mass vertical fluxes and mixing: when the coefficient is enhanced with respect to background conditions, oxygen or other nutrients can spread in the water column and to different lake layers. The flux F_S of a substance with concentration C_S in the lake can be described using the Fick's law, once K_V is known:

$$F_S = K_V \frac{dC_S}{dz} \qquad (3)$$

For waters stratified by temperature, mixing can be enhanced if $K_V > D_T$, where $D_T = 10^{-7}$ m^2 s^{-1} is the molecular temperature diffusivity. However, when $K_V \approx D_T$ dissolved substances will spread very slowly at the molecular level only.

In Eq. 2, Γ is a parameter representing the efficiency of the mixing and provides an estimate of how much energy is converted to mixing (b) with respect to the dissipated energy ε. Laboratory and experimental observations suggest $\Gamma \leq 0.2$ (Ivey et al. 2008; Bouffard and Boegman 2013) for wind-generated turbulence. However, for biogenic mixing the value for Γ is still not known. Several conditions and parameters affect the biomixing process, and thus Γ, such as the species of organisms concerned, their size, concentration, swimming behavior, and the environmental conditions such as the stratification strength and the background turbulence dissipation level. If swimming organisms do not efficiently mix the water column, creating small water disturbances, Γ would be too small and K_V does not increase.

Kunze et al. (2006) measured for the first time ε generated by the vertical migration of a population of krill (organism's length $l_{OR} = 1$–2 cm) in Saanich Inlet (Canada). Observed dissipation rates of TKE from biogenic inputs peaked between

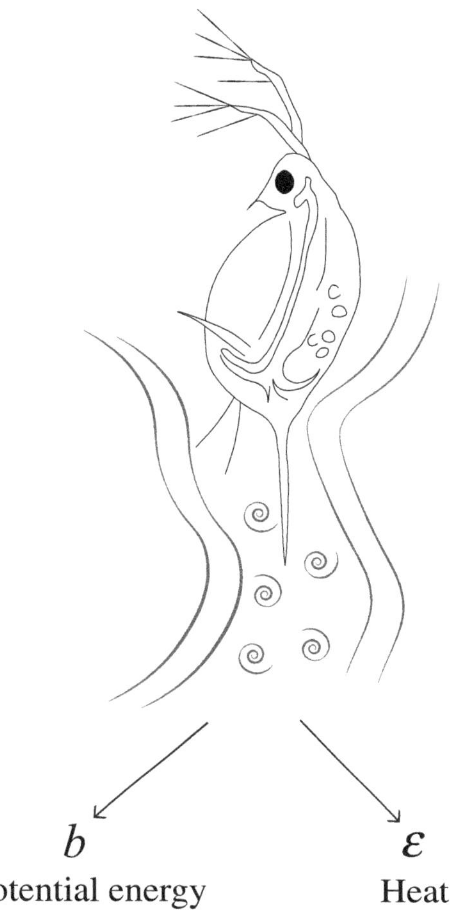

Fig. 2. Schematic of the partition of turbulent kinetic energy imparted by a swimmer (*Daphnia* spp.). The continuous line depicts the wake left by the swimmer, while the eddies are the turbulent instabilities created within the wake that can be a source of TKE. The source energy is converted into potential energy (b), increasing the mixing, and into heat as energy is dissipated (ε) due to water molecular viscosity.

10^{-4} W Kg^{-1} and 10^{-5} W Kg^{-1}, compared to typical background level of 10^{-9} W Kg^{-1}. Dissipation spanned five orders of magnitude, suggesting an important krill biomixing contribution as much as mixing from wind and tides. High concentration, and associated multi-body hydrodynamic interactions, probably played an important role, despite weak wind forcing and the strong stratification gradient. The estimated eddy diffusivity from Eq. 2, assuming $\Gamma = 0.2$, ranged between 2×10^{-1} and 2×10^{-2} m^2 s^{-1}, an increase of five orders of magnitude when compared to the daily-averaged level. However, elevated TKE rates were observed by Kunze only for a few minutes during the migration, indicating that the source of turbulence is not constant in time, as was later observed by Rousseau et al. (2010). Rippeth et al. (2007) drew the same conclusions and did not observe such important increases in turbulence from their measurements of TKE dissipation rates in stratified coastal waters of the UK.

Table 1. Main biomixing studies in the literature classified by type of study. For the different kind of analyzed organisms and swimming behaviors, we reported the main results for generated turbulence and mixing. Gray-shaded rows show the few biomixing observations for freshwater zooplankton.

Reference	Type of study	Organism (size)	Swimming behavior	Average ε (W kg^{-1})	Mixing
Huntley and Zhou (2004)	Model	Euphausiids-Whales	Aggregated	10^{-5}	-
Kunze et al. (2006)	Field (ocean)	Krill (1–2 cm)	Aggregated	10^{-5}–10^{-4}	$K_V = 2 \times 10^{-1}$–2×10^{-2} m^2 s^{-1} (with $\Gamma = 0.2$)
Rippeth et al. (2007)	Field (ocean)	Krill	Aggregated	No enhancement	-
Gregg and Horne (2009)	Field (ocean)	Nekton	School	10^{-6}–10^{-5}	No enhancement
Rousseau et al. (2010)	Field (ocean)	Euphausiids	Aggregated	$<10^{-8}$	$K_V \sim 10^{-5}$ m^2 s^{-1} (with $\Gamma = 0.2$)
Thiffeault and Childress (2010)	Model	Krill	Aggregated	$\sim 10^{-6}$	-
Lorke and Probst (2010)	Field (lake)	Perch	Aggregated	3×10^{-9}–10^{-8}	-
Leshansky and Pismen (2010)	Model	Small zooplankton	Aggregated	2×10^{-7}	-
Kunze (2011)	Model	Small zooplankton	Aggregated	10^{-9} (assumption)	$K_V = 2 \times 10^{-7}$ m^2 s^{-1}
Noss and Lorke (2012)	Laboratory	*Daphnia magna* (4 mm)	Tethered on a filament	8×10^{-7} (max: 2×10^{-5})	-
			Freely swimming	2×10^{-6} (max: 3×10^{-4})	$K_V \sim 10^{-5}$ m^2 s^{-1}
Noss and Lorke (2014)	Laboratory	*Daphnia magna* (3 mm)	Aggregated	-	$K_V \sim 10^{-9}$ m^2 s^{-1}
Wagner et al. (2014)	Model	Small zooplankton	Single organism	-	$\Gamma \sim 0.03$
Dean et al. (2015)	Model	Krill	Aggregated	10^{-6}–10^{-7} (highest concentration)	-
Wang and Ardekani (2015)	Model	Small zooplankton	Aggregated	-	$K_V \sim 10^{-6}$ m^2 s^{-1}
Tanaka et al. (2017)	Laboratory	Sardine	Aggregated	2.3×10^{-4}	$K_V \sim 10^{-2}$–10^{-1} m^2 s^{-1} $\Gamma = 0.02$–0.08

Other ocean studies estimated dissipation rates ε and eddy diffusivity K_V through laboratory experiments and models. A summary is presented in Table 1. These studies show that mixing by krill is not feasible (Rousseau et al. 2010) and only possible with high concentrations (Kunze et al. 2006; Dean et al. 2015) but other vertical migrators, such as copepods or other small zooplankton, may still be able to enhance ocean mixing (Huntley and Zhou 2004; Katija 2012). Direct comparisons of dissipation ε, between current models in the literature and field measurements, is not always possible because microstructure profilers, such as the one used by Kunze et al. (2006), are not able to sample turbulence near the organism's body, providing smaller turbulence dissipations than those estimated from models. Finally, the quantification of biomixing, as done by Kunze et al. (2006), must not rely only on the estimation of dissipation rates (ε) and on the assumption that $\Gamma = 0.2$ (Visser 2007*a,b*; Subramanian 2010) but must also be based on direct assessment of Γ and K_V in Eq. 2.

Biomixing in lakes

Biomixing observations in lakes are very limited. So far, the only experimental biomixing study in a lake was conducted by Lorke and Probst (2010) for perch (*Perca fluviatilis*), while the first investigations of zooplankton-generated mixing were carried out under controlled laboratory conditions for *Daphnia* only. *Daphnia* is a very common zooplanktonic genus in lakes, with body lengths approximately between 1 mm and 3 mm. Organisms within this genus often undertake diel vertical migration (DVM), ascending at dusk toward the food-rich surface layer to forage on phytoplankton, and sinking back at dawn into deeper, aphotic waters (Ringelberg 1999). DVM is mainly adopted as a predator-avoidance mechanism but other migratory drivers, such as UV exposure or temperature, may play a role (Williamson et al. 2011). Migrations can last anywhere from minutes to a few hours, and their magnitude differs among lakes and between seasons (Ringelberg 2010).

Noss and Lorke (2012) conducted the first laboratory study of dissipation rates (ε) of TKE produced by *Daphnia*. By using a particle image velocimetry (PIV) technique combined with laser-induced fluorescence, they could estimate some energetic parameters of the planktonic organism swimming in different configurations with a density gradient typical of the thermocline ($N = 0.07$ s^{-1}). TKE dissipation rates (ε) and diffusion coefficient (K_V) were estimated considering the water volume influenced by the organism while swimming, which is usually larger than the organism size. Estimated average dissipation was 2×10^{-6} W kg^{-1} with a maximum of 3×10^{-4} W kg^{-1}, in accordance with results from Huntley and Zhou (2004)'s model. Eddy diffusivity was enhanced in the organism vicinity ($K_V \sim 10^{-5}$ m^2 s^{-1}) and was two orders of magnitude bigger than the molecular heat diffusivity ($D_T \sim 10^{-7}$ m^2 s^{-1}), indicating the potential for an impact on temperature gradients in lakes. However, during the experiment, K_V was not measured in the whole tank, therefore it is not certain whether the zooplankton could have affected mixing on scales larger than the organism size. Moreover, the impact of the re-stratification was not evaluated and no conclusion can be drawn about the mixing efficiency Γ.

Later Noss and Lorke (2014) studied the same organism in different swimming configurations and quantified mixing via the diffusion of a fluorescent dye (Rhodamine 6G) injected into a stratified water tank ($N = 0.08$ s^{-1}). *Daphnia* (max. concentration $\sim$ 4 org. L^{-1}) were forced to vertically migrate generating a global diffusivity in the tank as low as 10^{-9} m^2 s^{-1}. Even when swimming in aggregations, *Daphnia* had a small impact on dissolved substances or gases, whose molecular diffusivity D_G is 10^{-9} m^2 s^{-1}. This result differs however from the previous study, because it provides the diffusion coefficient affected at larger scales, while Noss and Lorke (2012) measured the diffusivity in the near vicinity of a single organism only. For *Daphnia*, at organism-scale dissipation ε and mixing can be enhanced, but when K_V is assessed over the effective and larger volume influenced by *Daphnia* migration, the impact on mixing is negligible if compared to wind-induced mixing. To affect temperature stratification in lakes, *Daphnia* aggregation must be able to increase K_V above $D_T = 10^{-7}$ m^2 s^{-1}.

Wilhelmus and Dabiri (2014) later performed another laboratory experiment in an unstratified tank to analyze the fluid instabilities and mixing induced by *Artemia salina*, a small zooplanktonic species ($l_{OR} = 5$mm) that lives in saline lakes. During the vertical migration, induced artificially with a laser, collective swimming dynamics from different organisms created a large downward jet. The length of the generated eddies near its boundary was considerably larger ($l \sim 1$ cm) than a single organism. Their measurements clearly show that swimmers, when present at high concentration, can deliver kinetic energy at scales bigger than the single organism's length with a possible impact on mixing. However, the lack of a stable stratification did not allow the estimation of the real migration effect on mixing after buoyancy restores the initial density gradient: displaced water parcels and properties can return to their initial position with no effect on mixing if swimmers are not sufficiently efficient.

Physics-based models can also be used to evaluate biogenic mixing for lakes. Kunze (2011) estimated the eddy diffusivity coefficient from simple physical considerations and by assuming that each organism can transport a water volume comparable to its size as it swims in a dense aggregation. Kunze (2011) found that the apparent diffusivity depends on the organism concentration C and for *Daphnia* with $C = 100$ org. L^{-1}, the resulting diffusivity is $K_V = 1.7 \times 10^{-7}$ m^2 s^{-1}, suggesting a negligible enhancement in mixing. More importantly, the model does not consider any re-stratification effect and is not suitable for small zooplankton, such as for *Daphnia*, because it assumes that the organism Reynolds number $\mathrm{Re} = U \cdot l_{OR}/\nu < 1$, where U is the organism's speed and ν the kinematic water viscosity.

Laboratory experiments show that $Re \sim 30$–80 for *Daphnia* (Noss and Lorke 2014; Wickramarathna et al. 2014). Furthermore, inertial forces neglected by the model, can further enhance mixing (Noss and Lorke 2014). Another simple and similar approach was previously proposed by Leshansky and Pismen (2010). In their model, swimmers can disperse the turbulent local flow as a function of the school concentration C, the turbulent dissipation ε, the size l of the produced hydrodynamic instabilities, and speed U. By assuming that for a *Daphnia* swarm, $C = 100$ org. L^{-1}, $\varepsilon = 10^{-9}$ W kg^{-1}, $U = 30$ mm/s and $l = l_{OR} = 1$ mm (Gries et al. 1999; Wickramarathna et al. 2014), the diffusion coefficient is 4×10^{-7} m^2 s^{-1}. Diffusivity increases to $K_V = 10^{-5}$ m^2 s^{-1} when $C = 10{,}000$ org. L^{-1}. Estimated coefficients from these models provide a lower bound of mixing and generally suggest that zooplankton may not be able to alter vertical temperature stratification, since $K_V \approx D_T$.

Wagner et al. (2014) provided instead an estimation of mixing in terms of its efficiency Γ (Eq. 2). In their model, each organism is considered very small and swimming in a stable stratified fluid. For a single vertically migrating zooplankton $\Gamma \sim 0.03$, but it may achieve unity depending on the organism's length, swimming mode, and stratification. The model suggests that biomixing seems a feasible mechanism but does not provide any information about the eddy diffusion coefficient K_V. Moreover, the model is more suitable for micro-organisms and does not consider any influence of the zooplankton packaging density C, which may be the main boosting factor for the mixing.

Finally, Wang and Ardekani (2015) numerically resolved the flow field influenced by an aggregation of interacting swimmers in a stratified medium in the intermediate Reynolds number regime. The model is particularly suitable to model small zooplankton and provide a complete description of biomixing. Simulations were performed with a small number of swimmers and aggregations corresponding to very high densities of $C = 10{,}000$ org. L^{-1} to provide an upper-bound for mixing. In particular, organism swimming behavior was modeled as a "squirmer" (Lighthill 1952; Blake 1971) and controlled by a parameter β which scales with the

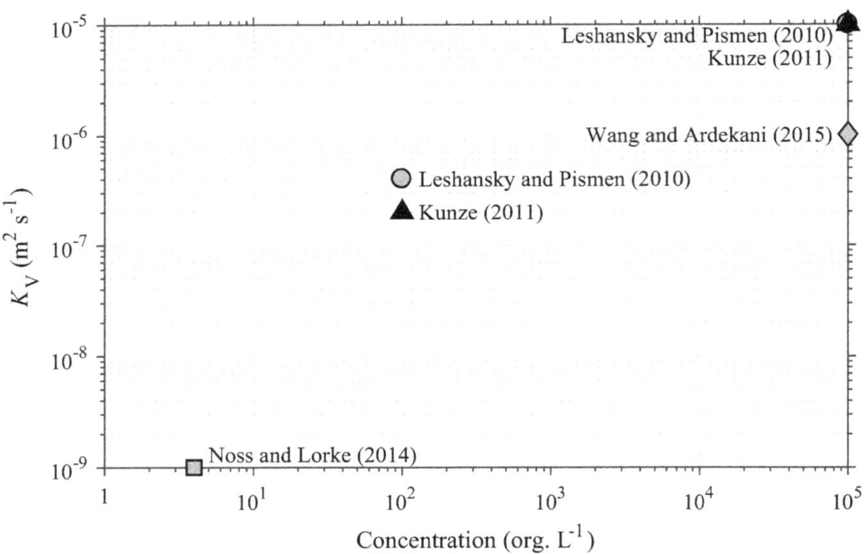

Fig. 3. Eddy diffusivity K_V as a function of zooplankton concentration C from numerical simulations and laboratory experiments.

organism's size l_{OR}, velocity U and fluid generated vorticity; for *Daphnia* $\beta = 1$ (Wickramarathna et al. 2014; Wickramarathna 2016). From this model, the estimated mixing efficiency Γ for *Daphnia* was 0.01, and eddy diffusivity K_V was as low as 2×10^{-7} m^2 s^{-1} but for a very strong density stratification with $N = 1.9$ s^{-1}. However, for a weaker but more realistic stratification, the numerical model by Wang and Ardekani (2015) showed that swimmers were less efficient ($\Gamma = 3 \cdot 10^{-4}$) but generate a higher diffusivity, with $K_V = 10^{-6}$ m^2 s^{-1}. Change of swimming trajectories, vertical orientation as well as organism buoyancy can further enhance these values (Wang and Ardekani 2015).

The need for field studies

Biomixing studies for oceans cannot be used to draw conclusions for lakes because oceans are physico-chemically different to freshwater bodies, and because marine planktonic organisms are more diverse and potentially larger than their freshwater counterparts (Hessen and Kaartvedt 2014), and biomixing is an organism-dependent mechanism. The few studies in the literature for freshwater zooplankton collectively yield differing conclusions about the role of biomixing (Fig. 3). Numerical simulations by Wang and Ardekani (2015) show that biomixing by *Daphnia* is a feasible process when the zooplankton concentration is as high as $C = 10{,}000$ org. L^{-1}. On the other hand, the experimental study by Noss and Lorke (2014) suggests that mixing is negligible with a smaller concentration of organisms from the same genus (4 org. L^{-1}). In the two studies K_V varies by three orders of magnitude, while the concentration C covers four orders of magnitude. Zooplankton abundance depends on both biotic and abiotic environmental conditions; their density in lakes can vary greatly and can be substantially higher than that used in the experiment by Noss and Lorke (2014), especially during the DVM (George and Hewitt 1999; Straile and Adrian 2000; Hembre and Megard 2003; Talling 2003).

Zooplankton aggregation density is important and may have emergent effects on biomixing: higher concentrations can enable interactions of wakes originating from single organisms and enhance shear and mixing in the same fashion as observed by Wilhelmus and Dabiri (2014). The form of the relationship between zooplankton density and biomixing is currently not known e.g., there may be a concentration threshold over which biomixing is enhanced. In addition, numerical simulations currently simplify taxonomic variability in biomixing potential e.g., Kunze (2011) and Wang and Ardekani (2015) describe all the zooplanktonic species with general models, while in reality zooplankton species swim in different ways, and species-specific models may be more suitable to model *Daphnia* and to describe their particular swimming behavior (Jiang and Kiørboe 2011). These interactions, taking place in a real environment, between individuals from multiple species may be stochastic and challenging to describe mathematically. However, community-level effects may be observable in the field.

Field observations are needed to understand the feasibility of biomixing by freshwater zooplankton communities generally, and *Daphnia* specifically, for several reasons. With field studies, it is possible to overcome limitations arising from laboratory experiments under controlled conditions. In the laboratory, diel vertical migration cycles are artificially simulated by alternating light and dark periods with LED panels or using laser beams with a constant intensity. These methods trigger the zooplankton primary phototaxis, which is the movement toward or away from a light beam. *Daphnia* DVM in the field is instead triggered by the secondary phototactic behavior, which is the reaction due to the rate of change in light intensity, usually peaking at dusk and sunrise only (Ringelberg 1999, 2010). These two different behavioral responses also explain why the DVM does not occur during the day or at night and therefore zooplankton responses in lab tanks may be very different from those in the field. Without field observation, it is not known whether the difference in the DVM trigger can affect *Daphnia* swimming responses and, thus, biomixing. Moreover, it is not certain how laser beams, used to fluoresce the fluid in the tank, impact upon zooplankton migration behavior. The use of artificial light, generated by LEDs in Noss and Lorke (2014) to trigger the migration, may explain why only 16% of the organisms into the tank moved and why some of them remained at the tank top or bottom. Field sampling allows the study of organisms in their natural environment without altering the behavior, potentially increasing the realism of biomixing estimates. Field studies also allow understanding the zooplankton concentration during the DVM, compared to the daily zooplankton densities in the lake. Finally, lakes are populated by variable abundances of zooplanktonic species (species of *Daphnia*, *Bosmina*, *Cyclops*, etc.) and other migrators that can interact with *Daphnia*. Other species can affect *Daphnia* density and force them to frequently change their swimming direction in the migrating layer, which could affect the vertical mixing. Such species interactions cannot be easily reproduced in lab experiments, and they may be difficult to address numerically with models. However, field studies would allow us to construct empirical relationships between abundance and biomixing for communities of different compositions, against which to test developing theoretical expectations.

Migration frequently acts as an avoidance mechanism from visual predators such as larval or juvenile fish (Ringelberg 1999; De Robertis 2002; Waya 2004). The presence of chemical substances released by predators, such as kairomones, and sensed by zooplankton, affect DVM leading to increased migration amplitude or faster swimming reactions (Loose and Dawidowicz 1994; Dodson et al. 1997; Ringelberg 1999, 2010). These behavioral responses can increase the size of the generated instabilities and may increase the vertical diffusion K_V. Moreover, food in lab experiments is usually absent and its availability in real lakes, such as a surface or deep chlorophyll maxima, may be another key factor affecting migration amplitude (Dodson et al. 1997; Ringelberg 1999; Rinke et al. 2007). Tank size, light distribution, temperature, and other features of the environment can also change the swimming behavior and limit the swimming reaction (Buchanan et al. 1982; Dodson et al. 1997). Field studies are needed to confirm whether results from experiments under simplified conditions and numerical models are applicable to biomixing mechanisms in complex natural environments. Only field measurements can tell us which lakes are, and are not, prone to such effects so that we can make generalizations about the importance of biomixing.

Challenges for future field investigations

Field investigation should be performed on vertical migrators during the DVM of zooplankton. *Daphnia* are a good candidate to develop our understanding of freshwater biomixing because (1) they are a very common and abundant migrating species in lakes. (2) Despite their smaller size, dissipation rates of kinetic energy are higher for *Daphnia* compared to theoretical estimates for other zooplanktonic species due to their unique swimming mode (Wickramarathna et al. 2014). (3) Finally, they have been studied in the lab, therefore field studies can be used to validate numerical models and compare experimental results under very controlled conditions.

In particular, the DVM can be directly studied both through zooplankton collection and analysis and indirectly via acoustic devices such as ADCPs or echo sounders, allowing a higher spatial and temporal resolution (Lorke et al. 2004; Rinke et al. 2007; Huber et al. 2011). These instruments are usually employed to measure current velocities in three dimensions and to infer turbulence levels as well. The backscatter strength (BS) or amplitude of the scattered wave provided by ADCPs can be used as a proxy for the zooplanktonic concentration and to estimate zooplankton velocities. Higher values of BS indicate higher zooplankton abundance while lower values usually indicate a lack of scatterers in the water. Recent studies by Huber et al. (2011) and Lorke et al. (2004) suggest that ADCPs can be calibrated against the zooplankton concentrations estimated by more traditional means, allowing continuous estimation of their abundance in the water column. However, these devices do not directly provide any information about the zooplankton abundance, size or taxonomy, but they can be used to track their displacement, to understand the timing of the migration and the part of the water column they inhabit during the day.

A first step in assessing biomixing in the field is to measure TKE dissipation rates ε. Generated turbulence during the DVM in lakes can be measured with microstructure profilers which are nowadays normally employed in sampling TKE dissipation rates. In particular, turbulence should be sampled before and after the DVM, to characterize the background turbulence condition without migrators, and during the zooplankton ascent. The duration of observations depends on the time scale of biomixing and measurements should continue for the whole migration duration to understand whether turbulence is patchy and short-lived or energy production by zooplankton is a regular process. Vertical migrators usually swim unsteadily (Noss and Lorke 2012) but asynchronous motions of organisms in the migrating layer may lead to quasi-stationary conditions of turbulence production. If turbulence is enhanced during the migration, this is an indication that energy is generated by zooplankton but, alone, this is not a sufficient proof of biologically-generated mixing. Available energy (m in Eq. 1) can be dissipated as heat with no changes in the potential energy b. However, if no turbulence is observed, zooplankton DVM is not a feasible mechanism for mixing water. Eddy diffusivity K_V can also be inferred from turbulence measurements by using parametrization of Eq. 2, but attention must be paid to the models used because the underlying hypotheses of the mixing parametrizations may not be applicable to biomixing.

If turbulence is generated during the DVM, the next natural step would be to directly measure mixing efficiency Γ or eddy diffusivity K_V via tracer injections (Wüest et al. 1996; Goudsmit et al. 1997; Wain et al. 2013) to measure the effect of the DVM on the eddy diffusivity K_V. This assessment should rely on measurements and comparison of diffusion before and during the DVM. The duration of tracer sampling should continue until after the migration is completed, and longer than the dissipation measurements. This allows understanding of how tracer diffusion is affected over longer time scales, when stratification restores the initial water column density structure affected by the zooplankton migration.

Attempts to study biomixing in the field can however pose important challenges. For example, zooplankton may avoid plankton nets (Brinton 1967; Harris et al. 2000) but disturbance can be limited by using nets with mouth-reducing cones or by reducing the towing speed (UNESCO 1968). The same avoidance mechanisms might be adopted toward free-falling probes (Benoit-Bird et al. 2010; Ross 2014) however, probes are usually designed to avoid any forward disturbances while sampling turbulence. Moreover, turbulence probes may not be able to resolve turbulence produced by a single organism: generated fluid structures from a single individual are generally smaller than the instrument spatial resolution or the turbulence signal may be contaminated by noise.

Zooplankton spatial heterogeneity is another important issue relevant to the role of biomixing in the field. If biologically generated mixing is sampled in the field, results of the measurements may depend on the chosen location within the lake interior because of horizontal zooplankton patchiness (Thackeray at al. 2004; Blukacz et al. 2009). Turbulence profile collection should therefore be coupled with ADCP measurements to continuously measure zooplankton concentration. ADCPs with multiple beams, bottom-mounted in different lake locations, or surveys with a boat-mounted ADCP, allow understanding of vertical and horizontal variations in abundance in the migrating layer and during the DVM. Vertical distribution before DVM and also horizontal patchiness and temporal variation in zooplankton concentration in the migrating layer are relevant to the spatio-temporal dynamics of biomixing. These dynamics can only be observed in the natural environment.

Conclusions

In this paper, we presented an overview of existing studies of turbulence and mixing generated by small zooplankton in lakes. Lake research currently yields mixed conclusions about the feasibility of biomixing, generally showing that small zooplankton can generate turbulence but different levels of mixing depending on the type of study and on the zooplankton abundance. Field studies are needed to overcome limitations arising from lab studies and to confirm the importance of biomixing in complex natural environments such as lakes, and without altering the behavior of the animals generating the biomixing under study.

References

Benoit-Bird, K. J., M. A. Moline, O. M. Schofield, I. C. Robbins, and C. M. Waluk. 2010. Zooplankton avoidance of a profiled open-path fluorometer. J. Plankton Res. **32**: 1413–1419. doi:10.1093/plankt/fbq053

Blake, J. R. 1971. A spherical envelope approach to ciliary propulsion. J. Fluid Mech. **46**: 199. doi:10.1017/S002211207100048X

Blukacz, E. A., B. J. Shuter, and W. G. Sprules. 2009. Towards understanding the relationship between wind conditions and plankton patchiness. Limnol. Oceanogr. **54**: 1530–1540. doi:10.4319/lo.2009.54.5.1530

Bouffard, D., and L. Boegman. 2012. Basin-scale internal waves, p. 102–107. *In* L. Bengtsson, R. W. Herschy, and R. W. Fairbridge [eds.], Encyclopedia of lakes and reservoirs. Springer Netherlands.

Bouffard, D., and L. Boegman. 2013. A diapycnal diffusivity model for stratified environmental flows. Dyn. Atmos. Oceans **61–62**: 14–34. doi:10.1016/j.dynatmoce.2013.02.002

Brinton, E. 1967. Vertical migration and avoidance capability of euphausiids in the California Current. Limnol. Oceanogr. **12**: 451–483. doi:10.4319/lo.1967.12.3.0451

Buchanan, C., B. Goldberg and R. Mccartney. 1982. A laboratory method for studying zooplankton swimming behaviors. Hydrobiologia **89**: 77–89. doi:10.1007/BF00008635

De Robertis, A. 2002. Size-dependent visual predation risk and the timing of vertical migration: An optimization model. Limnol. Oceanogr. **47**: 925–933. doi:10.4319/lo.2002.47.4.0925

Dean, C., A. Soloviev, A. Hirons, T. Frank, and J. Wood. 2015. Biomixing due to diel vertical migrations of zooplankton: Comparison of computational fluid dynamics model with observations. Ocean Model. **98**: 51–64. doi:10.1016/j.ocemod.2015.12.002

Dodson, S. I., S. Ryan, R. Tollrian, and W. Lampert. 1997. Individual swimming behavior of Daphnia: Effects of food, light and container size in four clones. J. Plankton Res. **19**: 1537–1552. doi:10.1093/plankt/19.10.1537

Fischer, H. B., E. J. List, R. C. Y. Koh, J. Imberger, and N. H. Brooks. 1979. Mixing in inland and coastal waters, v. 114. Academic Press.

George, D. G., and D. P. Hewitt. 1999. The influence of year-to-year variations in winter weather on the dynamics of Daphnia and Eudiaptomus in Esthwaite Water, Cumbria. Funct. Ecol. **13**: 45–54. doi:10.1046/j.1365-2435.1999.00007.x

Goudsmit, G., F. Peeters, M. Gloor, and A. Wüest. 1997. Boundary versus internal diapycnal mixing in stratified natural waters. J. Geophys. Res. **102**: 27903–27914. doi:10.1029/97JC01861

Gregg, M. C., and J. K. Horne. 2009. Turbulence, acoustic backscatter, and pelagic nekton in Monterey Bay. J. Phys. Oceanogr. **39**: 1097–1114. doi:10.1175/2008JPO4033.1

Gries, T., K. Jöhnk, D. Fields, and J. R. Strickler. 1999. Size and structure of "footprints" produced by Daphnia: Impact of animal size and density gradients. J. Plankton Res. **21**: 509–523. doi:10.1093/plankt/21.3.509

Harris, R., P. Wiebe, J. Lenz, H. R. Skjoldal, and M. Huntley. 2000. Zooplankton methodology manual. Academic Press.

Hembre, L. K., and R. O. Megard. 2003. Seasonal and diel patchiness of a Daphnia population: An acoustic analysis. Limnol. Oceanogr. **48**: 2221–2233. doi:10.4319/lo.2003.48.6.2221

Hessen, D. O., and S. Kaartvedt. 2014. Top-down cascades in lakes and oceans: Different perspectives but same story? J. Plankton Res. **36**: 914–924. doi:10.1093/plankt/fbu040

Huber, A. M. R., F. Peeters, and A. Lorke. 2011. Active and passive vertical motion of zooplankton in a lake. Limnol. Oceanogr. **56**: 695–706. doi:10.4319/lo.2011.56.2.0695

Huntley, M. E., and M. Zhou. 2004. Influence of animals on turbulence in the sea. Mar. Ecol. Prog. Ser. **273**: 65–79. doi:10.3354/meps273065

Ivey, G. N., and J. Imberger. 1991. On the nature of turbulence in a stratified fluid. Part I: The energetics of mixing. J. Phys. Oceanogr. **21**: 650–658. doi:10.1175/1520-0485(1991)021 < 0650:OTNOTI>2.0.CO;2

Ivey, G. N., K. B. Winters, and J. R. Koseff. 2008. Density stratification, turbulence, but how much mixing? Annu. Rev. Fluid Mech. **40**: 169–184. doi:10.1146/annurev.fluid.39.050905.110314

Jiang, H., and T. Kiørboe. 2011. The fluid dynamics of swimming by jumping in copepods. J. R. Soc. Interface **8**: 1090–1103. doi:10.1098/rsif.2010.0481

Jonas, T., A. Stips, W. Eugster, and A. Wuest. 2003. Observations of a quasi shear-free lacustrine convective boundary layer: Stratification and its implications on turbulence. J. Geophys. Res. **108**: 1–15. doi:10.1029/2002JC001440

Katija, K. 2012. Biogenic inputs to ocean mixing. J. Exp. Biol. **215**(Pt 6): 1040–1049. doi:10.1242/jeb.059279

Kunze, E. 2011. Fluid mixing by swimming organisms in the low-Reynolds-number limit. J. Mar. Res. **69**: 591–601. doi:10.1357/002224011799849435

Kunze, E., J. F. Dower, I. Beveridge, R. Dewey, and K. P. Bartlett. 2006. Observations of biologically generated turbulence in a coastal inlet. Science **313**: 1768–1770. doi:10.1126/science.1129378

Leshansky, A. M., and L. M. Pismen. 2010. Do small swimmers mix the ocean? Phys. Rev. E **82**: 25301. doi:10.1103/PhysRevE.82.025301

Lighthill, M. J. 1952. On the squirming motion of nearly spherical defomable bodies through liquids at very small reynolds numbers. Commun. Pure Appl. Math. **5**: 109–118. doi:10.1002/cpa.3160050201

Loose, C. J., and P. Dawidowicz. 1994. Trade-offs in diel vertical migration by zooplankton: The costs of predator avoidance. Ecology **75**: 2255–2263. doi:10.2307/1940881

Lorke, A., D. F. Mcginnis, P. Spaak, and A. Wüest. 2004. Acoustic observations of zooplankton in lakes using a Doppler current profiler. Freshw. Biol. **49**: 1280–1292. doi:10.1111/j.1365-2427.2004.01267.x

Lorke, A., and W. N. Probst. 2010. In situ measurements of turbulence in fish shoals. Limnol. Oceanogr. **55**: 354–364. doi:10.4319/lo.2010.55.1.0354

MacIntyre, S., K. M. Flynn, R. Jellison, and J. R. Romero. 1999. Boundary mixing and nutrient fluxes in Mono Lake, California. Limnol. Oceanogr. **3**: 512–529. doi:10.4319/lo.1999.44.3.0512

Noss, C., and A. Lorke. 2012. Zooplankton induced currents and fluxes in stratified waters. Water Qual. Res. J. Can. **47**: 276. doi:10.2166/wqrjc.2012.135

Noss, C., and A. Lorke. 2014. Direct observation of biomixing by vertically migrating zooplankton. Limnol. Oceanogr. **59**: 724–732. doi:10.4319/lo.2014.59.3.0724

Osborn, T. R. 1980. Estimates of the local rate of vertical diffusion from dissipation measurements. J. Phys. Oceanogr. **10**: 83–89. doi:10.1175/1520-0485(1980)010<0083:EOTLRO>2.0.CO;2

Pujiana, K., J. N. Moum, W. D. Smyth, and S. J. Warner. 2015. Distinguishing ichthyogenic turbulence from geophysical turbulence. J. Geophys. Res. Oceans **120**: 3792–3804. doi:10.1002/2014JC010659

Ringelberg, J. 1999. The photobehaviour of Daphnia spp. as a model to explain diel vertical migration in zooplankton. Biol. Rev. Camb. Philos. Soc. **74**: 397–423. doi:10.1017/S0006323199005381

Ringelberg, J. 2010. Diel vertical migration of Zooplankton in Lakes and Oceans (Springer). Springer Netherlands.

Rinke, K., I. Hübner, T. Petzoldt, S. Rolinski, M. König-Rinke, J. Post, A. Lorke, and J. Benndorf. 2007. How internal waves influence the vertical distribution of zooplankton. Freshw. Biol. **52**: 137–144. doi:10.1111/j.1365-2427.2006.01687.x

Rippeth, T. P., J. C. Gascoigne, J. A. M. Green, M. E. Inall, M. R. Palmer, J. H. Simpson, and P. J. Wiles. 2007. Turbulent Dissipation of Coastal Seas, a response to "Observations of Biologically Generated Turbulence in a Coastal Inlet." Science Electronic Letters.

Ross, T. 2014. A video-plankton and microstructure profiler for the exploration of in situ connections between zooplankton and turbulence. Deep-Sea Res. Part I **89**: 1–10. doi:10.1016/j.dsr.2014.04.003

Rousseau, S., E. Kunze, R. Dewey, K. Bartlett, and J. Dower. 2010. On turbulence production by swimming marine organisms in the open ocean and coastal waters. J. Phys. Oceanogr. **40**: 2107–2121. doi:10.1175/2010JPO4415.1

Straile, D., and R. Adrian. 2000. The North Atlantic Oscillation and plankton dynamics in two European lakes-two variations on a general theme. Glob. Chang. Biol. **6**: 663–670. doi:10.1046/j.1365-2486.2000.00350.x

Subramanian, G. 2010. Viscosity-enhanced bio-mixing of the oceans. Curr. Sci. **98**: 1103–1108.

Talling, J. F. 2003. Phytoplankton-zooplankton seasonal timing and the "clear-water phase" in some English lakes. Freshw. Biol. **48**: 39–52. doi:10.1046/j.1365-2427.2003.00968.x

Tanaka, M., T. Nagai, T. Okada, and H. Yamazaki. 2017. Measurement of sardine-generated turbulence in a large tank. Mar. Ecol. Prog. Ser. **571**: 207–220. doi.org/10.3354/meps12098 doi:10.3354/meps12098

Thackeray, S. J., D. G. George, R. I., Jones, and I. J. Winfield. 2004. Quantitative analysis of the importance of wind-induced circulation for the spatial structuring of planktonic populations. Freshw. Biol. **49**: 1091–1102. doi:10.1111/j.1365-2427.2004.01252.x

Thiffeault, J. L., and S. Childress. 2010. Stirring by swimming bodies. Phys. Lett. A **374**: 3487–3490. doi:10.1016/j.physleta.2010.06.043

Thorpe, S. A. 2005. The Turbulent Ocean. Cambridge University Press.

UNESCO. 1968. Zooplankton sampling. UNESCO.

Visser, A. W. 2007*a*. Biomixing of the Oceans? Science (New York, N.Y.) **316**: 838–839. doi:10.1126/science.1141272

Visser, A. W. 2007*b*. Visser's response to Kunze (2006). Science **318**: 1239–1239. doi:10.1126/science.318.5854.1239b

Wagner, G. L., W. R. Young, and E. Lauga. 2014. Mixing by microorganisms in stratified fluids. J. Mar. Res. **72**: 47–72. doi:10.1357/002224014813758940

Wain, D., M. Kohn, J. Scanlon, and C. Rehmann. 2013. Internal wave driven transport of fluid away from the boundary of a lake. Limnol. Oceanogr. **58**: 429–442. doi:10.4319/lo.2013.58.2.0429

Wang, S., and A. M. Ardekani. 2015. Biogenic mixing induced by intermediate Reynolds number swimming in stratified fluids. Sci. Rep. **5**: 17448. doi:10.1038/srep17448

Waya, R. 2004. Diel vertical migration of zooplankton in the Tanzanian waters of Lake Victoria. Tanzan. J. Sci. **30**: 123–124. doi:10.4314/tjs.v30i1.18394

Wickramarathna, L. N. 2016. Kinematics and energetics of swimming zooplankton. Universität Koblenz-Landau.

Wickramarathna, L. N., C. Noss, and A. Lorke. 2014. Hydrodynamic trails produced by Daphnia: Size and energetics. PloS One **9**. doi:10.1371/journal.pone.0092383

Wilhelmus, M. M., and J. O. Dabiri. 2014. Observations of large-scale fluid transport by laser-guided plankton aggregations. Phys. Fluids **26**: 101302. doi:10.1063/1.4895655

Williamson, C. E., J. M. Fischer, S. M. Bollens, E. P. Overholt, and J. K. Breckenridge, 2011. Towards a more comprehensive theory of zooplankton diel vertical migration: Integrating ultraviolet radiation and water transparency into the biotic paradigm. Limnol. Oceanogr. **56**: 1603–1623. doi:10.4319/lo.2011.56.5.1603

Wüest, A., D. C. Van Senden, J. Imberger, G. Piepke, M. Gloor, A. Wijest, G. Piepke, and M. Gloor. 1996. Comparison of diapycnal diffusivity measured by tracer and microstructure techniques. Dyn. Atmos. Oceans **24**: 27–39. doi:10.1016/0377-0265(95)00408-4

Wüest, A., and A. Lorke. 2003. Small scale hydrodynamics in lakes. Annu. Rev. Fluid Mech. **35**: 373–412. doi:10.1146/annurev.fluid.35.101101.161220

Acknowledgments

We thank the Editors and two anonymous reviewers for their valuable feedback and suggestions on the manuscript. Funding for this work was provided by a UK Royal Society Research Grant (Y0106WAIN) and an EU Marie Curie Career Integration Grant (PCIG14-GA-2013-630917) awarded to D. J. Wain.

16

Preliminary estimates of the contribution of Arctic nitrogen fixation to the global nitrogen budget

Rachel E. Sipler,[1]* *Donglai Gong*,[1] *Steven E. Baer*,[1,2] *Marta P. Sanderson*,[1] *Quinn N. Roberts*,[1] *Margaret R. Mulholland*,[3] *Deborah A. Bronk*[1]

[1]The Virginia Institute of Marine Science, College of William & Mary, Gloucester Point, Virginia; [2]Bigelow Laboratory for Ocean Sciences, East Boothbay, Maine; [3]Department of Ocean, Earth and Atmospheric Sciences, Old Dominion University, Norfolk, Virginia

Scientific Significance Statement

Nitrogen fixation is an important source of "new" nitrogen to both freshwater and marine ecosystems where it supports biological production and balances aquatic nitrogen budgets. Although freshwater nitrogen fixation is relatively common at high latitudes, water column marine nitrogen fixation has historically been considered to be a mostly warm water, oligotrophic process. Here, we provide evidence to show that marine nitrogen fixation could be an important source of nitrogen to the seasonally nitrogen-limited Arctic Ocean, which could have far reaching implications for primary productivity and biogeochemical budgets in this rapidly changing ecosystem.

Abstract

Dinitrogen (N_2) fixation is the source of all biologically available nitrogen on earth, and its presence or absence impacts net primary production and global biogeochemical cycles. Here, we report rates of 3.5–17.2 nmol N L^{-1} d^{-1} in the ice-free coastal Alaskan Arctic to show that N_2 fixation in the Arctic Ocean may be an important source of nitrogen to a seasonally nitrogen-limited system. If widespread in surface waters over ice-free shelves throughout the Arctic, N_2 fixation could contribute up to 3.5 Tg N yr^{-1} to the Arctic nitrogen budget. At these rates, N_2 fixation occurring in ice-free summer waters would offset up to 27.1% of the Arctic denitrification deficit and contribute an additional 2.7% to N_2 fixation globally, making it an important consideration in the current debate of whether nitrogen in the global ocean is in steady state. Additional investigations of high-latitude marine diazotrophic physiology are required to refine these N_2 fixation estimates.

Net primary productivity in the Arctic Ocean is projected to increase in response to higher temperatures and consequent ice melt (Arrigo et al. 2012). Any increase in productivity, however, will require sufficient nitrogen to support it (Tremblay and Gagnon 2009). While several studies have speculated on changes to the Arctic nitrogen cycle and resultant effects on the primary producers that form the base of the food chain (Tremblay et al. 2008; Popova et al. 2012), the role of dinitrogen (N_2) fixation in the Arctic Ocean has received comparatively little attention.

Historically, marine N_2 fixation has been considered a warm-water process that occurs mainly in subtropical oligotrophic gyres and tropical seas, where it provides a new source of nitrogen to the system (Sohm et al. 2011). Recent work has shown, however, that N_2 fixers (Moisander et al.

*Correspondence: rachelsipler@gmail.com

Author Contribution Statement: DAB and RES designed the experi-ment. RES and MPS performed the experiments. RES, DAB, MPS, SEB, QNR, and MRM analyzed the data. DG determined the area and dura-tion of ice-free regions. RES and DG calculated the potential contributions of N_2 fixation to local and global nitrogen budgets. RES and DAB wrote the paper with input from all authors.

Table 1. Site location and physical characteristics.

Station	Longitude (N)	Latitude (W)	Water column depth (m)	Sample collection depth (m)	Time started (local)	Temperature (°C)	Salinity
Marine							
1	71°20′40″	156°41′25″	17.1	4	19:10	5.0	30.5
2	71°20′40″	156°41′25″	17.5	8	15:16	4.7	30.4
3	71°20′36″	156°39′30″	8.0	2	20:38	4.6	30.4
4	71°20′40″	156°41′25″	17.5	8	23:57	4.5	30.4
8	71°19′02″	156°50′18″	ND	8	10:30	ND	30.4
Estuarine							
5	71°7′8″	156°54′24″	ND	Surface	15:53	6.7	12.0
6	71°7′45″	157°1′10″	ND	Surface	16:04	6.2	15.0
7	71°8′45″	157°3′45″	ND	Surface	16:18	6.0	20.1

Marine stations were located in the coastal Chukchi Sea (2.5 km northwest of Barrow, Alaska) and estuarine stations were in Walakpa Bay (20 km southwest of Barrow, Alaska). The local time of day when incubations was started is presented based on as 24-h clock.

2010; Blais et al. 2012; Díez et al. 2012; Fernández-Méndez et al. 2016) and N_2 fixation (Blais et al. 2012; Mulholland et al. 2012) can occur at lower temperatures and higher latitudes than once thought and in areas where nitrogen is not limiting (Hamersley et al. 2011; Sohm et al. 2011) shifting the paradigm of where marine N_2 fixation occurs.

The *nifH* gene encodes for the nitrogenase enzymes, which is responsible for N_2 fixation processes. Targeted *nifH* gene and broad metagenomic assessments have revealed the genetic capability for N_2 fixation in a variety of polar microbial communities. For example, *nifH* or species known to possess it have been detected in sea ice (e.g., Bowman et al. 2014; Fernández-Méndez et al. 2016), polar lakes (e.g., Toetz 1961; Shtarkman et al. 2013), the Arctic Ocean (e.g., Blais et al. 2012; Díez et al. 2012; Fernández-Méndez et al. 2016), glaciers (e.g., Telling et al. 2011), and polar soils (e.g., Solheim et al. 1996; Dickson 2000). These investigations provide compelling evidence that diverse polar microbial communities have the genes required to fix nitrogen. Fewer studies provide actual rates of marine N_2 fixation that can be used to estimate the amount of nitrogen contributed to global nitrogen budgets (e.g., Knowles and Wishart 1977; Haines et al. 1981; Gihring et al. 2010; Blais et al. 2012).

The goal of this study was to determine the rate of N_2 fixation in the surface mixed layer of the coastal Alaskan Arctic and to estimate how much nitrogen this process could contribute to local and global nitrogen budgets. These objectives were achieved using ^{15}N tracer techniques to measure rates of N_2 fixation during the summer under ice-free conditions at marine and estuarine sites in the Chukchi Sea, near Barrow, Alaska. The data presented here, and reported in Blais et al. (2012), were extrapolated to provide rough estimates of the potential range of fixed nitrogen contributed by Arctic marine N_2 fixation.

Methods

N_2 fixation

Field samples were collected under ice-free conditions 15–20 August 2011, in the coastal waters of the Chukchi Sea (salinity of 30) and Walakpa Bay (salinity of 12–20) near Barrow, Alaska. Rates of N_2 fixation were measured using the ^{15}N bubble addition method (Montoya et al. 1996). Gas-tight 1L glass KIMAX™ media bottles (model # 611001000) capped with Wheaton™ black open-top caps with gray butyl septa (model # 240680) were used for all incubations. The glass media bottles were acid washed (10% HCl), rinsed four times with high purity water (18.2 MΩ cm^{-1}) and combusted at 500°C for 4 h. The caps were submersed in a saltwater brine (approximately salinity of 60) for ~ 30 d, to condition them, and then acid washed (10% HCl) and rinsed with copious amounts of high purity water.

Seawater was collected using a low-pressure submersible electric pump (Johnson Pump model #16004) powered by a portable generator. Triplicate bottles were filled with seawater, capped and the ambient air bubbles were removed. The bottles were then amended with 1.2 mL (at a ratio of 1 mL of gas per 1L of seawater) of enriched (> 99%) $^{15}N_2$ gas purchased from Cambridge Isotope Laboratories (lot #11-10077). Samples were incubated for 24 h in environmental chambers located at the Barrow Arctic Research Center. To mimic in situ conditions, the temperature was set at 4.5°C and light levels to 50 μmol m^{-2} s^{-1} of light on a 24 h light cycle. No control incubation was done without added tracer gas. Sample locations, collection depths, and environmental conditions are presented in Table 1.

After 24 h, the incubations were terminated by filtering through 3.0 μm silver filters or pre-combusted (450°C for 2 h) GF/F filters with a nominal pore size of 0.7 μm. Filters were placed in sterile microcentrifuge tubes and stored frozen at −20°C. Prior to analysis, filters were thawed and dried at 40°C overnight. Based on cell counts by flow cytometry, approximately 61% of the bacterial cells were retained by the GF/F (described in more detail in Baer et al. 2017). Isotopic measurements for ^{15}N fixation rates were analyzed on a Europa GEO 20/20 mass spectrometer with an ANCA-SL autosampler. The atom % enrichments for all N_2 fixation samples ranged from 0.3691 to 0.4463 and sample mass ranged from 9.25 μg N to 37.25 μg N. N_2 fixation rates were calculated using a mixing model (Montoya et al. 1996) and are reported as the mean ± the standard error for each site ($n = 3$) or grouped as marine ($n = 15$, five sites) or estuarine ($n = 9$, three sites) systems. The atom % enrichment of the particulate samples was measured against an atmospheric standard. They were not corrected for any variations in the natural abundance of the particulate material at the start of the incubation. This is a potential issue because particulate nitrogen has been shown to be enriched in ^{15}N in some regions of the coastal Arctic; in the summer of 2012, $\delta^{15}N$ values of 6.66‰ and 7.78‰ were reported at coastal sites in the Chukchi to the west of the sample sites from this study (Yu et al. 2014). These $\delta^{15}N$ values would not negate our observations.

We note that the bubble method has been shown to underestimate N_2 fixation based on gas solubility and may have biases against smaller symbiotic diazotrophs (Mohr et al. 2010; Großkopf et al. 2012). However, the overall magnitude of this underestimation is dependent upon the duration of the incubation and the temperature of the water in which the bubble is dissolved (Mohr et al. 2010). Culture experiments show that 75% of $^{15}N_2$ gas had reached equilibrium after 24 h at considerably higher temperatures (28°C) using the bubble method, and result in only a 15% underestimation in N_2 fixation rates (Mohr et al. 2010). Unlike the bubble method, the direct injection method (Mohr et al. 2010) requires a 10% dilution of the microbial community with filtered $^{15}N_2$ supersaturated seawater. Hypothesizing that fixation rates and diazotrophic abundance in the Arctic samples would be low, the bubble method was used to ensure that we did not dilute the community and therefore the rates of N_2 fixation. There have been no studies that report the bubble method overestimating N_2 fixation, therefore, the values reported here should be considered conservative estimates of N_2 fixation rates for the coastal Chukchi Sea.

We also note that contaminants ($^{15}NH_4^+$, $^{15}NO_3^-/NO_2^-$) have been measured in some $^{15}N_2$ gas stocks (Dabundo et al. 2014). If contaminated $^{15}N_2$ is used, measured N_2 fixation rates may be inflated due to the uptake of the contaminant and not fixation of the added $^{15}N_2$ gas. Dabundo et al. (2014) compared the level of ^{15}N gas contamination from three commonly used manufacturers of $^{15}N_2$. The $^{15}N_2$ gas used in our study was purchased from Cambridge Isotope Laboratories, which had the lowest level of contamination of all sources tested. The highest levels of contamination reported for Cambridge Isotope Laboratories, $^{15}N_2$ gas were not substantial enough to negate the rates we observed in the Arctic and would have accounted for a negligible proportion ($\leq$ 0.3%) of the rates observed.

Nutrients

Ammonium concentrations were analyzed using the colorimetric phenol-hypochlorite method (Koroleff 1983). Concentrations of nitrate, nitrite, and phosphate were measured using a Lachat QuikChem 8500 autoanalyzer (Parsons et al. 1984). Nutrient data are reported as the mean ± the standard error of triplicate samples (Table 2).

Spatial and temporal calculations

A first order regional and pan-Arctic estimate of the amount of nitrogen that may be released via marine N_2 fixation was calculated using the total volume in the upper 50 m of the ice-free water column, integrated over the summer season (June–September), then multiplying that volume by the range of N_2 fixation rates observed in this and the Blais et al. (2012) studies. The extrapolation of the total nitrogen fixed under ice-free conditions between June and September is done for three domains: western Arctic shelves, pan Arctic continental shelves, and the entire Arctic Ocean. The western Arctic shelf is specified because the N_2 fixation rates reported here and in Blais et al. (2012), encompass the area between Baffin Bay on the northeastern boundary and the Chukchi Sea on the northwestern boundary (Fig. 1). For the purpose of this study, the western Arctic shelf was defined as bodies of water in the longitudinal range between 175°E and 45°W with water column depths shallower than 200 m and deeper than 1 m. Generally, this covers the Chukchi Sea, Beaufort shelves, Canadian Arctic Archipelago, the shelves surrounding Baffin Bay, and northern portion of the Labrador Sea. The Arctic continental shelves domain was defined as bodies of water that are not rivers or lakes north of latitude 64°N and with a water column depth shallower than 200 m and deeper than 1 m. The Arctic Ocean domain was defined as bodies of water that are not rivers or lakes north of latitude 64°N and with a water column depth of 1 m or greater. The 1 m shallow depth cut off was implemented to eliminate regions that are strongly wave-influenced or very shallow tidal systems as measurements used in this study were not made in those types of environments.

For the volumetric calculations of seasonally ice-free waters, we used the 30 arcsec resolution IBCAO3 bathymetric data (Jakobsson et al. 2012) and Multisensor Analyzed Sea Ice Extent (MASIE) Northern Hemisphere data published by the National Snow and Ice Data Center (Fetterer et al. 2010). The MASIE sea ice cover data has a constant horizontal resolution of 4 km, with each grid covering an area of 16 km^2. The climatological open water extent is calculated using MASIE data from 2006 to 2016.

Table 2. Chemical and N_2 fixation rate data.

Station	NH_4^+ (μmol N L^{-1})	NO_3^- (μmol N L^{-1})	PO_4^- (μmol P L^{-1})	N_2 fixation rate (nmol N L^{-1} d^{-1}) Whole	> 3 μm	> 3 μm: Whole (%)
Marine						
1	1.55 ± 0.07	0.45 ± 0.00	0.55 ± 0.01	5.7 ± 1.7	1.6 ± 0.3	34
2	0.59 ± 0.00	0.32 ± 0.00	0.51 ± 0.00	17.2 ± 7.0	12.6 ± 5.0	84
3	0.47 ± 0.02	0.33 ± 0.00	0.47 ± 0.00	3.5 ± 0.2	1.1 ± 0.2	31
4	0.59 ± 0.03	0.29 ± 0.00	0.55 ± 0.00	4.4 ± 1.2	1.7 ± 0.5	40
8	0.22 ± 0.00	0.30 ± 0.00	0.39 ± 0.00	7.7 ± 2.1	1.9 ± 0.4	25
Average	***0.68 ± 0.12***	***0.34 ± 0.02***	***0.49 ± 0.01***	***7.7 ± 1.8***	***3.8 ± 1.5***	***43***
Estuarine						
5	0.14 ± 0.01	0.22 ± 0.01	0.17 ± 0.00	4.9 ± 0.4	2.6 ± 0.5	52
6	0.11 ± 0.00	0.59 ± 0.00	0.17 ± 0.01	5.4 ± 0.3	5.1 ± 0.6	96
7	0.10 ± 0.00	0.21 ± 0.00	0.45 ± 0.00	5.6 ± 0.6	3.1 ± 0.2	57
Average	***0.12 ± 0.01***	***0.34 ± 0.06***	***0.22 ± 0.05***	***5.3 ± 0.2***	***3.6 ± 0.4***	***68***

Station numbers correspond to site locations described in Table 1. Samples were collected 15–20 August 2011. Data represent the mean ± the standard error of triplicate samples.

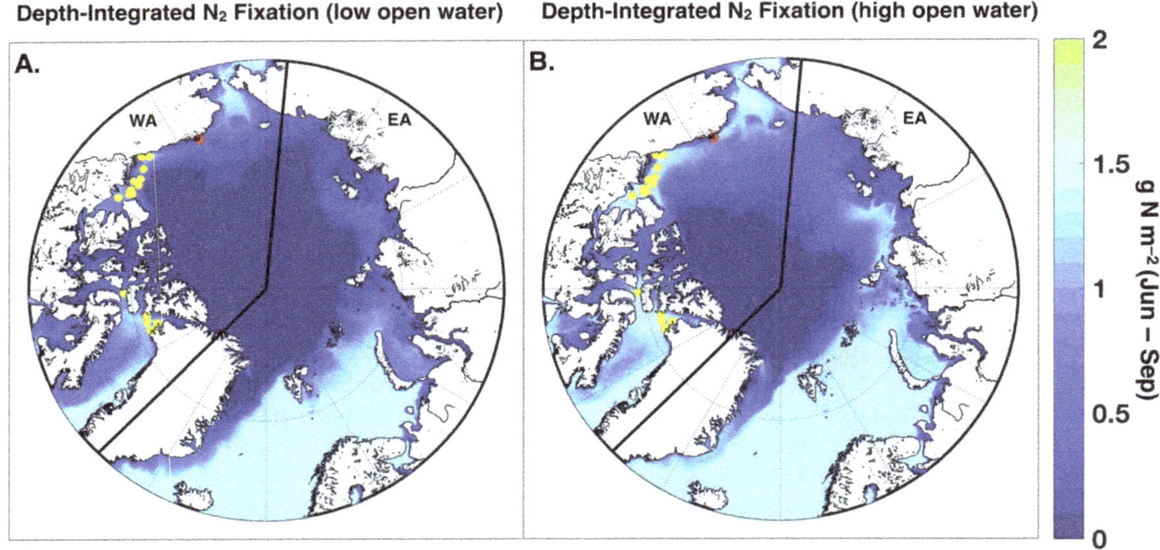

Fig. 1. Summer depth-integrated contribution in the Arctic Ocean under low and high open water conditions. The maps depict the time averaged, depth-integrated nitrogen fixed per unit area during the summer months in the Arctic Ocean. The calculation is based on the number of open water days for each location in the Arctic Ocean and the integration is down to a maximum depth of 50 m which includes both the upper mixed layer as well as the seasonal pynocline. The maps depict the amount of nitrogen contributed to the Arctic Ocean between June and September under (A) low (1 standard deviation below the mean condition) and (B) high (1 standard deviation above the mean condition) open water scenarios. The MASIE ice extent data used in the calculation is from June to September (2006–2016). A solid black line bisecting the maps delineates the Western Arctic (WA) longitudinal range used in this study. The remaining Eastern Arctic sector is identified as EA. Sites listed in Table 1 are depicted by red circles and sites sampled by Blais et al. (2012) are depicted by yellow circles. These points provide the spatial range of sites with measured N_2-fixation rates and the data used in the extrapolations presented in Table 3.

For each of the three domains that we described above (western Arctic shelves, pan-Arctic shelves, and whole Arctic Ocean), the total amount of nitrogen fixed each year was calculated by integrating, in space and time, the measured N_2 fixation rates over the water column depth down to a maximum depth of 50 m for ice-free days from 01 June to 30 September. These restrictions in time and depth are based on the conditions under which the observations were made (Blais et al. 2012; this study) and our assumptions about the depth range over which the measurements may be valid. All available measurements were collected under ice-free conditions, within the seasonal surface mixed layer. For the purpose of this extrapolation, we have assumed that rates are constant above 50 m depth because the goal was to determine the maximum contribution potential. Admittedly this may be a source of error if the profile of N_2 fixation rates is a function of depth. Recent *nifH* studies indicate that the majority of Arctic marine diazotrophs are heterotrophic (~92%; Fernández-Méndez et al. 2016) and thus may not be directly dependent on light for energy. More studies are required to understand the physiological limitations of Arctic pelagic diazotrophs.

To estimate the potential range of uncertainty introduced by variability in seasonal sea ice cover from year to year, we also calculated the standard deviation of the extent of the seasonal open water area from 2006 to 2016. The standard deviation of the sea ice cover map is then added and subtracted from the mean sea ice cover map, then integrated with the measured N_2 fixation rates to obtain estimates of total fixed nitrogen for both "high" and "low" open water years. Maps of the high (mean + standard deviation) and low (mean − standard deviation) open water days for the entire Arctic are shown in Fig. 1.

Table 3. Contribution of marine N_2 fixation to local and global nitrogen budgets.

	Ice-free shelves western Arctic	Ice-free shelves entire Arctic
$Tg\ N\ yr^{-1}$	$\leq 1.2 \pm 0.3$	$\leq 3.5 \pm 0.7$
% Annual global N_2 fixation	≤ 0.9	≤ 2.7
% Arctic denitrification deficit	0.01–9.3	0.02–27.1

Amount of nitrogen contributed by Arctic N_2 fixation and its potential proportional contributions to annual global N_2 fixation (estimated at 130 $Tg\ N\ yr^{-1}$; Eugster and Gruber 2012) and the Arctic denitrification deficit (estimated at 13 $Tg\ N\ yr^{-1}$; Chang and Devol 2009) based on two scenarios from June to September: (1) N_2 fixation limited to ice-free shelves in the western Arctic Ocean (1.74 million km^2) and (2) N_2 fixation present on all ice-free shelves in the entire Arctic Ocean (4.91 million km^2). The Arctic Ocean is defined as bodies of water that are not rivers or lakes north of 64°N with a water column depth of 1 m or greater. Continental shelves are defined as bodies of water that are not rivers or lakes between 1 m and 200 m in depth; the western Arctic shelf is defined in the longitude range between 175°E and 45°W with water between 1 m and 200 m in depth. The total integrated depth was 50 m or less depending on total water column depth. The ranges reported here are based on the range of marine N_2 fixation rates (0.01–17.15 $nmol\ N\ L^{-1}\ d^{-1}$) reported by Blais et al. (2012) and this study.

Results and discussion

Active N_2 fixation was found in the coastal Chukchi Sea near Barrow, Alaska with average rates of 7.7 ± 1.8 $nmol\ N\ L^{-1}\ d^{-1}$ and 5.3 ± 0.2 $nmol\ N\ L^{-1}\ d^{-1}$ at marine and estuarine sites, respectively (Table 2). A comparison of the > 3 μm and whole water (> 0.7 μm) size fractions revealed that approximately 43% of marine and 68% estuarine diazotrophs were < 3 μm in size (Table 2). The rates measured in the Chukchi Sea are higher than rates (0.01 ± 0.01 $nmol\ N\ L^{-1}\ d^{-1}$ to 4.45 ± 0.23 $nmol\ N\ L^{-1}\ d^{-1}$) measured in the Canadian Arctic (Blais et al. 2012). The N_2 fixation rates reported here and in Blais et al. (2012), encompass the area between Baffin Bay on the northeastern boundary and the Chukchi Sea on the northwestern boundary. Assuming uniform N_2 fixation rates in the upper 50 m of the water column, representing the surface mixed layer and the seasonal pycnocline, we estimate that N_2 fixation on the western Arctic shelves can provide 0.01–0.89 $mmol\ N\ m^{-2}\ d^{-1}$, which would contribute up to 1.2 $Tg\ N\ yr^{-1}$ (Table 3). However, most Arctic coastal regions fall within the comparatively wide ranges of temperature (−1.2°C to 19.4°C), salinity (4.4–32.4), and nutrient concentrations (< 0.03–6.92 μmol nitrate L^{-1} and 0.17–1.25 μmol phosphate L^{-1}) where N_2 fixation was observed in the western Arctic (Blais et al. 2012; this study), which suggests that like the conditions under which it was found, N_2 fixation may be more widespread in the Arctic Ocean. Indeed recent investigations using *nifH* gene sequencing found a diverse array of diazotrophs throughout the central Arctic and Eurasian basin (Fernández-Méndez et al. 2016).

The Arctic Ocean covers 20% of the continental shelf area in the global ocean. If N_2 fixation occurs across all Arctic shelves, the total amount of N_2 fixed would increase to 3.5 $Tg\ N\ yr^{-1}$ or 2.7% of global N_2 fixation, based on the current estimate of global marine N_2 fixation of 130 $Tg\ N\ yr^{-1}$ (Table 3; Eugster and Gruber 2012). Taking this analysis to the extreme, if N_2 fixation is widespread across the entire ice-free Arctic, Arctic marine N_2 fixation could contribute as much as 9.2 $Tg\ N\ yr^{-1}$ or 7.1% of global marine N_2 fixation in the ice-free summer months.

We note that the estimates described here are based on *net* N_2 fixation rates. Numerous studies have shown that some diazotrophs release upwards of 50% of recently fixed nitrogen as ammonium or dissolved organic nitrogen (e.g., Glibert and Bronk 1994; Benavides et al. 2013). Therefore, the amount of nitrogen released into the ice-free Arctic Ocean is likely more than our reported rates imply.

We also estimated the variability introduced by variation in annual sea ice cover during the summer months. For the entire Arctic Ocean, using the upper range of the observed N_2 fixation rates (17.2 $nmol\ N\ L^{-1}\ d^{-1}$), the 1 sigma anomaly in total fixed N_2 is 1.7 $Tg\ N\ yr^{-1}$ or 18% of the estimate based on average summer sea ice cover. For just the Arctic shelf regions, the 1 sigma anomaly is 0.71 $Tg\ N\ yr^{-1}$ or 21% of the average, and for the western Arctic only shelf regions, the 1 sigma anomaly is 0.26 $Tg\ N\ yr^{-1}$ or 22% of the average.

The presence of active N_2 fixation in the Arctic is relevant to outstanding questions of whether the nitrogen budget of the Arctic is balanced. Although some estimates suggest that the annual Arctic nitrogen budget is balanced (Torres-Valdés et al. 2013), measured rates of denitrification on the Arctic shelves are high and produce an estimated nitrogen deficit of 13 Tg N yr^{-1} (Chang and Devol 2009), with indirect rate calculations generating higher estimates (Mills et al. 2015). This implies that there is a currently unquantified source of nitrogen to the Arctic Ocean. Although some studies have concluded that N_2 fixation is too small to offset any appreciable portion of denitrification (Blais et al. 2012; Mills et al. 2015), our data suggests otherwise. Comparing the estimated denitrification deficit with our N_2 fixation estimates described above, N_2 fixation within the ice-free western Arctic would offset up to 9.3% of the estimated 13 Tg N yr^{-1} Arctic denitrification deficit (Chang and Devol 2009). This increases to 27.1% if N_2 fixation is occurring across all ice-free Arctic shelves. Again, taking the extrapolation to the extreme, 70.7% of the Arctic denitrification deficit could be offset by nitrogen fixation if it is active across the entire seasonally ice-free Arctic Ocean (Table 3).

If we consider just the Chukchi shelf, N_2 fixation could offset 40% of the denitrification based on direct measurements of denitrification (average measured rate of 0.96 mmol N m^{-2} d^{-1}; Devol et al. 1997; Chang and Devol 2009) with the average N_2 fixation rates observed at our marine sites (0.39 ± 0.09 mmol N m^{-2} d^{-1}). Although we constrain the depth used for calculations to 50 m based on current measurements (Blais et al. 2012; this study), which are limited to the surface mixed layer, N_2 fixation has now been found at sub-euphotic depths (Hamersley et al. 2011; Jayakumar et al. 2012; Rahav et al. 2013). Sub-euphotic N_2 fixation would greatly increase the potential contribution of N_2 fixation to the Arctic as would sea-ice and under-ice N_2 fixation (Fernández-Méndez et al. 2016). We have also not included N_2 fixation in Arctic marine sediments in our estimates, which would add a small (0.001–0.02 mmol N m^{-2} d^{-1}) additional source of fixed nitrogen (Knowles and Wishart 1977; Haines et al. 1981; Gihring et al. 2010). These additional sources would further increase the Arctic's potential as an important source of newly fixed nitrogen and underscore the need for more comprehensive studies investigating the rate of N_2 fixation throughout the Arctic Ocean over broader spatial and temporal scales.

From a global change perspective, the most pressing question is whether N_2 fixation has always occurred in the Arctic or whether it is an emerging phenomenon. The Chukchi Sea has warmed by 0.6°C and the ice-free period has lengthened by 11 d over the last decade (Stroeve et al. 2014). Earlier assumptions that N_2 fixation was restricted to warm and nutrient impoverished euphotic waters have limited water-column N_2 fixation studies in high-latitude marine systems. However, more than 50 yr ago N_2 fixation rate measurements were made in the lakes, ponds, and coastal ocean around Barrow, Alaska. Though active N_2 fixation was observed in inland waters, no N_2 fixation was detected at any of the estuarine or marine sites (Toetz 1961), including the locations where we found active N_2 fixation in this study. The Toetz (1961) study followed ^{15}N tracer methods described by Dugdale et al. (1959), which differed from the method used in our study. Briefly, the Toetz study aerated samples with an 80 : 20 helium : oxygen mixture under a reduced pressure of 0.8 atm. $^{15}N_2$ gas (95% enriched) was then added to bring the pressure within the sample back up to 1 atm. Any sample with an atom % enrichment higher than 0.386 indicated significant N_2 fixation. With such limited data and a difference in methodological approaches, we cannot say definitively that N_2 fixation is new to the Arctic. However, long-term increases in N_2 fixation have been correlated to changes in climate (Sherwood et al. 2014), and the trend towards greater nitrogen limitation in the region makes N_2 fixation more likely.

If N_2 fixation is increasing in the Arctic, cascading changes in the current nitrogen, carbon, and phosphorus cycles may be significant. Current and projected estimates of carbon sequestration in the Arctic are based on the observations that nitrogen, typically nitrate, limits primary production in this region (Tremblay and Gagnon 2009; Popova et al. 2012). As a source of new nitrogen, N_2 fixation could enhance sequestration of carbon dioxide and should be considered in future estimates. Additionally, the Arctic Ocean is a source of phosphorus to the North Atlantic where it supports primary production and N_2 fixation (Yamamoto-Kawai et al. 2006; Torres-Valdés et al. 2013). If Arctic N_2 fixation is providing an additional source of nitrogen, diazotrophy, and the new biomass that it supports could draw down dissolved phosphorus within the Arctic and ultimately reduce the amount of phosphorus reaching the North Atlantic, potentially altering productivity within that region. Increases in Arctic N_2 fixation could thus profoundly affect areal patterns of productivity within both of these oceans. Decreases in soluble phosphorus, in the absence of nitrate, have been observed in regions of the Canadian Arctic (Simpson et al. 2008; Tremblay et al. 2008), providing evidence that this may already be occurring. Clearly, more studies of N_2 fixation in Arctic waters are needed. The extrapolations provided in this study highlight the potential contributions of Arctic N_2 fixation and the need to better define its current contribution and to determine how this may change in the future.

References

Arrigo, K. R., and others. 2012. Massive phytoplankton blooms under Arctic sea ice. Science **336**: 1408. doi: 10.1126/science.1215065

Baer, S. E., R. E. Sipler, Q. N. Roberts, P. L. Yager, M. E. Frischer, and D. A. Bronk. 2017. Seasonal nitrogen uptake and regeneration in the western coastal Arctic. Limnol. Oceanogr. doi:10.1002/lno.10580

Benavides, M., D. A. Bronk, N. S. Agawin, M. D. Pérez-Hernández, A. Hernández-Guerra, and J. Arístegui. 2013. Longitudinal variability of size-fractionated N2 fixation and DON release rates along 24.5° N in the subtropical North Atlantic. J. Geophys. Res. Oceans **118**: 3406–3415. doi:10.1002/jgrc.20253

Blais, M., J. É. Tremblay, A. D. Jungblut, J. Gagnon, J. Martin, M. Thaler, and C. Lovejoy. 2012. Nitrogen fixation and identification of potential diazotrophs in the Canadian Arctic. Global Biogeochem. Cycles **26**: GB3022. doi:10.1029/2011GB004096

Bowman, J. S., C. T. Berthiaume, E. V. Armbrust, and J. W. Deming. 2014. The genetic potential for key biogeochemical processes in Arctic frost flowers and young sea ice revealed by metagenomic analysis. FEMS Microbiol. Ecol. **89**: 376–387. doi:10.1111/1574-6941.12331

Chang, B. X., and A. H. Devol. 2009. Seasonal and spatial patterns of sedimentary denitrification rates in the Chukchi Sea. Deep-Sea Res. Part II Top. Stud. Oceanogr. **56**: 1339–1350. doi:10.1016/j.dsr2.2008.10.024

Dabundo, R., M. F. Lehmann, L. Treibergs, C. R. Tobias, M. A. Altabet, P. H. Moisander, and J. Granger. 2014. The contamination of commercial 15N2 gas stocks with 15N–labeled nitrate and ammonium and consequences for nitrogen fixation measurements. PLoS One **9**: e110335. doi:10.1371/journal.pone.0110335

Devol, A. H., L. A. Codispoti, and J. P. Christensen. 1997. Summer and winter denitrification rates in western Arctic shelf sediments. Cont. Shelf Res. **17**: 1029–1050. doi: 10.1016/S0278-4343(97)00003-4

Dickson, L. G. 2000. Constraints to nitrogen fixation by cryptogamic crusts in a polar desert ecosystem, Devon Island, NWT, Canada. Arct. Antarct. Alp. Res. **32**: 40–45. doi:10.2307/1552408

Díez, B., B. Bergman, C. Pedrós-Alió, M. Antó, and P. Snoeijs. 2012. High cyanobacterial nifH gene diversity in Arctic seawater and sea ice brine. Environ. Microbiol. Rep. **4**: 360–366. doi:10.1111/j.1758-2229.2012.00343.x

Dugdale, R., V. Dugdale, J. Neess, and J. Goering. 1959. Nitrogen fixation in lakes. Science **130**: 859–860. doi: 10.1126/science.130.3379.859

Eugster, O., and N. Gruber. 2012. A probabilistic estimate of global marine N-fixation and denitrification. Global Biogeochem. Cycles **26**: GB4013. doi:10.1029/2012GB004300

Fernández-Méndez, M., K. A. Turk-Kubo, P. L. Buttigieg, J. Z. Rapp, T. Krumpen, J. P. Zehr, and A. Boetius. 2016. Diazotroph diversity in the sea ice, melt ponds, and surface waters of the Eurasian Basin of the central Arctic Ocean. Front. Microbiol. **7**: 1884. doi:10.3389/fmicb.2016.01884

Fetterer, F., M. Savoie, S. Helfrich, and P. Clemente-Colón. 2010. Multisensor analyzed sea ice extent-northern hemisphere (MAISE-NH). National Snow Ice Data Center.

Gihring, T. M., G. Lavik, M. M. Kuypers, and J. E. Kostka. 2010. Direct determination of nitrogen cycling rates and pathways in Arctic fjord sediments (Svalbard, Norway). Limnol. Oceanogr. **55**: 740–752. doi:10.4319/lo.2009.55.2.0740

Glibert, P. M., and D. A. Bronk. 1994. Release of dissolved organic nitrogen by marine diazotrophic cyanobacteria, Trichodesmium Spp. Appl. Environ. Microbiol. **60**: 3996–4000.

Großkopf, T., and others. 2012. Doubling of marine dinitrogen-fixation rates based on direct measurements. Nature **488**: 361–364. doi:10.1038/nature11338

Haines, J. R., R. M. Atlas, R. P. Griffiths, and R. Y. Morita. 1981. Denitrification and nitrogen fixation in Alaskan continental shelf sediments. Appl. Environ. Microbiol. **41**: 412–421.

Hamersley, M. R., K. A. Turk, A. Leinweber, N. Gruber, J. P. Zehr, T. Gunderson, and D. G. Capone. 2011. Nitrogen fixation within the water column associated with two hypoxic basins in the Southern California Bight. Aquat. Microb. Ecol. **63**: 193. doi:10.3354/ame01494

Jakobsson, M., and others. 2012. The international bathymetric chart of the Arctic Ocean (IBCAO) version 3.0. Geophys. Res. Lett. **39**: L12609. doi:10.1029/2012GL052219

Jayakumar, A., M. M. Al-Rshaidat, B. B. Ward, and M. R. Mulholland. 2012. Diversity, distribution, and expression of diazotroph nifH genes in oxygen-deficient waters of the Arabian Sea. FEMS Microbiol. Ecol. **82**: 597–606. doi: 10.1111/j.1574-6941.2012.01430.x

Knowles, R., and C. Wishart. 1977. Nitrogen fixation in Arctic marine sediments: Effect of oil and hydrocarbon fractions. Environ. Pollut. **13**: 133–149. doi:10.1016/0013-9327(77)90098-2

Koroleff, F. 1983. Determination of nutrients, p. 125–187. *In* K. Grasshoff, M. Eberhardt, and K. Kremling [eds.], Methods of seawater analysis. Verlag Chemie.

Mills, M. M., and others. 2015. Impacts of low phytoplankton NO_3^-: PO_4^{3-} utilization ratios over the Chukchi Shelf, Arctic Ocean. Deep-Sea Res. Part II Top. Stud. Oceanogr. **188**: 105–121.

Mohr, W., T. Grosskopf, D. W. Wallace, and J. Laroche. 2010. Methodological underestimation of oceanic nitrogen fixation rates. PLoS One **5**: e12583. doi:10.1371/journal.pone.0012583

Moisander, P. H., R. A. Beinart, I. Hewson, A. E. White, K. S. Johnson, C. A. Carlson, J. P. Montoya, and J. P. Zehr. 2010. Unicellular cyanobacterial distributions broaden the oceanic N_2 fixation domain. Science **327**: 1512–1514. doi: 10.1126/science.1185468

Montoya, J. P., M. Voss, P. Kahler, and D. G. Capone. 1996. A simple, high-precision, high-sensitivity tracer assay for N (inf2) fixation. Appl. Environ. Microbiol. **62**: 986–993.

Mulholland, M. R., and others. 2012. Rates of dinitrogen fixation and the abundance of diazotrophs in North

American coastal waters between Cape Hatteras and Georges Bank. Limnol. Oceanogr. **57**: 1067–1083. doi:10.4319/lo.2012.57.4.1067

Parsons, T. R., Y. Maita, and C. Lalli [eds.]. 1984. A manual of chemical and biological methods for seawater analysis. Pergamon Press.

Popova, E. E., and others. 2012. What controls primary production in the Arctic Ocean? Results from an intercomparison of five general circulation models with biogeochemistry. J. Geophys. Res. Oceans **117**: C00D12. doi:10.1029/2011JC007112

Rahav, E., E. Bar-Zeev, S. Ohayon, H. Elifantz, N. Belkin, B. Herut, M. R. Mulholland, and I. Berman-Frank. 2013. Dinitrogen fixation in aphotic oxygenated marine environments. Front. Microbiol. **4**: 227. doi:10.3389/fmicb.2013.00227

Sherwood, O. A., T. P. Guilderson, F. C. Batista, J. T. Schiff, and M. D. Mccarthy. 2014. Increasing subtropical North Pacific Ocean nitrogen fixation since the Little Ice Age. Nature **505**: 78–81. doi:10.1038/nature12784

Shtarkman, Y. M., Z. A. Koçer, R. Edgar, R. S. Veerapaneni, T. D'Elia, P. F. Morris, and S. O. Rogers. 2013. Subglacial Lake Vostok (Antarctica) accretion ice contains a diverse set of sequences from aquatic, marine and sediment-inhabiting bacteria and eukarya. PLoS One **8**: e67221. doi:10.1371/journal.pone.0067221

Simpson, K. G., J. É. Tremblay, Y. Gratton, and N. M. Price. 2008. An annual study of inorganic and organic nitrogen and phosphorus and silicic acid in the southeastern Beaufort Sea. J. Geophys. Res. Oceans **113**: C07016. doi:10.1029/2007JC004462

Sohm, J. A., E. A. Webb, and D. G. Capone. 2011. Emerging patterns of marine nitrogen fixation. Nat. Rev. Microbiol. **9**: 499–508. doi:10.1038/nrmicro2594

Solheim, B., A. Endal, and H. Vigstad. 1996. Nitrogen fixation in Arctic vegetation and soils from Svalbard, Norway. Polar Biol. **16**: 35–40. doi:10.1007/BF02388733

Stroeve, J., T. Markus, L. Boisvert, J. Miller, and A. Barrett. 2014. Changes in Arctic melt season and implications for sea ice loss. Geophys. Res. Lett. **41**: 1216–1225. doi:10.1002/2013GL058951

Telling, J., A. M. Anesio, M. Tranter, T. Irvine-Fynn, A. Hodson, C. Butler, and J. Wadham. 2011. Nitrogen fixation on Arctic glaciers, Svalbard. J. Geophys. Res. Biogeosci. **116**: G03039. doi:10.1029/2010JG001632

Toetz, D. W. 1961. Nitrogen fixation in natural waters in the Alaskan Arctic. Masters thesis. Univ. of Wisconsin.

Torres-Valdés, S., and others. 2013. Export of nutrients from the Arctic Ocean. J. Geophys. Res. Oceans **118**: 1625–1644. doi:10.1002/jgrc.20063

Tremblay, J. É., K. Simpson, J. Martin, L. Miller, Y. Gratton, D. Barber, and N. M. Price. 2008. Vertical stability and the annual dynamics of nutrients and chlorophyll fluorescence in the coastal, southeast Beaufort Sea. J. Geophys. Res. Oceans **113**: C07S90. doi:10.1029/2007JC004547

Tremblay, J.-E., and J. Gagnon. 2009. The effects of irradiance and nutrient supply on the productivity of Arctic waters: A perspective on climate change, p. 73–93. *In* J .C. J. Nihoul, and A. G. Kostianoy [eds.], Influence of climate change on the changing Arctic and sub-Arctic conditions. Springer, Dordrecht, Netherlands, doi:10.1007/978-1-4020-9460-6_7.

Yamamoto-Kawai, M., E. Carmack, and F. Mclaughlin. 2006. Nitrogen balance and Arctic throughflow. Nature **443**: 43–43. doi:10.1038/443043a

Yu, X., J. Lei, X. Yao, J. Zhu, and X. Jin. 2014. The composition and origination of particles from surface water in the Chukchi Sea, Arctic Ocean. Adv. Polar Sci. **25**: 147–154. doi:10.13679/j.advps.2014.3.00147

Acknowledgments

We thank L. Killberg-Thoreson and T. Connelly for lab and field assistance, and CH2MHill, UMIAQ science support staff and Barrow Whaling Captains Association for facilitating field work. This work is supported by NSF grant ARC-0909839 to DAB and ARC-1504307 to RES. This paper is Contribution No. 3648 of the Virginia Institute of Marine Science, College of William & Mary.

17

Mangrove outwelling is a significant source of oceanic exchangeable organic carbon

James Z. Sippo,[1,2]* *Damien T. Maher,*[1,2] *Douglas R. Tait,*[1,2] *Sergio Ruiz-Halpern,*[1] *Christian J. Sanders,*[1,2] *Isaac R. Santos*[1,2]
[1]School of Environment, Science and Engineering, Southern Cross University, Lismore, 2480 Australia; [2]National Marine Science Centre, Southern Cross University, Coffs Harbour, New South Wales 2450, Australia

Scientific Significance Statement

Exchangeable dissolved organic carbon (EDOC) consists of volatile and semi volatile organic compounds which are globally ubiquitous carbon species and play an important role in atmospheric chemistry. However, no studies have quantified EDOC outwelling from mangrove forests, in spite of these systems contributing > 10% of the terrestrially derived dissolved organic carbon (DOC) to the ocean. By measuring EDOC and DOC outwelling from a mangrove tidal creek and concentrations in mangrove forests over a large latitudinal gradient (26°), we found that EDOC makes up ~13% of the DOC pool, and that mangroves may export 3.1 Tg C yr^{-1} of EDOC to the coastal ocean. These results highlight that outwelling from mangroves may be a major, yet previously overlooked source of EDOC to the ocean.

Abstract

Exchangeable dissolved organic carbon (EDOC) makes up a significant proportion of the oceanic dissolved organic carbon (DOC) pool, yet EDOC sources to the coastal ocean are poorly constrained. We measured the exchange of EDOC and concentrations of EDOC and DOC in mangrove waters over a 26° latitudinal gradient. A clear latitudinal trend was observed, with the highest EDOC concentrations in the tropics. EDOC exports to the coastal ocean were 4.7 ± 1.9 mmol m^{-2} d^{-1}, equivalent to 11% of DOC exports (42.1 ± 6.7 mmol m^{-2} d^{-1}). Pore-water and groundwater exchange were minor sources of EDOC. EDOC concentrations were equal to 13% ± 4% of DOC concentrations. Based on previous global DOC export estimates, and our EDOC : DOC ratios, mangroves outwell 3.1 Tg C yr^{-1} as EDOC, equivalent to ~60% of the global EDOC flux from the ocean to the atmosphere. However, seasonality of mangrove EDOC cycling requires further research.

Exchangeable dissolved organic carbon (EDOC) consists of volatile organic compounds (VOCs) and semi-volatile organic compounds (SOCs) which are ubiquitous carbon species that play a major role in global tropospheric chemistry and the global carbon cycle (Fehsenfeld et al. 1992; Fuentes et al. 2000). Sources of VOCs and SOCs can be both anthropogenic and biogenic. Anthropogenic VOC and SOC production occurs during fossil fuel combustion, industrial chemical production, and waste processing (Fuentes et al. 2000). Biogenic VOC and SOC emissions from terrestrial vegetation accounts for the vast majority of the inputs to the atmosphere (Guenther et al. 1995; Laothawornkitkul et al. 2009). The atmospheric exchange of marine biogenic VOCs and SOCs has also been identified as both an important sink

*Correspondence: j.sippo.11@student.scu.edu.au

Author Contribution Statement: IS, DM, and SR-H came up with the research question and designed the study approach. DT led the field survey for five of the six sites and DM and SR-H led the field survey for one site. All authors helped with data analysis and writing of the manuscript. JS helped to collect the data, analyzed the data, and wrote the manuscript.

Table 1. Site description identifying the predominant mangrove species at the six study sites in Australia.

Site	Description	Latitude	Longitude	Mangrove species	Dominant species
Darwin	Sadgroves Creek within Darwin Harbour, Northern Territory	−12.441°	130.860°	36	*Avicennia marina, Aegiceras corniculatum, Bruguiera gymnorrhiza* and *Ceriops australis*
Hinchinbrook Island	Coral Creek on Hinchinbrook Island, Great Barrier Reef (GBR), Queensland	−18.244°	146.228°	29	*Rhizophora stylosa*
Seventeen Seventy	Tom's Creek near Seventeen Seventy, Queensland	−24.190°	151.876°	13	*A. marina* and *R. stylosa*
Jacobs Well	Unnamed tidal creek in southern Moreton Bay, Queensland	−27.783°	153.375°	7	*A. marina, A. corniculatum, B. gymnorrhiza* and *R. stylosa.*
Newcastle	Dunns Creek near Newcastle, New South Wales	−32.850°	151.767°	3	*A. marina*
Barwon Heads	An unnamed tidal creek on Barwon River, south west Melbourne, Victoria	−38.262°	144.496°	1	*A. marina*

for atmospheric hydroxyl radicals, and a major source of secondary organic aerosols to the atmosphere (Gantt et al. 2009; Carpenter et al. 2012).

The exchange of VOCs and SOCs occurs between oceans and atmosphere where the ocean can act as both a source and as sink of anthropogenic and biogenic VOC and SOC compounds (Dachs et al. 2005; Ruiz-Halpern et al. 2010; Gonzalez-Gaya et al. 2016). The estimated VOC evasion from the ocean surface to the atmosphere is 4.9–5 Tg C a^{-1} (Guenther et al. 1995; Ehhalt 2001). However, these estimates are based largely on the exchange of major compounds and may not account for a significant fraction of VOCs and SOCs present in the atmosphere or ocean (Goldstein and Galbally 2007; Ruiz-Halpern et al. 2014*a*). For example, Park et al. (2013) found that the 10 major compounds measured in most VOC assessments accounted for only 63% of total VOCs in a forested valley. Based on a modeling study by Jurado et al. (2008) the deposition of exchangeable organic carbon to the ocean could be as high as 187 Tg C a^{-1}.

The problem of measuring VOCs and SOCs collectively is comparable to that of estimating the pool of dissolved organic carbon (DOC) from the sum of the concentrations of the individual compounds (Ruiz-Halpern et al. 2014*a*). The solution adopted by Dachs et al. (2005) is to collectively measure VOCs and SOCs as EDOC if measured in water, and as gaseous organic carbon (GOC) if measured in air. This approach is comparable to that of conventional DOC analysis to operationally quantify these compounds (Ruiz-Halpern et al. 2014*a*) and provides a total measure of carbon that is exchangeable between the air and water.

EDOC makes up a significant proportion (30–67%) of the oceanic DOC pool, which is one of the earth's largest reactive carbon reservoirs (Dachs et al. 2005). It is well documented that the coastal interface generally exhibits high rates of carbon fixation, respiration, and transformation (Walsh 1991; Mackenzie et al. 1998). Further, DOC and VOC can have significantly higher concentrations in coastal waters than in the open ocean (Dachs et al. 2005; Ruiz-Halpern et al. 2014*b*). Mangrove forests have also been shown to produce significant levels of isoprene, the main biogenic VOC constituent of terrestrial vegetation (Barr et al. 2003; Exton et al. 2015). However, there are no studies examining the lateral exchange of EDOC at the land-ocean interface.

Mangrove forests are recognized as the largest store of carbon in the coastal zone and are amongst the most productive ecosystems on earth (Bouillon 2011; Breithaupt et al. 2012). Mangrove forests contribute > 10% of terrestrially derived DOC flux to the ocean (Dittmar et al. 2006). This contribution does not include EDOC because the analytical methods used to measure DOC normally acidify samples and purge inorganic carbon prior to measurement of DOC. This procedure also removes VOCs and SOCs from water samples. As a result, previous DOC exchange estimates in mangroves (e.g., Bouillon et al. 2007; Maher et al. 2013) have not taken into account the contribution of EDOC.

Here, we report EDOC and DOC observations in six mangrove creeks along a latitudinal gradient to assess whether mangrove forests export EDOC to the coastal ocean, and whether VOCs make a significant contribution to the mangrove DOC pool. We also investigate whether pore-water exchange is a major source of EDOC to surface mangrove waters.

Methods

Study sites

The study was undertaken in six pristine mangrove tidal creeks on the northern, eastern, and southern Australian coastlines over a latitudinal range from 12°S to 38°S (Table 1;

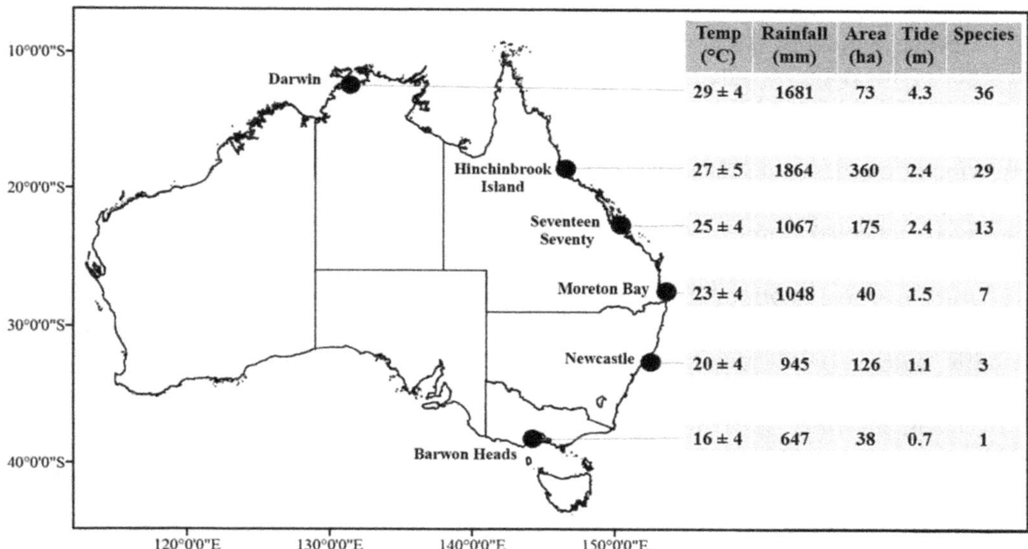

Fig. 1. The six study sites were in tidal creeks of pristine mangrove forests over a 26° latitudinal gradient in Australia. Site characteristics are included where Temp is average annual air temperature (°C), Rainfall is average annual rainfall (mm), Area is catchment area of tidal creeks (ha) and Tide is tidal range (m).

Fig. 1). All study sites had low lying topography with maximum elevations below 5.5 m Australian Height Datum and no obvious source of riverine input. Therefore, the creeks received no apparent freshwater input other than direct rainfall and it is assumed that any changes in water chemistry are related to processes occurring within the mangrove intertidal flat rather than influences from upstream freshwater inputs. This study builds on recent investigations at the same sites focusing on pore-water exchange (Tait et al. 2016), alkalinity and dissolved inorganic carbon exchange (Sippo et al. 2016), nitrous oxide fluxes (Maher et al. 2016), trace metal fluxes (Holloway et al. 2016), and system scale carbon stock estimates (Sanders et al. 2016).

Sampling and analysis

Surface-water EDOC and DOC concentrations, were measured at hourly intervals over 25 h periods (i.e., two tidal cycles) near the mouth of the tidal creek at all sites. At the Moreton Bay site, a more intensive sampling campaign was conducted to quantify exchange with the coastal ocean and air water fluxes. EDOC, GOC, and DOC concentrations were measured hourly over three 25-h periods for the duration of a spring-neap-spring cycle ($\sim$15 days). All six sites were sampled during their respective dry season in 2014, i.e., during the winter for the three low latitude sites and during the summer for the three high latitude sites.

VOC is operationally defined as EDOC in the water phase and as GOC in the air phase (Dachs et al. 2005). EDOC concentrations were measured from 1 L of water added to an acid cleaned bottle following Dachs et al. (2005) and Ruiz-Halpern et al. (2010). Briefly, water samples were purged by recirculating air in the dead volume of the sample bottle and Teflon tubing. The gas stream containing the exchangeable VOC was redissolved in duplicate precombusted 40 mL borosilicate vials containing acidified ($pH < 2$ to ensure no inorganic carbon adsorption) milliQ water. The air stream returned to the sampling bottle in a closed loop. Exchange of EDOC from the water sample was enhanced through continuous bubbling via a glass frit in the sample bottle. Samples were then stored in the dark at 4°C until analyses.

GOC concentrations were measured as GOC H'^{-1} (that is the equilibrium concentration of GOC), therefore, avoiding the need to calculate the Henry's law constant (H') for each compound, which are unknown (*see* Dachs et al. 2005 and Ruiz-Halpern et al. 2010 for a detailed explanation). At the Moreton Bay Site, samples were taken at 3–4 h intervals and interpolated to hourly intervals near the mouth of the tidal creek. No GOC H^{-1} data was available for the other five sites. GOC H^{-1} concentrations were measured by equilibrating acidified milliQ with the atmosphere. Atmospheric air was pumped through duplicate precomubusted 40 mL borosilicate vials containing acidified milliQ water for 25 min using an oil-free diaphragm pump. DOC samples were collected with a 40-mL sample rinsed polypropylene syringe, and filtered through GFF filters into precombusted borosilicate vials.

The samples were kept cool and in the dark until analysis using a TOC analyser (TOC-VCPH; Shimadzu). The vials were analyzed as for DOC (NPOC method, Spyres et al. 2000) but with the purging step eliminated. Following this analysis, the same samples were analyzed for DOC and TIC to check for contamination. If DOC and/or inorganic carbon contamination was detected, it was subtracted from the VOC measurement. DOC reference standards (2 μmol L^{-1} and 44 μmol L^{-1}; Hansell Lab, University of Miami) were run throughout each analysis (CV = 1.5%). Methods for measuring estuarine water depth, water temperature, wind speed, salinity, and current velocity methods and data are reported in a companion paper (Sippo et al. 2016).

Pore-water samples were collected from ~12 sites within the catchment of each tidal creek. Samples were taken by digging bores with a hand auger to a depth of ~30 cm below the water table. Water within the bore was purged three times using a peristaltic pump. EDOC and DOC samples were then taken from the bottom of the bore into 1 L Schott bottles (EDOC), or by using a syringe with the sample filtered through GFF filters into precombusted 40 mL borosilicate vials (DOC). Samples were treated and analyzed as described earlier for surface-water samples.

Atmospheric evasion

Air-water VOC flux values were calculated according to:

$$F=k(\text{EDOC}-\text{GOC } H'^{-1}) \quad (1)$$

where F is the flux value (mmol m^{-2} d^{-1}), k is the gas transfer velocity (m d^{-1}), EDOC and GOC H^{-1} are the exchangeable organic carbon concentration in the water, and the atmospheric equilibrium concentration, respectively. Positive values indicate a flux from water to the atmosphere while negative values indicate a flux from the atmosphere to water.

The gas transfer velocity (k) was calculated using a combination of empirically derived k models which integrate wind speed, current velocity, and depth, which was developed in tidal mangrove forests using a deliberate gas tracer experiment (Ho et al. 2014) and used in several recent mangrove gas flux papers (Call et al. 2015; Maher et al. 2016; Sippo et al. 2016, Rosentreter et al. 2016). The equation merges the reaeration coefficient determined by O'Connor and Dobbins (1958) using depth and current velocity, with the wind-driven k parameterization of Raymond and Cole (2001):

$$k_{600} = \left(1.58e^{0.3u}\right) + \left(1.539v^{0.5}h^{-0.5}\right) \quad (2)$$

where k_{600} is the gas transfer velocity (cm h^{-1}) normalized to a Schmidt number of 600, u is the wind speed at a height of 10 m (m s^{-1}), v is the water velocity (cm s^{-1}), and h is water depth (m). k_{600} values were corrected to the in situ Schmidt number assuming a molecular weight of VOC of 120 g mol^{-1} (*see* Dachs et al. 2005 for further details). Evasion was calculated only for the Moreton Bay system because atmospheric GOC H'^{-1} data were unavailable for the other five mangrove creeks.

Outwelling estimates

Oceanic exchange rates were calculated in mmol m^{-2} day^{-1} at Moreton Bay where atmospheric evasion rates were also available. A digital elevation model (DEM) of the creek catchment was constructed with LiDAR data (1 m resolution ± 0.2 m elevation accuracy). Change of water volume within the catchment was calculated using the DEM and ARC GIS Hydrology toolbox (Maher et al. 2013). Concentrations of EDOC and DOC were multiplied by the change in water volume over the hourly sampling interval, integrated over two tidal cycles and normalized to catchment area and the duration of the study to give an export rate in mmol m^{-2} (catchment area) d^{-1}.

Results and discussion

Drivers of EDOC concentrations

Surface-water EDOC concentrations averaged 34 μM (range = 14–82 μM) across the six sites, with the highest concentrations occurring at lower latitude sites (Fig. 2). Pore-water EDOC concentrations were similar to surface waters, averaging 35 (range = 11–154) μM across the six sites, decreasing at higher latitudes. These average concentrations are similar to EDOC concentrations observed in the open ocean (30 ± 6 μM in spring and ~36 μM in fall) (Dachs et al. 2005), although with much higher variability. The mean surface-water DOC concentration was 279 μM (range = 84–478 μM) across the sites while the mean DOC concentration in pore water was 1155 μM (range = 332–3430 μM) and were 2- to 13-fold higher than those in respective surface waters at the six sites (Fig. 2). No clear tidal trends were observed at five out of the six sites, so average concentrations only are reported in Fig. 2.

We anticipated that pore-water inputs would be a major driver of mangrove EDOC exports, similar to DOC and DIC (Bouillon et al. 2007, Maher et al. 2013), and alkalinity exports (Sippo et al. 2016). Despite significant pore-water exchange reported at the sites (2.1–35.5 cm d^{-1}, Tait et al. 2016) and consistent inverse relationships between water depth and EDOC concentrations (Fig. 4), mean pore-water EDOC concentrations were lower than surface waters at five of the six sites (Fig. 2). Therefore, pore-water exchange is unlikely to be a major source of surface-water EDOC in mangroves.

Temperature and light have been identified as environmental controls of biogenic hydrocarbon emissions from plant leaves (Guenther et al. 1995; Fuentes et al. 2000; Laothawornkitkul et al. 2009). However, temperature did not follow a clear trend with latitude due to the seasonality of the study (i.e., the three low latitude sites were sampled in the winter and the three high latitude sites were sampled in the summer), explaining the weak correlation between temperature and EDOC (Fig. 3). Although this is the first estimate of EDOC dynamics in mangroves, clearly the lack of

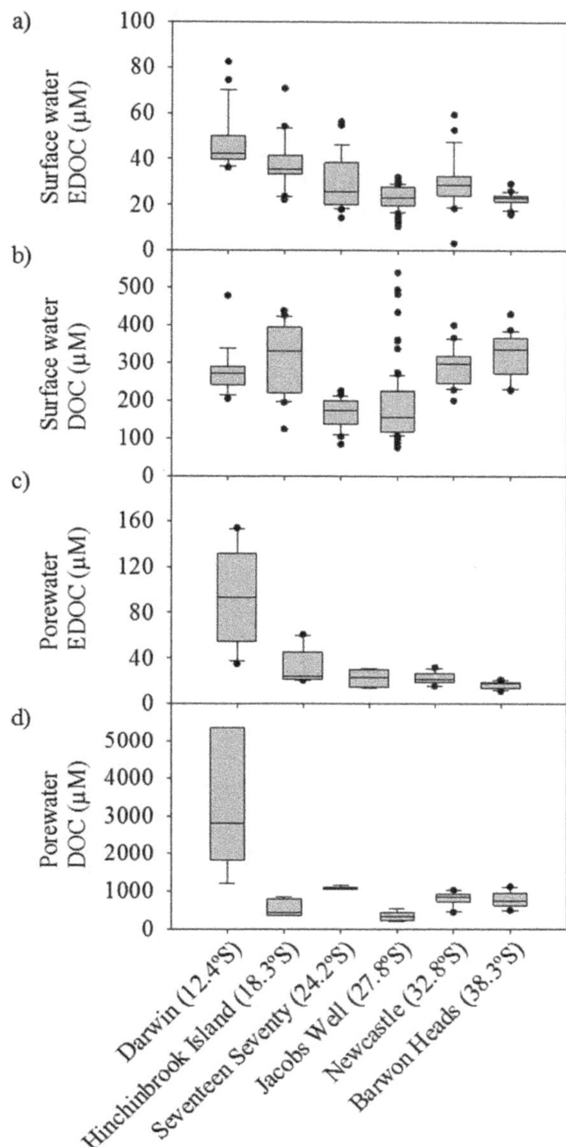

Fig. 2. Box plots of surface-water EDOC and DOC in surface waters and pore waters of six Australian mangrove dominated tidal creeks, ordered from low latitude on the left to high latitude on the right. The central horizontal line represents the median, the box represents the upper and lower quartiles, and the whiskers represent the maximum and minimum values excluding outliers, i.e., black dots. Sample size > 25 at each site.

seasonal data must be considered when interpreting these findings. However, mean temperatures range 29–16°C over the 26° latitude, which is a much greater difference than the variability of mean annual minimum and maximum temperatures at each site (Fig. 1).

Above ground plant biomass had the highest positive correlation to EDOC ($r^2 = 0.91$) and offers the best explanation to the latitudinal trend in EDOC concentrations (Fig. 3). Average annual temperatures also strongly correlated with EDOC concentrations ($r^2 = 0.89$), yet this is likely because

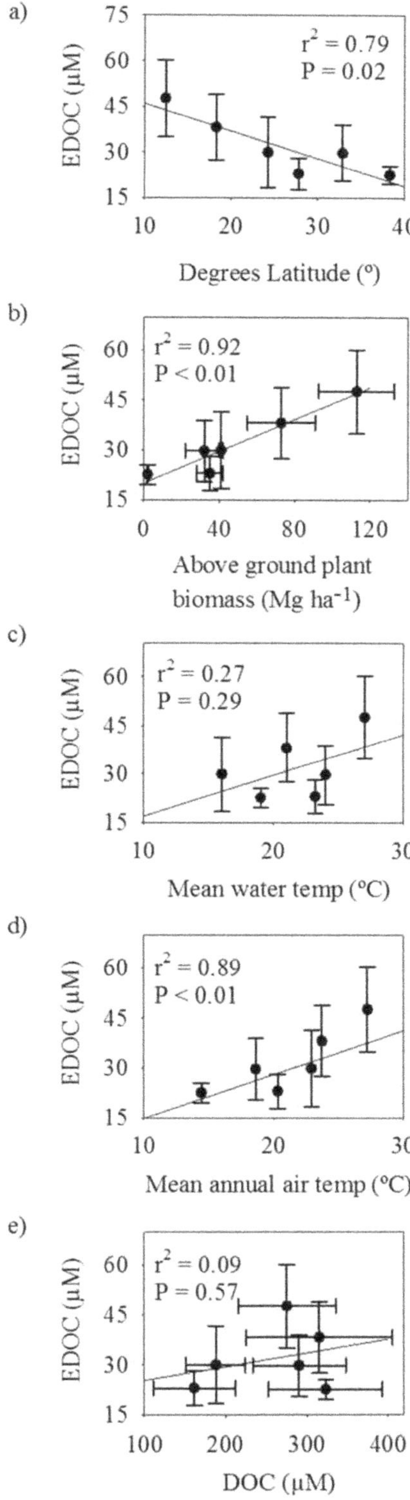

Fig. 3. Correlations of mean EDOC concentrations ($n > 25$ at each site) at the six study sites vs. (**a**) degrees latitude, (**b**) above ground plant biomass* (**c**) mean water temperature, (**d**) mean annual air temperature, and (**e**) DOC concentrations ($n > 25$ at each site). *Above ground plant biomass data is sourced from Sanders et al. (2016). Error bars indicate standard error (SE).

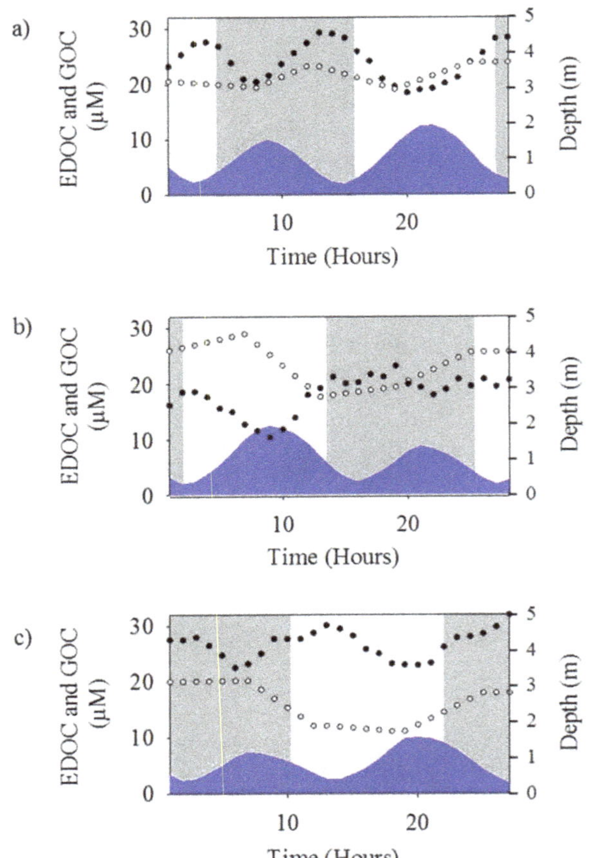

Fig. 4. Moreton Bay time series of EDOC (black) and GOC (white) at (**a**) Spring tide, (**b**) Mid tide, and (**c**) Neap tide. Blue represents water depth and gray bars represent night time.

temperature is a significant driver of plant biomass (Hutchison et al. 2014; Sanders et al. 2016). Vegetation is the largest source of VOCs to the atmosphere (Guenther et al. 1995; Ehhalt et al. 2001) and VOCs are emitted from all parts of a plant (Laothawornkitkul et al. 2009). We suggest that leaching of EDOC primarily through mangrove roots and submerged pneumatophores is the major source of EDOC to creek waters, and, therefore, water column concentrations and exchange rates are coupled to biomass. However, while mangrove biomass does not vary over seasons, the specific mechanism of EDOC release has not been constrained and may be seasonal, e.g., due to seasonal variability in productivity and respiration. Further, processes not considered by this study such as EDOC production by phytoplankton or benthic microalgae may also be controlling EDOC concentrations in mangrove creek waters.

EDOC exchange with the coastal ocean and atmosphere

At Moreton Bay where atmospheric GOC concentrations were available, the EDOC flux between water and air averaged 0.6 mmol m^{-2} d^{-1}. The creek ranged from a sink of atmospheric EDOC at mid tide (−4.0 mmol m^{-2} d^{-1}) to a source at neap tide (5.3 mmol m^{-2} d^{-1}). Overall, the net exchange of EDOC between mangrove water and air was a small source to the atmosphere (0.6 ± 4.7 mmol m^{-2} d^{-1}). However, there was an oscillation between the source and sink of EDOC to the atmosphere over individual tidal cycles and over the spring neap cycle (Fig. 4). Water depth was correlated to EDOC concentrations during spring tide ($r^2 = 0.74$), mid tide ($r^2 = 0.39$), and neap tide ($r^2 = 0.42$), where low tides had the highest flux from water to air throughout the study. Observations in a subarctic fjord, a Mediterranean bay and the Antarctic were similar in both magnitude and direction, with EDOC flux oscillating between production and uptake (Ruiz-Halpern et al. 2014*a*,*b*).

Our study provides a first estimate of EDOC exchange between pristine mangrove forests and the coastal ocean and atmosphere. Mean lateral EDOC export to the coastal ocean was 4.7 ± 1.9 mmol m^{-2} d^{-1}, equivalent to 11.6% ± 6.1% of the calculated DOC exports (Table 2) and ~6% of DIC exports (83 mmol m^{-2} d^{-1}) calculated from the same creek by Sippo et al. (2016). Our estimates of DOC exports (average 42.1 ± 6.7 mmol m^{-2} d^{-1}) were at the high end of previous estimates for mangroves (e.g., *see* Adame and Lovelock, 2011). The ratio of EDOC to DOC concentrations observed across the six sites showed low variability (EDOC = 13% ± 4% of DOC) with a trend of increasing ratio with decreasing latitude.

Table 2. Carbon exchange between mangrove tidal creek waters and the atmosphere and coastal ocean over spring to neap tidal cycles at Moreton Bay.

Tide	Air-water EDOC flux (mmol m^{-2} d^{-1})	Tidal EDOC exchange (mmol m^{-2} d^{-1})	Net EDOC exchange (mmol m^{-2} d^{-1})	DOC exchange (mmol m^{-2} d^{-1})	EDOC % of DOC exchange
Spring	0.5	6.8	7.3	37.0	18.4
Mid	−4.0	3.2	−0.8	49.7	6.4
Neap	5.3	4.0	9.3	39.8	10.1
Average	**0.6 ± 4.7**	**4.7 ± 1.9**	**5.3 ± 5.3**	**42.2 ± 6.7**	**11.6 ± 6.1**

Net EDOC exchange is calculated as the sum of air-water EDOC flux and tidal EDOC exchange. So the sum of columns 1 and 2 = column 3. The last column, EDOC % of DOC, is calculated as tidal EDOC exchange/DOC exchange × 100.

There was no significant relationship between the spring-neap cycle and EDOC exports. However, the largest net export of EDOC from mangrove forests occurred during the neap tide (Table 2). This is similar to the results of a study by Call et al. (2015) at the same study site, who found that pore water enriched in CO_2 and CH_4 preferentially seeps into surface waters at neap tides. Further, water volumes are lowest in the creek at neap tides which would increase the influence that autochthonous EDOC production has on surface water concentrations.

Global implications

Global mangrove forests may export 24 ± 21 Tg C of DOC to the ocean annually (Bouillon et al. 2008). The methods used to measure DOC and estimate these exports generally exclude VOC. Our results indicate that EDOC contributes 13% ± 4% of the DOC pool within mangrove waters. If our results are representative of mangroves globally, an upscaling based on the EDOC to DOC ratio reveals that EDOC may account for an additional outwelling of 3.1 ± 1.0 Tg C yr^{-1} from mangroves. This value is equivalent to ~60% of the global VOC flux from the open ocean to the atmosphere (Guenther et al. 1995; Ehhalt 2001) and ~2% of the modeled atmospheric EDOC deposition to the ocean (Jurado et al. 2008). If mangrove outwelling accounts for 10% of terrestrial DOC in the ocean as suggested by Dittmar et al. (2006), our estimated mangrove EDOC exports could account for 1.3–2.5% of the total terrestrial DOC input to the ocean. This brings the total DOC exports from mangroves (including EDOC) to 11.3–12.5% of the terrestrially-derived DOC inputs to the ocean.

Our investigation further supports the increasing evidence on the disproportionally large role that mangroves play in the exchange of carbon between land and ocean. We conclude that (1) mangrove biomass is a major control of EDOC concentrations over a latitudinal gradient, (2) EDOC accounts for 13% ± 4% of DOC in mangroves, (3) pore-water exchange is not a major source of EDOC to mangrove creeks, and (4) outwelling from mangroves may be a significant, but previously overlooked source of EDOC to the ocean. Seasonality of EDOC production and exchange was not investigated in this study. Further research on the seasonal variability of EDOC within mangroves is required to further constrain our initial estimates.

References

Adame, M. F., & Lovelock, C. E. 2011. Carbon and nutrient exchange of mangrove forests with the coastal ocean. Hydrobiologia, **663**: 23–50. doi:10.1007/s10750-010-0554-7

Barr, J.G., J.D. Fuentes, N. Wang, Y. Edmonds, J.C. Zieman, B.P. Hayden, D.L. Childers. 2003. Red mangroves emit hydrocarbons. Southeast. Nat. **2**: 499–510.

Bouillon, S. 2011. Carbon Cycle: Storage beneath mangroves.Nat. Geosci, **4**: 282–283. doi:10.1038/ngeo1130

Bouillon, S. and others 2008. Mangrove production and carbon sinks: A revision of global budget estimates. Global Biogeochem. Cyles **22**. doi:10.1029/2007gb003052

Bouillon, S., Dehairs, F., Velimirov, B., Abril, G., & Borges, A. V. 2007. Dynamics of organic and inorganic carbon across contiguous mangrove and seagrass systems (Gazi Bay, Kenya). J. Geophys. Res.: Biogeosci. **112**: doi:10.1029/2006JG000325

Breithaupt, J. L., Smoak, J. M., Smith, T. J., Sanders, C. J., & Hoare, A. 2012. Organic carbon burial rates in mangrove sediments: Strengthening the global budget. Global Biogeochem. Cycles, **26**: 475–488. doi:10.1029/2012GB004375

Call, M. and others 2015. Spatial and temporal variability of carbon dioxide and methane fluxes over semi-diurnal and spring-neap-spring timescales in a mangrove creek. Geochim. Cosmochim. Acta, **150**: 211–225. doi:10.1016/j.gca.2014.11.023

Carpenter, L. J., Archer, S. D., & Beale, R. 2012. Ocean-atmosphere trace gas exchange. Chem. Soc. Rev. **41**: 6473–6506. doi:10.1039/C2CS35121H

Dachs, J., Calleja, M. L., Duarte, C. M., del Vento, S., Turpin, B., Polidori, A., G. J. Herndl., Agustí, S. 2005. High atmosphere-ocean exchange of organic carbon in the NE subtropical Atlantic. Geophys. Res. Lett. **32**: doi:10.1029/2005GL023799

Dittmar, T., Hertkorn, N., Kattner, G., & Lara, R. J. 2006. Mangroves, a major source of dissolved organic carbon to the oceans. Global Biogeochem. Cycles **20**: doi:10.1029/2005gb002570

Ehhalt D., and others. 2001. Atmospheric chemistry and greenhouse gases, p. 881. In J. T. Houghton, Y. Ding, D. J. Griggs, M. Noguer, P. J. van der Linden, X. Dai, K. Maskell, and C. A. Johnson [eds.], Climate change 2001: The scientific basis. Contribution of Working Group I to the Third Assessment Report of the Intergovernmental Panel on Climate Change. Cambridge University Press.

Exton, D. A., McGenity, T. J., Steinke, M., Smith, D. J. and Suggett, D. J. 2015. Uncovering the volatile nature of tropical coastal marine ecosystems in a changing world. Glob. Change Biol. **21**: 1383–1394. doi:10.1111/gcb.12764

Fehsenfeld, F. and others 1992. Emissions of volatile organic compounds from vegetation and the implications for atmospheric chemistry. Global Biogeochem. Cycles **6**: 389–430. doi:10.1029/92GB02125

Fuentes, J. D. and others 2000. Biogenic Hydrocarbons in the Atmospheric Boundary Layer: A Review. Bull. Am. Meteorol. Soc. **81**: 1537–1575. doi:10.1175/1520-0477(2000)081 < 1537:BHITAB>2.3.CO2

Gantt, B., Meskhidze, N., & Kamykowski, D. 2009. A new physically-based quantification of marine isoprene and primary organic aerosol emissions. Atmos.Chem. Phys. **9**: 4915–4927. doi:10.5194/acp-9-4915-2009

Goldstein, A. H., & Galbally, I. E. 2007. Known and Unexplored Organic Constituents in the Earth's Atmosphere. Environ. Sci. Technol. **41**: 1514–1521. doi:10.1021/es072476p

Gonzalez-Gaya, B. and others 2016. High atmosphere-ocean exchange of semivolatile aromatic hydrocarbons. Nat. Geosci. **9**: 438–442. doi:10.1038/ngeo2714

Guenther, A. and others 1995. A global model of natural volatile organic compound emissions. J. Geophys. Res.: Atmos. **100**: 8873–8892. doi:10.1029/94JD02950

Ho, D. T., Ferrón, S., Engel, V. C., Larsen, L. G., & Barr, J. G. 2014. Air-water gas exchange and CO_2 flux in a mangrove-dominated estuary. Geophys. Res. Lett. **41**: 2013GL058785. doi:10.1002/2013GL058785

Holloway, C. J. and others 2016. Manganese and iron release from mangrove porewaters: A significant component of oceanic budgets? Mar. Chem. **184**: 43–52. doi:10.1016/j.marchem.2016.05.013

Hutchison, J., Manica, A., Swetnam, R., Balmford, A., & Spalding, M. 2014. Predicting Global Patterns in Mangrove Forest Biomass. Conserv. Lett. **7**: 233–240. doi: 10.1111/conl.12060

Jurado, E., Dachs, J., Duarte, C. M., & Simó, R. 2008. Atmospheric deposition of organic and black carbon to the global oceans. Atmos. Environ. **42**: 7931–7939. doi: 10.1016/j.atmosenv.2008.07.029

Laothawornkitkul, J., Taylor, J. E., Paul, N. D., & Hewitt, C. N. 2009. Biogenic volatile organic compounds in the Earth system. New Phytol. **183**: 27–51. doi:10.1111/j.1469-8137.2009.02859.x

Mackenzie, F.T., Lerman, A., Ver, L.M.B., 1998. Role of continental margin in the global carbon balance during the past three centuries. Geology. **26**: 423–426. doi:10.1130/0091

Maher, D., Santos, I., Golsby-Smith, L., Gleeson, J., & Eyre, B. 2013. Groundwater-derived dissolved inorganic and organic carbon exports from a mangrove tidal creek: The missing mangrove carbon sink? Limnol. Oceanog. **58**: 475–488. doi:10.4319/lo.2013.58.2.0475

Maher, D. T., Sippo, J. Z., Tait, D. R., Holloway, C., & Santos, I. R. 2016. Pristine mangrove creek waters are a sink of nitrous oxide. Sci. Rep. **6**: 25701. doi:10.1038/srep25701

O'Connor, D. J., & Dobbins, W. E. 1958. Mechanism of reaeration in natural streams. Trans. Am. Soc. Civ. Eng. **123**: 641–666.

Park, J.-H., Goldstein, A. H., Timkovsky, J., Fares, S., Weber, R., Karlik, J., & Holzinger, R. 2013. Active atmosphere-ecosystem exchange of the vast majority of detected volatile organic compounds. Science. **341**: 643–647. doi: 10.1126/science.1235053

Raymond, P., & Cole, J. 2001. Gas exchange in rivers and estuaries: Choosing a gas transfer velocity. Estuaries, **24**: 312–317. doi:10.2307/1352954

Rosentreter, J. A., D. T. Maher, D. T. Ho, M. Call, J. G. Barr, and B. D. Eyre. 2016. Spatial and temporal variability of CO_2 and CH_4 gas transfer velocities and quantification of the CH_4 microbubble flux in mangrove dominated estuaries. Limnol. Oceanogr. doi: 10.1002/lno.10444

Ruiz-Halpern, S., Sejr, M. K., Duarte, C. M., Krause-Jensen, D., Dalsgaard, T., Dachs, J., & Rysgaard, S. 2010. Air-water exchange and vertical profiles of organic carbon in a subarctic fjord. Limnol. Oceanogr. **55**: 1733–1740. doi: 10.4319/lo.2010.55.4.1733

Ruiz-Halpern, S., Calleja, M. L., Dachs, J., Del Vento, S., Pastor, M., Palmer, M., Agusti, S. and Duarte, C. M. 2014a. Ocean–atmosphere exchange of organic carbon and CO_2 surrounding the Antarctic Peninsula. Biogeosciences. **11**: 2755–2770. doi:10.5194/bg-11-2755-2014

Ruiz-Halpern, S., Vaquer-Sunyer, R., & Duarte, C. M. 2014b. Annual benthic metabolism and organic carbon fluxes in a semi-enclosed Mediterranean bay dominated by the macroalgae Caulerpa prolifera. Front. Mar. Sci. **1**: 2296–7745. doi:10.3389/fmars.2014.00067

Sanders, C. J., D. T. Maher, D. R. Tait, D. Williams, C. Holloway, J. Z. Sippo, and I. R. Santos 2016. Are global mangrove carbon stocks driven by rainfall?, J. Geophys. Res.: Biogeosci,. **121**: 2600–2609, doi:10.1002/2016JG003510

Sippo, J. Z., Maher, D. T., Tait, D. R., Holloway, C., & Santos, I. R. 2016. Are mangroves drivers or buffers of coastal acidification? Insights from alkalinity and dissolved inorganic carbon export estimates across a latitudinal transect. Global Biogeochem. Cycles. **30**: 753–766. doi:10.1002/2015GB005324

Spyres, G., Nimmo, M., Worsfold, P. J., Achterberg, E. P., and Miller, A. E. J. 2000. Determination of dissolved organic carbon in seawater using high temperature catalytic oxidation techniques. Trends Analyt. Chem. **19**: 498–506. doi: 10.1016/S0165-9936(00)00022-4

Tait, D. R., Maher, D. T., Macklin, P. A., & Santos, I. R. 2016. Mangrove pore water exchange across a latitudinal gradient. Geophys. Res. Lett. **43**: 3334–3341. doi:10.1002/2016GL068289

Walsh, J.J., 1991. Importance of continental margins in the marine biogeochemical cycling of carbon and nitrogen. Nature. **350**: 53–55. doi:10.1038/350053a0

Acknowledgments

We would like to thank Ceylena Holloway, Paul Macklin, Darren Williams, and a number of members from the Centre for Coastal Biogeochemistry for data collection at Moreton Bay. James Z. Sippo acknowledges support through an AINSE Postgraduate Research Award. This project was funded by the Australian Research Council (DE140101733, DE150100581, DE160100443, DP150103286, LE120100156 and LE140100083) and the Hermon Slade Foundation.

Nutritional support of inland aquatic food webs by aged carbon and organic matter

Amber R. Bellamy, *James E. Bauer*

Aquatic Biogeochemistry Laboratory, Department of Evolution, Ecology and Organismal Biology, Ohio State University, Columbus, Ohio

Scientific Significance Statement

A large proportion of the carbon and organic matter in streams, rivers, and lakes originates from terrestrial ecosystems. Much of this terrestrially derived material has been stored and aged in watersheds for up to geologically relevant timescales before being mobilized to aquatic settings. It is generally assumed that aged materials contribute little to aquatic food webs compared to modern sources. Here, we show that aged materials can contribute measurably to consumer nutrition in many inland water food webs, either by augmenting or displacing modern contemporary sources of carbon and organic matter to aquatic autotrophs and heterotrophs. This demonstrates that linkages between modern aquatic food webs and geological carbon and organic matter reservoirs are widespread in inland water ecosystems.

Abstract

Aged (typically tens to thousands of years old) forms of non-living carbon (C) and organic matter (OM) predominate in many inland water ecosystems. Advances in the methodologies used to measure natural abundance radiocarbon (^{14}C) have led to increased use of natural ^{14}C as both a source and age tracer in aquatic ecosystem and food web studies. Here, we review (1) $\Delta^{14}C$ values and ages of C and OM typically found in different inland water systems, (2) the mechanisms through which these materials enter inland water ecosystems, and (3) all available ^{14}C data on aquatic consumers across a range of inland water ecosystem types. Using $\Delta^{14}C$ values of aquatic consumers and their potential nutritional resources, we estimate contributions of aged C and OM to aquatic consumer biomass. We conclude that in nearly every case, one or more forms of aged C and/or OM contribute to aquatic consumer nutrition in inland water ecosystems.

Growing evidence indicates that aged and even ancient forms of carbon (C) and organic matter (OM) dominate in most inland waters (Raymond and Bauer 2001; Hossler and Bauer 2013; Spencer et al. 2014*a*). The sizes of Earth's geologically aged C and OM reservoirs are orders of magnitude larger than modern to moderate-aged reservoirs (Hedges 1992). These aged C and OM sources are continuously mobilized to inland water (hereafter referred to as "aquatic") ecosystems by various mechanisms (Butman et al. 2015; Marwick et al. 2015). However, the extent to which aged forms of C and OM contribute to consumer nutrition and food webs, and the factors controlling these contributions, are poorly understood and have not been accounted for in the vast majority of studies, leading to potentially major gaps in our conceptual and quantitative models of C and energy flow in aquatic ecosystems (Guillemette et al. 2017).

The goals of this review and synthesis are to (1) evaluate the literature on potential sources and inputs of C and OM to inland water ecosystems using natural abundance ^{14}C as a tracer, (2) present representative $\Delta^{14}C$ values and ages of particulate and dissolved organic C (POC and DOC, respectively) and dissolved

*Correspondence: bellamy.41@osu.edu

Author Contribution Statement: A.R.B. and J.E.B. contributed equally to the writing of this manuscript.

inorganic C (DIC) pools in streams, rivers, and lakes potentially available to aquatic consumers, (3) assess the $\Delta^{14}C$ values and apparent ages of aquatic consumer organisms in various systems, and (4) estimate aged C contributions to aquatic metazoan consumer biomass and nutrition.

Background

Stable isotopes as tracers of aquatic C and OM nutritional sources

Natural abundance stable isotopes (e.g., ^{13}C, ^{15}N, and ^{2}H) have been used extensively for evaluating OM utilization by aquatic consumers and have been shown to be far more quantitative than, e.g., classical gut content analysis (Peterson and Fry 1987; Junger and Planas 1994; Leberfinger et al. 2011). However, in most aquatic systems $\delta^{13}C$ and $\delta^{15}N$ have relatively small dynamic ranges (tens of ‰ at most), and differentiation of multiple dietary and nutritional sources having overlapping $\delta^{13}C$ and $\delta^{15}N$ signatures can be challenging and non-definitive (Phillips and Gregg 2003). Stable isotope ratios of H ($\delta^{2}H$ or δD) have a much larger dynamic range than $\delta^{13}C$ and $\delta^{15}N$ in natural systems ($\geq \sim 100$‰), and when used in conjunction with $\delta^{13}C$ and $\delta^{15}N$ may provide greater (i.e., isotopically three-dimensional) differentiation of dietary and nutritional contributions (Deines et al. 2009; Cole et al. 2011; Tanentzap et al. 2017). $\delta^{2}H$ has been increasingly used to distinguish between allochthonous and autochthonous sources of OM in aquatic systems because terrestrial vegetation can be enriched in ^{2}H by ~ 100‰ or more over aquatic vegetation (Doucett et al. 2007; Finlay et al. 2010). A major limitation of using stable isotopes in aquatic dietary and nutritional studies is that they cannot easily differentiate C and OM from newly formed contemporary (i.e., modern) sources and those originating from mobilization of far more abundant aged (i.e., 10^{2} to $\geq 10^{6}$ year timescales) sources. Knowledge of the ages of C and OM contributing to consumer nutrition is also important for reassessing the paradigm that young, recently produced materials are more—or even exclusively—biologically available and reactive compared to aged sources (Caraco et al. 2010; McCallister and del Giorgio 2012; Guillemette et al. 2017).

Natural ^{14}C as a source and age tracer of C and OM in aquatic systems

Natural abundance radiocarbon (^{14}C), the radioactive isotope of carbon ($t_{1/2} = 5568$ yr; Stuiver and Polach 1977), has at least a one to two order-of-magnitude greater dynamic range than $\delta^{13}C$, $\delta^{15}N$, and $\delta^{2}H$. Natural ("pre-bomb") $\Delta^{14}C$ values range over ~ 950‰ (i.e., from -1000‰ to ~ -50‰; McNichol and Aluwihare 2007; Taylor 2016) and are even greater (up to ~ 1900‰ range) when anthropogenic ^{14}C inputs such as thermonuclear weapons testing and nuclear reactors are considered (McNichol and Aluwihare 2007; Taylor 2016). In addition to being a highly sensitive source tracer, ^{14}C uniquely allows for determination of C and OM ages in both non-living and living aquatic C-containing components.

Natural abundance ^{14}C has historically been employed much less frequently than stable isotopes in aquatic and terrestrial food web studies. This was due in large part to the challenges in obtaining adequate sample C quantities, lengthy sample processing times, and analytical cost (McNichol and Aluwihare 2007). However, the advent of accelerator mass spectrometry (AMS) has resulted in increased use of natural ^{14}C measurements in ecosystem studies. Prior to the development and widespread use of AMS, ^{14}C analyses were conducted using β-decay methods such as gas proportional and liquid scintillation counting (McNichol and Aluwihare 2007; Taylor 2016). Decay counting requires grams of C and up to days of counting time per sample for an accurate assessment of ^{14}C content, whereas AMS methods require only tens to hundreds of micrograms of C and minutes or less analysis times (McNichol and Aluwihare 2007; Taylor 2016).

Most inland water food web studies utilizing natural ^{14}C have been conducted in subtropical, temperate, and subarctic North American and European systems, but a growing number are being conducted in Asian systems (Table 1). To our knowledge, no such studies have yet been carried out in desert (e.g., hypersaline), tropical, or arctic aquatic systems. While the use of natural ^{14}C for estimating ages of C and OM utilized in aquatic food webs is in principal relatively straightforward, the influence of 1950s–1960s nuclear weapons testing (i.e., the so-called "bomb" ^{14}C) must also be considered, especially for the mid-20th to early 21st century period (McNichol and Aluwihare 2007; Taylor 2016). In addition to the apparent ages of consumer organisms resulting from their utilization of non-living aged OM, live aquatic autotrophs consumed for nutrition may also possess apparent "age" due to fixation of aged DIC (as $CO_2(aq)$) commonly observed in inland waters (Ishikawa et al. 2013, 2014) and must be accounted for.

Variability in the ages of carbon and organic matter in inland waters

Modern-aged forms of carbon and organic matter in inland waters

Non-living C and OM in aquatic systems have been found to range from modern ($\Delta^{14}C \geq \sim 0$‰) to fossil ($\Delta^{14}C \leq -1000$‰) in age (Fig. 1; Supporting Information Table S1; Hedges et al. 1997; Raymond et al. 2004; Butman et al. 2012) and may be either autochthonous or allochthonous in origin. Living or recently living terrestrial vegetation is derived from modern-day CO_2 recently fixed from the atmosphere ($\Delta^{14}C \geq 0$‰; Fig. 1; Supporting Information Table S1; Garnett and Billett 2007; Gaudinski et al. 2009; Carbone et al. 2013). However, modern-aged terrestrial vegetation is a globally relatively small reservoir of organic C compared to moderately aged (i.e., that having $\Delta^{14}C < 0$‰ but > -1000‰) and fossil aged C reservoirs (i.e., those having $\Delta^{14}C = \sim -1000$‰; Figs. 1, 2; Supporting

Table 1. Inland water food web studies from the published literature that have employed natural abundance ^{14}C analyses of consumer organisms.

Study location	Study dates	Organisms studied	References*
North America			
Great Basin Lakes, U.S.A. and Canada	1950s	Fish, brine shrimp	1
Colville River/Beaufort Sea, Alaska, U.S.A.	1980	Fish, invertebrates, birds	2
Hudson River, New York, U.S.A.	2000–2001	Bacteria	3
Eastern Townships Lakes, Quebec, Canada	2004	Bacteria[†], cladocerans, copepods	4
Hudson River, New York, U.S.A.	2004–2005	Cladocerans and copepods	5
Everglades, Florida, U.S.A.	2006–2007	Fish	6
Lake Superior, U.S.A. and Canada	2009	Cladocerans and copepods	7
Pigeon River, Michigan, U.S.A.	2011	Ammocoetes (lamprey larva)	8
Clear Fork River, Ohio, U.S.A.	2011	Ammocoetes (lamprey larva)	8
Lake Superior, U.S.A. and Canada	2012	Cladocerans, copepods, mysids, amphipods, fish	9
Herbert River, Alaska, U.S.A.	2012	Macroinvertebrates and fish	10
Paint Creek, Ohio, U.S.A.	2012–2013	Macroinvertebrates	11
Muskingum River, Ohio, U.S.A.	2013	Freshwater mussels	12
Susquehanna River watershed, Pennsylvania, U.S.A.	2011–2014	Macroinvertebrates	13
Mohawk-Hudson, New York, U.S.A.	2014	Macroinvertebrates	14
Nyack Floodplain, Montana, U.S.A.	2013–2015	Stoneflies	15
Europe			
Trave River, Germany	2007–2010	Fish, crayfish	16
Weibe River, Lake Rosenfield‡, Germany	2009/2011	Mussels	17
Midtdalsbreen Glacier, Norway	2010/2014	Fish, chironomids, aquatic beetles	18
Lough Erne, Ireland	2011	Fish, cladocerans, copepods, mysids	19
Lakes Schwerin‡ and Ostorf‡, Germany	2011	Fish, eels, zebra mussels	20
Rivers Rebbe and Schloss Wilhelosthal[‡], Germany	2013	Fish	21
Asia			
Lake Biwa watershed, Japan	2006–2008	Macroinvertebrates and fish	22
Kanno River watershed, Japan	2011–2012	Macroinvertebrates	22

* References: 1- Broecker and Walton (1959), 2- Schell (1983), 3- McCallister et al. (2004), 4- McCallister and del Giorgio (2008, 2012), 5- Caraco et al. (2010), 6- Wang et al. (2014), 7- Zigah et al. (2012*a*), 8- Evans 2012, 9- Kruger et al. (2016), 10- Fellman et al. (2015), 11- Bellamy et al. (unpubl. data), 12- Weber et al. (2017), 13- Bellamy et al. (pers. comm.), 14- Bellamy et al. (2017), 15- DelVecchia et al. (2016), 16- Philipsen and Heinemeier (2013), 17- Fernandes et al. (2012), 18- Hagvar and Ohlson (2013) and Hagvar et al. (2016), 19- Keaveney et al. (2015), 20- Fernandes et al. (2013), 21- Fernandes et al. (2016), 22- Ishikawa et al. (2010, 2014, 2016).

[†] $\Delta^{14}C$ of bacterial respired CO_2 was measured as a proxy for bacterial biomass.

[‡] Due to small sample sizes from each of these lakes, data were grouped for Fig. 3 as "German Lakes."

Information Table S1). Potential sources of newly formed autochthonous OM in aquatic systems include macrophytes, benthic algae, phytoplankton, cyanobacteria (Allan and Castillo 2007) and young OM-utilizing heterotrophic bacteria (Hall et al. 2000), and biogenic methane (Fig. 1; Supporting Information Table S1; Chanton et al. 1995; Grey 2016).

Moderate- to fossil-aged forms of carbon and organic matter in inland waters

Non-living aged autochthonous and allochthonous OM sources to aquatic systems include aquatic sediments, terrestrial soils, thermogenic and biogenic methane, and sedimentary rocks that have been stored in watersheds, streams, and rivers for decades to many millions of years since their deposition (Fig. 1; Supporting Information Table S1; Hedges 1992; Copard et al. 2007; Battin et al. 2009). Soil OM from leaf litter, roots, and woody debris derived from recent primary production are typically modern- or near-modern (i.e., decadal or less) aged (Fig. 1; Supporting Information Table S1; Gaudinski et al. 2000; Trumbore 2009). Century to millennial aged soil OM typically arises from more highly degraded terrestrial vegetation (Figs. 1, 2; Supporting Information Table S1; Trumbore 2009; Schmidt et al. 2011) and depends on soil depth and horizon, OM reactivity, redox conditions, etc. (Trumbore 2000; Jenkinson et al. 2008; Schrumpf et al. 2013). The resuspension of aged fine soil and sedimentary POM can also lead to moderately aged suspended POM commonly observed in aquatic systems, especially in turbulent flowing waters (Bianchi and Bauer 2011; Hossler and Bauer 2013). Biogenic methane, depending on the age of its OM source (e.g., peat) supporting methanogenesis, can also be moderately to highly aged (Fig. 1; Supporting Information Table S1; Chanton et al. 1995; DelVecchia et al. 2016).

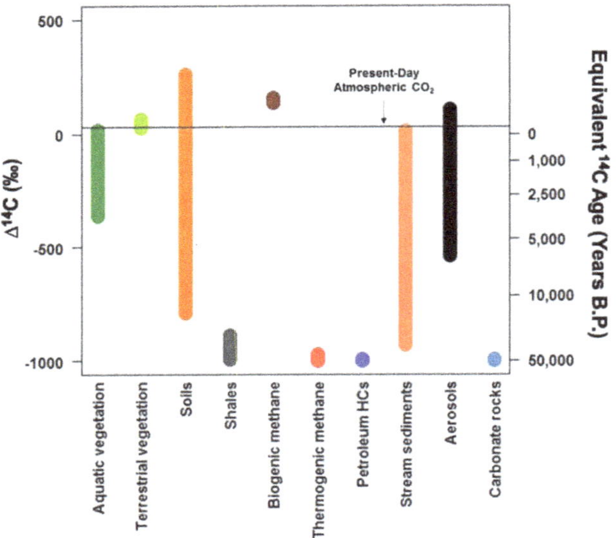

Fig. 1. Representative ranges of $\Delta^{14}C$ values and equivalent ^{14}C ages of potential carbon (C) and organic matter (OM) sources to various inland water ecosystems from the published literature. For details of the values and associated references, *see* Supporting Information Table S1.

Aquatic primary production in aquatic systems is typically assumed to be a source of modern- or near-modern aged OM (Zigah et al. 2011; Kruger et al. 2016). However, ^{14}C is often depleted in DIC and CO_2(aq) in inland waters relative to atmospheric CO_2 (known as the freshwater reservoir offset; Keaveney and Reimer 2012; Philipsen and Heinemeier 2013; Fernandes et al. 2016) and can lead to ^{14}C-depletion and apparent "age" in living autotrophic biomass. ^{14}C-depleted DIC and CO_2(aq) in aquatic systems can originate from fossil carbonaceous rock weathering and respiration of non-living, aged water column soil and sediment OM (Fig. 1; Supporting Information Table S1; Broecker and Walton 1959; Butman and Raymond 2011; Keaveney and Reimer 2012; Ishikawa et al. 2014). This ^{14}C-depleted living or recently living aquatic OM may thus serve as a source of "aged" nutrition to aquatic consumers (Broecker and Walton 1959; Ishikawa et al. 2013, 2014).

Ancient or fossil-aged ($\Delta^{14}C \leq -1000‰$) OM reservoirs derive from materials stored on geological timescales in sedimentary rocks (e.g., shales, kerogens, coal, petroleum, and thermogenic methane; Figs. 1, 2; Supporting Information Table S1). Fossil aged C and OM are far more abundant in global reservoirs than contemporary or moderately aged C and OM by several orders of magnitude (Fig. 2; Hedges 1992). Thus, mobilization of even small amounts of these abundant and highly aged materials has the potential to contribute significantly to the amounts and ^{14}C ages of C and OM in modern-day aquatic systems (Caraco et al. 2010; Hossler and Bauer 2013; Marwick et al. 2015).

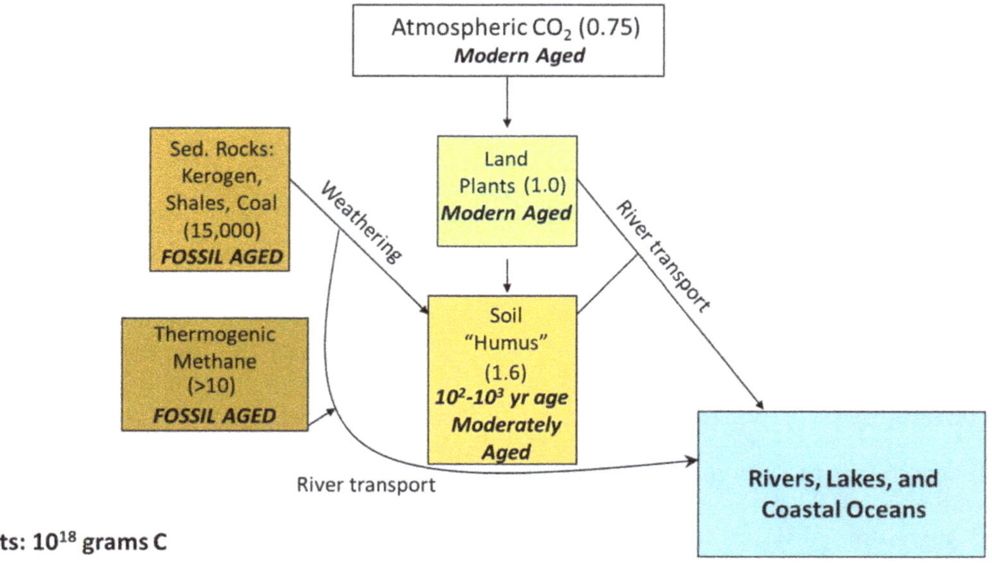

Fig. 2. Potential sources and ages of allochthonous carbon (C) and organic matter (OM) in aquatic systems, and their generalized transport pathways and global reservoir sizes. Values in parenthesis are 10^{18} grams C. Adapted from Hedges (1992) and Bauer and Bianchi (2011).

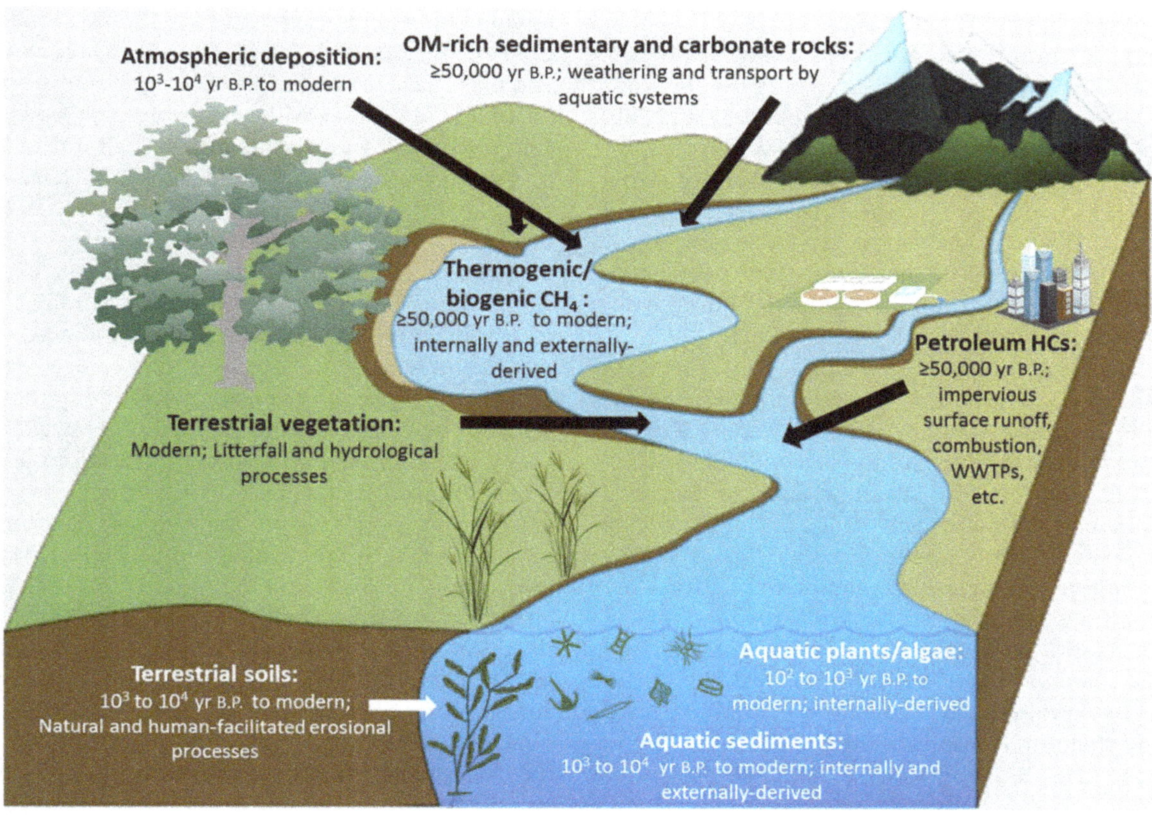

Fig. 3. Potential routes through which carbon (C) and organic matter (OM) of varying ages enter inland water ecosystems. Sources: Bellamy et al. (2017), Bellamy et al. (pers. comm.), Bellamy et al. unpubl. data, Chanton et al. (1995), Clark and Fritz (1997), Currie et al. (1997), Ishikawa et al. (2015), Pessenda et al. (1996), Petsch et al. (2001), and Pohlman et al. (2009), Wozniak et al. (2011, 2012*a,b*). Adapted from http://ian.umces.edu/.

Sources and routes of input of aged C and OM to inland water systems

Aquatic systems and terrestrial landscapes are closely linked, and their connectivity facilitates hydrologic inputs of different sources of C and OM of varying age (Fig. 3; Aufdenkampe et al. 2011). While inland waters cover a small fraction of the earth's surface (~ 3%; Raymond et al. 2013), C and OM yields from the surrounding landscape are large and can support ecosystem processes such as metabolism and secondary production, leading to inland waters being largely net heterotrophic (Cole et al. 2007; McCallister and del Giorgio 2012; Wilkinson et al. 2013). Geomorphology, lithology, climate (e.g., temperature, precipitation), and land use may all influence the forms and ages of C and OM mobilized to inland waters (Marwick et al. 2015).

Aged materials were historically not considered as significant components of aquatic C and OM pools, as they are known to be today (Hedges et al. 1986; Raymond and Bauer 2001). Human activities have dramatically further increased the mobilization of these materials from their storage reservoirs (Fig. 3; Regnier et al. 2013; Butman et al. 2015). Consequently, inputs of aged allochthonous materials to aquatic systems may impact aquatic food webs to an even greater extent as agriculture, urbanization, and other land use changes have expanded (Aufdenkampe et al. 2011; Butman et al. 2015; Fellman et al. 2015). In addition, assumptions of the biological recalcitrance of allochthonous OM based on molecular structure and age are also changing (McCallister and del Giorgio 2012; Fellman et al. 2014; Marín-Spiotta et al. 2014).

Extensive aging of C and OM results from long-term storage in forms (e.g., sorbed to mineral particles) and environments (e.g., suboxic and anoxic) that inhibit or prevent their degradation and utilization by heterotrophic bacteria and higher organisms (Salmon et al. 2000; Lützow et al. 2006; Kleber 2010). However, when aged OM is released from its points of storage and protection, it may become increasingly susceptible to degradation and utilization (Petsch et al. 2001; Schillawski and Petsch 2008; Kleber and Johnson 2010). Interactions between terrestrial and aquatic environments via the hydrologic cycle therefore likely (1) increase rates and extents of aged C and OM processing (Blair and Aller 2012; Marín-Spiotta et al. 2014), and (2) modify the routes through which C and OM of differing sources and ages enter streams, rivers, and lakes (Fig. 3).

Streams and rivers

Carbon and OM in streams and small rivers is typically dominated by allochthonous materials (Fig. 3) for two primary reasons. First, stream waters arise from surface and subsurface runoff in watersheds, and these waters are in intimate contact with both living and non-living terrestrial biomass, as well as with soils and their associated terrestrial OM (Hope et al. 1997; Aitkenhead et al. 1999; Boix-Fayos et al. 2009). Second, shading by riparian vegetation and canopy cover can limit autochthonous production in streams and small rivers (Vannote et al. 1980; Finlay 2001).

Both natural and anthropogenic controls play important roles in the sources and ages of C and OM entering streams and rivers (Fig. 3; Hossler and Bauer 2013; Marwick et al. 2015). Natural controls include hydrogeomorphology, lithology, and climate, all of which may show dramatic temporal and spatial variability (Hossler and Bauer 2012, 2013). For example, steep small mountainous rivers (SMRs)—typically underlain by ancient kerogen-rich sedimentary rocks—are responsible for transport and delivery of large amounts of moderately to fossil aged C and OM to streams, rivers, and ocean margins (Kao and Liu 1996; Leithold et al. 2006; Hilton et al. 2008). Low-gradient watersheds containing fossil carbonates and shales may also introduce aged C and OM to aquatic systems (Fig. 3; Longworth et al. 2007; Marwick et al. 2015).

Human activities in watersheds are increasingly important controls on the sources and ages of C and OM in streams and rivers, and land use change may alter the ages of materials transported to aquatic systems (Fig. 3; Butman et al. 2015). Agricultural activity and/or removal of riparian vegetation may lead to higher inputs of moderately aged soil C to streams and rivers by destabilizing soils and accelerating erosion (Longworth et al. 2007; Restrepo et al. 2015). While global increases in watershed disturbance have been concentrated in developed countries, the intensity and timing of land use change has been highly variable across terrestrial regions (Goldewijk 2001; Vörösmarty et al. 2010; Lambin and Meyfroidt 2011). These and related factors may lead to variability in the delivery of aged C and OM to specific aquatic systems. In urbanized regions, petroleum hydrocarbons in runoff, wastewater treatment plants, and aerosol deposition can all transport aged C and OM to streams and rivers (Fig. 3; Griffith et al. 2009; Wozniak et al. 2012*a*,*b*). Increasing stream water temperatures resulting from climate warming have also led to observed increases in aged C and OM inputs to streams and rivers, especially in high latitude regions associated with thawing permafrost soils and glaciers (Vonk et al. 2013; Hood et al. 2015).

Lakes

Depending upon their size, lakes are also integrated to varying extents with the terrestrial landscape and are influenced by many of the same factors that affect inputs of C and OM of different sources and ages to streams and rivers (Caraco and Cole 2004; Cole et al. 2007). However, in contrast to streams and rivers, lake basins can serve as significant storage reservoirs for imported C and OM from surrounding watersheds (Fig. 3; Cole et al. 2007; Tranvik et al. 2009). High stream flow may lead to increased delivery of younger-aged DOC (Fig. 3; Zigah et al. 2011, 2014; Spencer et al. 2014*b*). The surface area and volume of lakes further influence the sources and ages of C and OM that predominate in them (Wetzel 1990; Zigah et al. 2012*a*,*b*). Higher watershed : lake surface areas generally lead to greater inputs and amounts of allochthonous and aged C and OM in small lakes, whereas in large lakes modern-aged autochthonous production tends to be quantitatively more important (Zigah et al. 2012*a*; Wilkinson et al. 2013; Tanentzap et al. 2017). Aged C and OM may be more common in lakes in drainage basins having lithologies comprised of aged forms of C and OM, and in lakes at higher elevations and latitudes receiving glacial runoff and inputs from combustion-derived aerosol deposition (Broecker and Walton 1959; Keaveney and Reimer 2012; Spencer et al. 2014*a*; Keaveney et al. 2015).

Anthropogenic activities such as fossil fuel combustion may lead to increased inputs of aged C and OM to watersheds and aquatic water bodies (Fig. 3; Wozniak et al. 2012*a*,*b*; Iavorivska et al. 2017; Mahowald et al. 2017). Glacial meltwaters are also increasingly recognized sources of aged C and OM from both natural and anthropogenic aerosol deposition (Xu et al. 2009; Spencer et al. 2014*a*,*b*). The ages of C and OM both in, and transported by, aquatic systems can thus vary widely, and will depend on the sources and ages of C and OM mobilized to their POC, DOC, and DIC pools.

Ages of POC, DOC, and DIC pools in streams and rivers

The majority of studies have found POC to be more highly aged than DOC and DIC in streams and rivers (Fig. 4A; Supporting Information Table S2). This is thought to result from inputs of moderately to highly aged OM from fossil aged bedrock and accreted soils (Raymond and Bauer 2001*a*; Leithold et al. 2006; Blair et al. 2010; Marwick et al. 2015). The generally young ages of stream and river DOC (Hossler and Bauer 2013; Marwick et al. 2015; Fig. 4A; Supporting Information Table S2) may be attributed to significant contributions from leachates of fresh surface soil litter and root exudates (Aitkenhead-Peterson et al. 2003; Dodds et al. 2017). More highly aged riverine DOC is commonly associated with human activities such as agricultural, wastewater treatment plant, and petroleum inputs (Wang et al. 2012; Butman et al. 2015; Marwick et al. 2015). Glacial melt water also contributes moderately aged DOC to subarctic streams and rivers (Hood et al. 2009, 2015). With the exceptions of some agricultural and high relief watersheds, stream and river DIC is generally modern or near modern in age (Fig. 4A; Supporting Information Table S2), suggesting that it is predominantly derived from respiration of recently produced forms of OM (e.g., terrestrial and aquatic plants; Marwick et al. 2015). $\Delta^{14}C$ and $\delta^{13}C$ values of DIC are often more challenging to

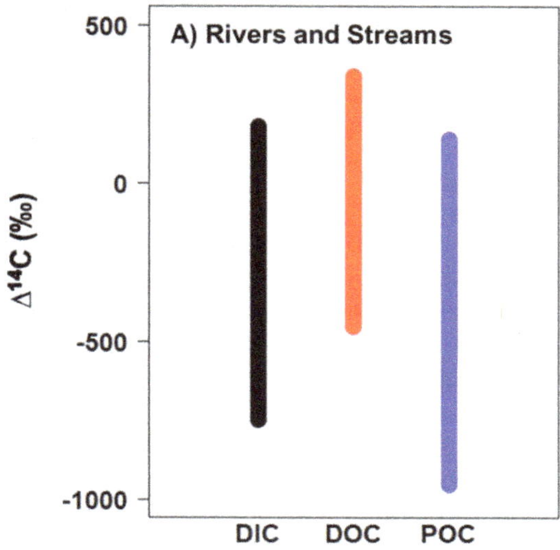

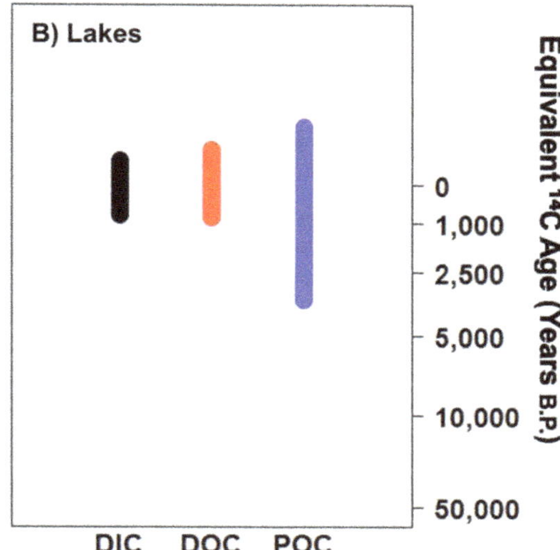

Fig. 4. Representative ranges of $\Delta^{14}C$ and equivalent ^{14}C ages of dissolved inorganic carbon (DIC), dissolved organic carbon (DOC), and particulate organic carbon (POC) from the published literature for various (**A**) streams and rivers and (**B**) lakes. For details on the values and associated studies, see Supporting Information Tables S2 and S3.

interpret than those of POC and DOC due to combined effects of respiration, dissolution of mineral carbonates, and both turbulent and diffusive atmospheric exchange (Finlay 2003; Raymond et al. 2004; Hossler and Bauer 2013).

Ages of POC, DOC, and DIC pools in lakes

Fewer studies have used natural ^{14}C in lakes than in streams and rivers, but this number has been gradually increasing. The $\Delta^{14}C$ values and ages of DIC, DOC, and POC in lakes studied to date are generally less variable than in streams and rivers (Fig. 4A,B; Supporting Information Tables S2, S3). Similar to streams and rivers, POC in lakes is generally more highly and variably aged than DOC or DIC (Fig. 4A,B; Supporting Information Tables S2, S3; Zigah et al. 2011, 2012*a*,*b*; McCallister and del Giorgio 2012; Fernandes et al. 2013; Keaveney et al. 2015). DOC and DIC in lakes are most often modern- to century-aged (Fig. 4A,B; Supporting Information Tables S2, S3) and thought to result from similar sources and processes as streams and rivers. Runoff from surrounding watersheds and inputs from groundwaters may also be important sources of aged DOC and DIC to lakes (Brady et al. 2009; Melymuk et al. 2014). We speculate that lake trophic status (i.e., nutrient loading) co-varies inversely with aquatic C and OM age, especially in fertilized agricultural and recently deforested watersheds.

$\Delta^{14}C$ *values and apparent ages of aquatic consumer organisms*

Utilization of aged C and OM by aquatic consumers

When aged C and OM is mobilized from its points of preservation and storage to aquatic systems, it may be assimilated by aquatic bacterial and metazoan consumers as particulate and/or dissolved OM (POM and DOM, respectively; note that POC and DOC refer to only the carbon-containing components of POM and DOM; Supporting Information Fig. S1). POM and DOM can contribute to consumer diet both directly and indirectly. Metazoan consumers may directly ingest aged POM (which also includes aquatic primary producers that have fixed aged aquatic DIC), as well as flocculated aged DOM (Supporting Information Fig. S1; Wallace et al. 1997; Kerner et al. 2003; Zigah et al. 2012*a*). Some soft-bodied organisms (e.g., dreissenid mussels) also have the ability to take up DOM directly (Supporting Information Fig. S1; Roditi et al. 2000; Baines et al. 2005). In addition to being consumers of non-living and aquatic OM, heterotrophic bacteria may also "repackage" aged DOM and POM into bacterial biomass that can then be consumed by higher consumers directly or by grazing on protozoan bacterivores (Supporting Information Fig. S1; Cherrier et al. 1999; McCallister et al. 2004; Berggren et al. 2010).

Bacterial utilization of aged C and OM in temperate and subtropical systems

It is well-documented that heterotrophic bacteria play an important role in the degradation and utilization of fossil aged petroleum hydrocarbons in many aquatic systems (Atlas 1981; Heitkamp and Johnson 1984; Ahad and Pakdel 2013). Bacteria, through their enzymatic and hydrolytic activities, are also known to mediate the mobilization of C and OM to inland waters (Guillemette et al. 2016). Natural ^{14}C measurements show that modern- to moderate-aged (i.e., non-fossil) OM contributes to heterotrophic bacterial biomass ($\Delta^{14}C$ range $-153‰$ to $214‰$; 1330 yr to modern-aged, respectively; McCallister et al. 2004) and/or respired $CO_2(aq)$ ($\Delta^{14}C$ range $-172‰$ to $94‰$; 1470 yr to modern-aged, respectively; McCallister and del Giorgio 2012) in temperate lake systems (Table 1). Fellman et al. (2014) reported that in the Kimberley region of Western Australia stream DOC was increasingly enriched in ^{14}C with increasing stream size. Use of a mass balance approach to estimate potential ^{14}C-enrichment of DOC from each stream suggested that one potential explanation for the ^{14}C enrichment was that heterotrophic bacteria selectively metabolized older, ^{14}C-depeleted DOC components (Fellman et al. 2014).

Fossil DOC leached from shales in watersheds may also enter streams and rivers and be utilized by heterotrophic bacteria. Petsch et al. (2001) estimated that 74–94% of the lipid C in bacteria growing on watershed shales was derived from fossil ($\geq$ 50,000 yr in age) OM. Rapid bacterial utilization of fossil DOC solubilized from shale OM (80% loss over 2-week incubation) further illustrates that highly aged forms of OM may be bioavailable once mobilized to aquatic systems (Schillawski and Petsch 2008).

Metazoan utilization of aged C and OM in temperate and subtropical systems

Invertebrates in inland waters play an important role in the processing of autochthonous and allochthonous C and OM through their feeding and metabolic activities (Anderson and Sedell 1979; Wallace et al. 2015; Tanentzap et al. 2017). They are also an important source of nutrition for higher consumers (e.g., vertebrates). Natural ^{14}C measurements have been conducted on both primary aquatic consumers as well as secondary consumers and predators (Table 1), and studies suggest that aged C and OM can be transferred to progressively higher trophic levels (Schell 1983; Wang et al. 2014; Hagvar et al. 2016).

Streams and rivers

Natural ^{14}C has been employed as a food web tracer in multiple stream and river systems (Table 1). ^{14}C-depletion in living consumer biomass may result from direct consumption and assimilation of (1) non-living aged OM (Schell 1983; Caraco et al. 2010; Hagvar and Ohlson 2013; Wang et al. 2014) and/or (2) living ^{14}C-depleted OM derived from fixation of ^{14}C-depleted DIC and $CO_2(aq)$ by aquatic primary producers (Figs. 1, 5; Supporting Information Table S1; Fernandes et al. 2012, 2016; Philipsen and Heinemeier 2013; Ishikawa et al. 2014).

Caraco et al. (2010) showed that cladoceran and copepod zooplankton from the Hudson River, New York, USA had a mean $\Delta^{14}C$ of $-240‰$ (equivalent age of $\sim$ 2200 yr B.P.) due to utilization of millennial-aged terrestrial soil OM (Fig. 5; Supporting Information Table S1). In this study, zooplankton mean $\Delta^{14}C$ was nearly 200‰ ($\sim$ 1790 yr) lower than average phytoplankton $\Delta^{14}C$ and about 300‰ ($\sim$ 2870 yr) lower than modern terrestrial plant biomass. Similar conclusions of aged OM assimilation by freshwater larval insect and mussel consumers were reached by studies in the Hudson-Mohawk watershed, New York and the Muskingum River, Ohio, respectively (Fig. 5; Weber et al. 2017; Bellamy et al. 2017). In the Hudson-Mohawk, aged soil OM contributions to macroinvertebrate nutrition were greater in streams with high vs. low agriculture watersheds, however, utilization of ^{14}C-depleted "aged" algae by macroinvertebrates could also not be ruled out (Bellamy et al. 2017). Recent $\Delta^{14}C$ measurements of stoneflies in the hyporheic zone of the Nyack floodplain on the Middle Fork of the Flathead River, Montana, USA revealed that they were highly aged (up to 6900 yr B.P.; Table 1; Fig. 5, Supporting Information Table S1), suggesting they utilized methanotrophic bacterial biomass derived from aged biogenic, and possibly thermogenic, methane (DelVecchia et al. 2016).

In the Kanno River, Japan, deforestation was found to lead to inputs and autotrophic fixation of ^{14}C-depleted DIC from fossil carbonate weathering, resulting in ^{14}C-depleted aquatic primary producer biomass (Ishikawa et al. 2016) (Fig. 5). Similar to the Hudson-Mohawk system, modern-aged forms of OM contributed more to macroinvertebrate biomass from Kanno River streams in watersheds having less disturbance (Ishikawa et al. 2016). These authors further suggest that during forest reestablishment, root biomass of *Cryptomeria japonica* (Japanese cedar) increased bacterial respiration of soil OM to CO_2, and consequently led to greater carbonic acid weathering of soil fossil carbonates and inputs of ^{14}C-depleted DIC to headwater streams (Ishikawa et al. 2016). Thus, various forms of watershed disturbance may facilitate the mobilization of aged C and OM into aquatic systems, and impact the ^{14}C content and apparent ages of consumers (Caraco et al. 2010; Ishikawa et al. 2016; Fig. 3).

Lakes and wetlands

Fewer studies using natural ^{14}C have been conducted in lake and wetland food webs than in streams and rivers (Table 1; Fig. 5, Supporting Information Table S1). In the lakes studied to date, fixation of aged DIC and $CO_2(aq)$ by primary producers appears to be the dominant pathway through which consumers assimilate aged C (Keaveney and Reimer 2012; Fernandes et al. 2013; Keaveney et al. 2015). In Eastern Townships lakes (Quebec, Canada), cladocerans and copepods were used as proxies for autochthonous primary production by assuming that zooplankton would select algal over detrital components of the POM pool (McCallister and del Giorgio 2008). Zooplankton $\Delta^{14}C$ values ranged from $-2‰$ to 40‰ (modern in age) and overlapped with the $\Delta^{14}C$ values of the DIC, thus appearing to confirm the validity of this assumption. Aged terrestrial OM was therefore deduced not to contribute significantly to zooplankton biomass in these lakes (Fig. 5; McCallister and del Giorgio 2008).

^{14}C-depletion ($\Delta^{14}C = -130‰$ to 33‰; 1120 yr to modern-aged) in vertebrates and molluscs in German lakes (Lakes Rosenfeld, Schwerin, Ostorf, and Schloss Wilhelosthal; Fig. 5) was also attributed to their consumption and utilization of ^{14}C-depleted algae ($\Delta^{14}C = -44‰$; 270 yr equivalent age) and

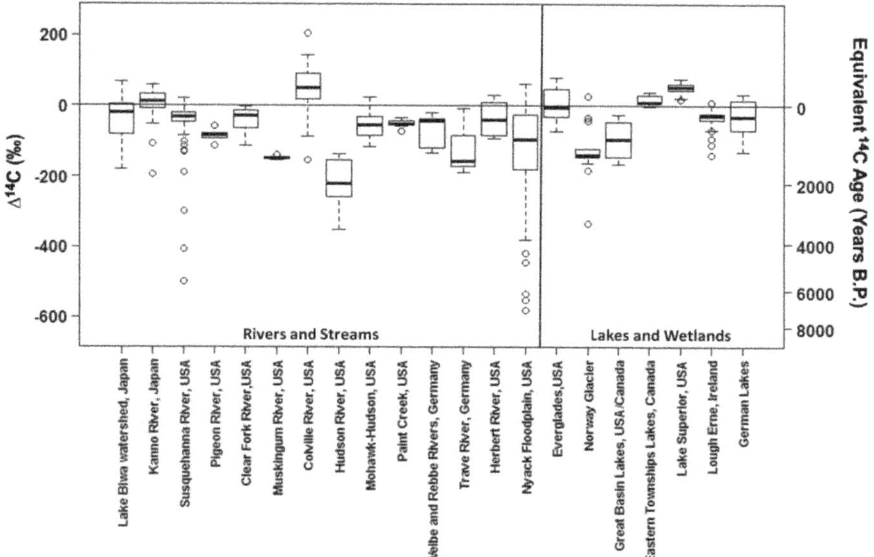

Fig. 5. $\Delta^{14}C$ values and equivalent ^{14}C ages for metazoan consumers of various inland waters from the published literature. *See* Table 1 for details of systems and organisms analyzed.

macrophytes ($\Delta^{14}C = -33‰$; 360 yr equivalent age) fixing ^{14}C-depleted DIC, which in Lake Schwerin, DIC was 240 yr in equivalent age ($\Delta^{14}C = -29‰$; Fernandes et al. 2013). However, mussel tissue collected from Lake Ostorf was more ^{14}C-depleted ($\Delta^{14}C = -130‰$; 1120 yr equivalent age) than fish and mussels collected from Lake Schwerin. It was speculated by the authors that ^{14}C-depleted, respired CO_2 from aged peat OM in the Lake Ostorf watershed contributed to lake DIC and phytoplankton (neither was measured directly) that served as a nutritional source to mussels there (Fernandes et al. 2013).

Extensive work in Lough Erne, Ireland has similarly shown algae to be ^{14}C-depleted (mean $\Delta^{14}C = -59 \pm 9‰$; 490 yr equivalent age), suggesting a significant freshwater reservoir offset, presumably from inputs of weathered fossil carbonates and/or aged soil CO_2 to lake DIC (Keaveney et al. 2015*a*,*b*). In contrast, some consumers in Lough Erne were enriched in ^{14}C (maximum invertebrate and fish $\Delta^{14}C$ values were 0‰ and 10‰, respectively, and modern-aged; Fig. 5), suggesting that modern terrestrial OM contributed to their biomass (Keaveney et al. 2015*a*). POC was more ^{14}C-depleted ($-122 \pm 61‰$; 1050 yr in age) than DIC or algae, suggesting that there was a moderately aged, but relatively non-utilized, source of OM in the lake. Copepods in Lough Erne apparently selected against significant amounts of this aged OM because their $\Delta^{14}C$ values (−50‰ to −4‰; 410 yr to modern-aged; Fig. 5) closely matched algae and DOC (depending on season), while cladocerans ($\Delta^{14}C = -138‰$ to $-67‰$; equivalent ages of 1190–560 yr B.P.; Keaveney et al. 2015) derived some of their biomass from moderately aged detrital OM.

Natural ^{14}C measurements of zooplankton and fish in the limited number of lakes studied to date suggest in the majority of cases that their utilization of aged C and OM is minimal (Fig. 5). Due to their greater watershed : lake surface area ratios, small lakes are more likely to receive relatively greater amounts of aged allochthonous materials than larger lakes (smaller watershed : lake surface areas) that contain relatively greater amounts of ^{14}C-enriched modern-aged autochthonous C and OM (Zigah et al. 2012*a*,*b*; Wilkinson et al. 2013) from autotrophic utilization of ^{14}C-enriched CO_2 from atmosphere-lake exchange. Studies in Lake Superior, the largest lake yet studied, showed that recently produced ^{14}C-enriched aquatic OM was the preferred source of nutrition to cladocerans and copepods (Fig. 5). Zooplankton $\Delta^{14}C$ values were slightly elevated ($\Delta^{14}C = 44–65‰$; modern-aged; Fig. 5) compared to POC and DOC ($\Delta^{14}C = 10 \pm 29‰$ and $38 \pm 21‰$, respectively; both modern-aged), and contributions of aged OM to the Lake Superior pelagic food web were minimal (Zigah et al. 2012*a*). More recent work in Lake Superior has shown that aged (940 yr equivalent age; $\Delta^{14}C = -117‰$; Li et al. 2013) sedimentary OM was assimilated by the benthic amphipod *Diaporeia*, which was more ^{14}C-depleted (by ~ 61‰, in some cases) relative to non-benthic organisms (e.g., zooplankton and fish; Kruger et al. 2016). However, young recently produced OM contributed more than aged sources even to benthic consumer biomass in Lake Superior. Examination of the relationships between zooplankton $\Delta^{14}C$ and hydrologic residence times (an indicator of lake size) suggested that allochthonous, and potentially aged resources, support more zooplankton biomass in smaller lakes (Zigah et al. 2012*a*). However, given that there have been limited studies utilizing ^{14}C in large lakes, we cannot definitively conclude that aged C and OM are not important to consumers in large lakes. Additionally, the composition of vegetation and soil in lake catchments will strongly influence the sources and ages of C and OM in lakes, regardless of size (Sobek et al. 2007; Tanentzap et al. 2017).

In contrast to most lake consumers, the ^{14}C contents of wetland consumers (fish and aquatic invertebrates) in the Florida Everglades, USA examined by Wang et al. (2014) suggest that they rely to varying extents on aged OM. Largemouth bass and sunfish collected from agriculturally impacted regions of the Florida Everglades were ^{14}C-depleted ($\Delta^{14}C$ range: $-52‰$ to $-22‰$; 430–180 yr equivalent age; Fig. 5). The authors assumed that OM derived from exposed peat deposits (estimated at $\sim$ 2000 yr old) from nearby agricultural activity was exported to Everglades wetlands. This aged OM was assimilated by aquatic invertebrates, and in turn consumed by fish (Wang et al. 2014). Wetlands differ from most lakes in a number of important ways. Their generally shallower nature, smaller volumes, and proximity to abundant aged sources (e.g., peat) of OM may lead to greater reliance of wetland consumers on aged sources of C and OM.

Bacterial utilization of aged C and OM in subarctic systems

The most commonly used method of assessing the age of OM utilized by aquatic bacteria in subarctic waters has been to measure changes in the amounts and $\Delta^{14}C$ values of DOC before and after a period (generally days to weeks) of dark incubation (Hood et al. 2009; Vonk et al. 2013). Isotopic mass balance between the amounts and $\Delta^{14}C$ values of DOC at the start and end of the incubation is then used to infer the $\Delta^{14}C$ of the utilized DOC, and by association, of the biomass of bacteria utilizing it. Laboratory incubation experiments using this approach with DOC from streams discharging to the Gulf of Alaska revealed that there was a significant positive correlation between DOC degradation and ^{14}C age (i.e., older DOC was preferentially utilized; Hood et al. 2009). Separate incubations using DOC from thawing, millennial-aged permafrost from Siberian Yedoma deposits showed that highly aged (24,100 and 21,700 yr equivalent ages in 2010 and 2011, respectively), ^{14}C-depleted ($-946 \pm 25‰$ and $-933‰$ for 2010 and 2011, respectively) DOC was utilized by stream heterotrophic bacteria from the Kolyama River, Siberia (Vonk et al. 2013). Thus, the information available to date from subarctic waters suggests that aged OM can be repackaged to bacterial biomass and potentially be available to higher consumer levels.

Metazoan utilization of aged C and OM in subarctic systems

Growing evidence suggests that aged C and OM from permafrost and melting glaciers can also contribute to higher aquatic consumers in downstream rivers and lakes in subarctic ecosystems (Schell 1983; Fellman et al. 2015). One of the earliest applications of natural ^{14}C in aquatic food web studies was conducted by Schell (1983) using β-decay counting in an Alaskan river-estuarine food web (Table 1; Fig. 5). Anadromous and riverine fish and oldsquaw ducks collected from the Colville River, Alaska, USA were significantly more ^{14}C-depleted ($\Delta^{14}C = -153‰$ to $-55‰$, 1330–460 yr equivalent age) than marine fish from the Beaufort Sea ($\Delta^{14}C = 17‰$ to $204‰$, modern-aged) due to the reliance of river invertebrate prey on millennial aged ($\sim$ 8700 yr B.P.) terrestrial peat (Schell 1983).

Recent studies in Norwegian subarctic systems found that tissues from predatory terrestrial invertebrates (spiders, harvestmen, and beetles) were moderately aged ($\Delta^{14}C = -130‰$ to $-41‰$; 1100–340 yr equivalent age), and gut content analysis revealed that aged chironomids ($\Delta^{14}C = -121‰$; 1040 yr equivalent age; Fig. 5) were a primary nutritional resource (Hagvar and Ohlson 2013). Hagvar and Ohlson (2013) further postulated that chironomid larvae that relied on aged, glacier-derived OM were then consumed by terrestrial invertebrate predators. In the same study, a predaceous diving beetle larva and adult collected from a nearby dam had equivalent ^{14}C ages of 1200 ($\Delta^{14}C = -139‰$) and 1130 yr ($\Delta^{14}C = -131‰$), respectively (Fig. 5). Diving beetles were also likely consuming aged chironomid larvae (Hagvar and Ohlson 2013). Subsequent studies in small lakes in the same region of Norway confirmed these earlier studies' original hypothesis that aged glacial C and OM was transferred through chironomid larvae before being transferred to terrestrial and aquatic predatory invertebrates (Hagvar et al. 2016).

Studies by Fellman et al. (2015) in the heavily glaciated Herbert River, Alaska, USA also provide evidence of direct assimilation of glacial meltwater-derived C and OM by aquatic consumers. In a non-glacial stream, $\Delta^{14}C$ values of macroinvertebrates and fish ranged from $-14‰$ to $52‰$ (115 yr to modern-aged), while in the glacially impacted upper and lower Herbert River, macroinvertebrates and fish had $\Delta^{14}C$ values as low as $-171‰$ (equivalent age of 1500 yr; Fellman et al. 2015; Fig. 5). Inputs of aged DOC from glaciers and thawing permafrost to inland waters are predicted to increase with increasing global temperatures (Hood et al. 2015; Spencer et al. 2015), and this aged OM could increasingly contribute to the secondary production of both bacteria and higher consumers (Fellman et al. 2015; Hagvar et al. 2016).

Estimates of aged C and OM utilization by aquatic consumers

The use of stable isotope mixing models in food web studies has provided important insights to the nutritional sources to consumers (Fry 2007; Moore and Semmens 2008; Layman et al. 2012). Consumers often rely on multiple nutritional sources, and use of isotopic values of organisms and their potential nutritional resources along with known isotopic fractionation factors in mixing models provide estimates of the relative contributions of the different potential nutritional sources to the consumer (Phillips and Gregg 2003; Solomon et al. 2011; Phillips et al. 2014). Early linear mixing models relied on a system of algebraic equations (Phillips and Gregg 2001), whereas more recent models employ a Bayesian approach and allow for improved incorporation of uncertainty (Phillips and Gregg 2003; Moore and Semmens 2008). Use of multiple (2–4) isotopes in a mixing model can additionally improve estimates of source contributions to consumer biomass by further reducing estimate uncertainty, especially when multiple isotopically unique potential nutritional sources are being considered (Finlay et al. 2010; Phillips et al. 2014). In addition, when natural ^{14}C is used in conjunction with stable isotopes in mixing models, estimates of the ages of potential nutritional resources to consumer biomass are also obtained.

In order to estimate maximum possible C contributions of varying ages to aquatic consumers in different aquatic systems where natural ^{14}C has been used (Table 1), we used the most ^{14}C-depleted values for consumer organisms from each study system shown in Fig. 5. For illustrative purposes, we employed a simple, linear two endmember mixing model approach with one fossil to millennial aged endmember (i.e., nutritional source) and one modern-aged endmember. Since we used only two potential nutritional sources, we employed the following mass balance equations to estimate the proportional contribution of each to the consumer (Phillips and Gregg 2001):

$$\Delta_{mix} = p_A * \Delta_A + p_B * \Delta_B \tag{1}$$

$$p_A + p_B = 1 \tag{2}$$

where *mix* represents the mixture (consumer), *A* and *B* each represent the two endmember nutritional sources, and *p* represents the proportional contribution of each nutritional source. Δ_A, Δ_B, and Δ_{mix} are the $\Delta^{14}C$ values of *A*, *B*, and the mixture, respectively.

Our calculations employed three different aged endmembers (*A*) in order to cover a realistic range of aged C and OM sources to aquatic systems and consumers. These consisted of: (1) a fossil-aged endmember (either carbonate or shale-derived OM; $\Delta^{14}C = -1000‰$, age $\geq$ 50,000 yr B.P.), (2) a moderately aged soil C and OM endmember ($\Delta^{14}C = -540‰$, age = 6240 yr B.P.; Hossler and Bauer (2012)), and (3) a 50 : 50 mixture of the moderately aged soil endmember ($\Delta^{14}C = -540‰$) and a young soil C and OM endmember ($\Delta^{14}C = 157‰$, modern in age; Hossler and Bauer 2012), yielding an endmember of $\Delta^{14}C = -192‰$ (age = 1710 yr B.P.). In their analysis of the $\Delta^{14}C$ and $\delta^{13}C$ values of POC, DOC, and DIC in eight northeast U.S. rivers, Hossler and Bauer (2012) determined that river OM was dominated by varying combinations of young and moderately aged soil endmembers. The results of the calculations using the three different aged endmembers are shown in Table 2.

Streams and rivers

Using the fossil endmember ($\Delta^{14}C = -1000‰$), our model estimates indicate that fossil C could potentially contribute maximums of 9–58% to consumer biomass in all streams and rivers included in our study (Table 2). When moderately aged, and a combination of moderately and young aged, soil endmembers were used, the maximum estimates of aged nutritional contributions to consumers increased correspondingly. Using the moderately aged soil C and OM endmember resulted in maximum nutritional contributions to consumers of this material ranging from 17% to > 100% (Table 2). Employing a combined moderately/young aged soil endmember yielded a maximum contribution to consumers of 48% to > 100% (Table 2). Estimates > 100% indicate that contributions from the moderately and combined moderately/young aged materials are not adequate to account for the entirety of aged consumer biomass. Therefore, utilization by consumers of more highly aged (i.e., greater amounts of fossil and/or moderately aged) materials are necessary to attain their observed $\Delta^{14}C$ values (Table 2).

Only some of the studies reviewed here (Caraco et al. 2010; Ishikawa et al. 2014, 2016; Fellman et al. 2015; DelVecchia et al. 2016) provided estimates of the contributions of aged C and OM to consumer biomass. Differences between our fossil endmember estimates and those of the studies of Fellman et al. (2015), Ishikawa et al. (2014, 2016), and DelVecchia et al. (2016) may be explained by the fact that the glacially derived C, periphyton C, and methane endmembers that each study used, respectively, were not as ^{14}C-depleted as our fossil C endmember. Use of soil C and OM endmembers (both moderately aged and a 50 : 50 mixture of moderately and young aged) in our model led to estimates closer to those calculated by Fellman et al. (2015) and Ishikawa et al. (2014, 2016). However, for DelVecchia et al. (2016), our moderately aged soil endmember was not old enough to explain their observed stonefly ages (*see* Supporting Information Section 2 for additional details regarding our vs. published estimates of aged nutritional contributions). Our estimates based on C and OM of moderately aged and 50 : 50 mixture of moderately and young aged soil may be more realistic because, as noted by Hossler and Bauer (2012), most rivers contain admixtures of moderately and young aged soil C and OM. This is consistent with our predicted contributions of the sources and ages of materials to consumer biomass for previous studies.

Lakes and wetlands

Our model indicates that maximum contributions of fossil, moderately aged, and a mixture of moderately/young aged soil C and OM to consumer biomass, were highly variable across all temperate and subtropical lake and wetland systems (Fig. 5). Maximum contributions of fossil, moderately aged, and mixed moderately/young aged material ranged from 0–17%, 0–31%, and 0–86%, respectively (Table 2). These estimates suggest that temperate and subtropical lake food webs may be less

Table 2. Maximum estimated contributions of fossil aged C and OM, moderately aged soil C and OM, and a 50 : 50 mixture of moderately aged and young aged soil C and OM endmembers to metazoan consumer biomass as calculated in the present study for lakes, wetlands, streams, and rivers in which organism $\Delta^{14}C$ was measured. Also provided are the original contributions of aged C and OM estimated in each of the studies.

System type	Fossil aged (%)*	Moderately aged soil (%)†	Moderately and young‡ aged soil (%)§	Original estimates (%)
Streams and rivers				
Lake Biwa watershed, Japan	18	33	93	41–62¶
Kanno River, Japan	20	36	>100‖	~45–90¶
Susquehanna River, Pennsylvania	50	92	>100‖	5–76#
Pigeon River, Michigan	11	21	58	ND
Clear Fork River, Ohio	11	21	58	ND
Muskingum River, Ohio	15	29	80	7–27¶
Colville River, Alaska	15	28	80	ND
Hudson River, New York	35	65	>100‖	21–57¶
Mohawk-Hudson, New York	11	21	60	0–93#
Weibe and Rebbe Rivers	13	24	69	ND
Trave River	19	35	97	ND
Herbert River, Alaska	9	17	48	25–36¶
Nyack Floodplain, Montana	58	>100§	>100‖	26.2–69.2¶
Lakes and wetlands				
Everglades, Florida	7	13	38	31¶
Norway Glacier	33	62	>100‖	ND
Great Basin Lakes	17	31	86	ND
Eastern Townships lakes, Quebec	0	0	0	ND
Lake Superior	0	0	0	3#
Lough Erne	14	26	72	ND
German Lakes	13	24	68	ND

* Fossil C and OM endmember $\Delta^{14}C = -1000‰$ (equivalent ^{14}C age ≥ 50,000 yr B.P.).

† The moderately aged soil material is defined as "passive aged" soil C and OM by Hossler and Bauer (2012). Estimated $\Delta^{14}C$ values for the passive soil C and OM pool were −538‰ (equivalent ^{14}C age = 6200 yr B.P.) for Inceptisols and −541‰ (equivalent ^{14}C age = 6260 yr B.P.) for Ultisols according to Hossler and Bauer (2012). We used the average of these two values for a passive aged soil OC endmember $\Delta^{14}C$ value of −540‰ (equivalent ^{14}C age = 6240 yr B.P.). See text for calculation details.

‡ The young aged soil material is defined as "slow turnover" soil C and OM ($\Delta^{14}C = 157‰$; modern-aged) by Hossler and Bauer (2012).

§ 50 : 50 mixture of moderately aged soil C and OM (mean $\Delta^{14}C = -540‰$; equivalent ^{14}C age = 6240 yr B.P.) and young aged soil C and OM ($\Delta^{14}C = 157‰$; modern-aged) pools, giving a mean $\Delta^{14}C$ value of −192‰ (mean equivalent ^{14}C age = 1710 yr B.P.). See text for calculation details.

‖ Estimates > 100% indicate that inputs of C and OM having $\Delta^{14}C < -192‰$ are required (i.e., lesser amounts of young aged soil material and greater amounts of moderately or fossil aged materials).

¶ Mean contribution.

\# Median contribution.

reliant on moderately and fossil aged C and OM, and more reliant on younger nutritional resources, than streams and rivers. In contrast, lentic food webs in agricultural watersheds, as well as those in subarctic regions, were supported by aged C and OM in the forms of soil-derived and glacial-derived material, respectively (Hagvar and Ohlson 2013; Hagvar et al. 2016; Wang et al. 2014). Estimates of aged C and OM contributions to consumer biomass were also not always provided for all lake studies (Table 1). However, discrepancies between estimates when provided and our estimates are likely due to differences in $\Delta^{14}C$ values and ages of the authors and our selected endmembers (*see* Supporting Information Section 2 for additional details).

Conclusions and future directions

The ages of nutritional resources utilized by aquatic consumers have generally not been evaluated in the vast majority of aquatic food web studies. However, in every inland water study that has so far measured natural abundance ^{14}C in consumers, consumer biomass is supported by one or more forms of aged C or OM, with the exception of the largest lakes (Table 2). Minimal contributions of aged C and OM to the food webs of very large lakes like the Lake Superior may be explained by their relatively low watershed : lake surface areas and the greater abundance and importance of modern-aged aquatic primary production to consumer nutrition (Zigah et al. 2012*a*; Kruger et al. 2016). Collectively these studies indicate that modern aquatic food webs rely not only on living or recently produced sources of nutrition, but are also supported by geologically aged reservoirs of C and OM.

The two primary mechanisms by which aged C can enter aquatic food webs are utilization of (1) aged OM directly (derived, e.g., from soils, sedimentary rocks, petroleum hydrocarbons, etc.) or bacterially "repackaged" biomass and (2) living or recently living biomass from photosynthetic fixation of aged DIC and CO_2(aq) by aquatic primary producers. While these mechanisms are independent from each other, both still represent contributions of moderately to geologically aged C to present-day aquatic food webs. Future studies using natural ^{14}C to assess contributions of aged C and OM to aquatic consumer biomass must better distinguish between the aged OM and aged DIC pathways as entry points of aged C to aquatic food webs. While both are sources of "aged" nutrition, the former represents remobilization of truly aged and previously sequestered OM sources while the latter represents recent photosynthetic fixation of dissolved CO_2 derived from some combination of (1) fossil carbonates mobilized via weathering and (2) respiration of aged OM. Use of novel methods including compound-specific ^{14}C analysis (Ishikawa et al. 2015; Kruger et al. 2016) and whole-system ^{13}C-DIC labeling (Cole et al. 2002; Pace et al. 2004) may help delineate these different mechanisms.

A number of important research questions remain to be addressed on the importance of the ages of C and OM contributing to aquatic food webs. These include: (1) does utilization of aged C and OM augment or displace modern-aged sources of nutrition?; (2) do contributions of aged C and OM to aquatic consumer nutrition alter food web community structure, ecosystem function, and biogeochemical cycling relative to aquatic systems that do not contain aged C and OM or utilize aged nutritional resources?; and (3) what are the dominant watershed- and larger-scale factors (e.g., changes in land use, hydrology, climate [i.e., temperature, rainfall, etc.]) influencing the mobilization of aged C and OM to aquatic systems and how might they alter food web nutrition and dynamics?

References

Ahad, J. M., and H. Pakdel. 2013. Direct evaluation of in situ biodegradation in Athabasca oil sands tailings ponds using natural abundance radiocarbon. Environ. Sci. Technol. **47**: 10214–10222. doi:10.1021/es402302z

Aiken, G. R., R. G. Spencer, R. G. Striegl, P. F. Schuster, and P. A. Raymond. 2014. Influences of glacier melt and permafrost thaw on the age of dissolved organic carbon in the Yukon River basin. Global Biogeochem. Cycles **28**: 525–537. doi:10.1002/2013GB004764

Aitkenhead, J. A., D. Hope, and M. F. Billett. 1999. The relationship between dissolved organic carbon in stream water and soil organic carbon pools at different spatial scales. Hydrol. Process. **13**: 1289–1302. doi:10.1002/(SICI)1099-1085(19990615)13:8 < 1289::AID-HYP766 > 3.0.CO;2-M

Aitkenhead-Peterson, J. A., W. H. McDowell, J. C. Neff, E. Stuart, and L. Robert. 2003. Sources, production, and regulation of allochthonous dissolved organic matter inputs to surface waters, p. 26–70. *In* S. E. G. Findlay and R. L. Sinsabaugh [eds.], Aquatic ecosystems: Interactivity of dissolved organic matter. Academic Press.

Allan, J. D., and M. M. Castillo. 2007. Stream ecology: Structure and function of running waters. Springer Netherlands Springer Science+ Business Media BV.

Anderson, N. H., and J. R. Sedell. 1979. Detritus processing by macroinvertebrates in stream ecosystems. Annu. Rev. Entomol. **24**: 351–377. doi:10.1146/annurev.en.24.010179.002031

Atlas, R. M. 1981. Microbial degradation of petroleum hydrocarbons: An environmental perspective. Microbiol. Rev. **45**: 180–209.

Aufdenkampe, A. K., E. Mayorga, P. A. Raymond, J. M. Melack, S. C. Doney, S. R. Alin, R. E. Aalto, and K. Yoo. 2011. Riverine coupling of biogeochemical cycles between land, oceans, and atmosphere. Front. Ecol. Environ. **9**: 53–60. doi:10.1890/100014

Baines, S. B., N. S. Fisher, and J. J. Cole. 2005. Uptake of dissolved organic matter (DOM) and its importance to metabolic requirements of the zebra mussel, *Dreissena polymorpha*. Limnol. Oceanogr. **50**: 36–47. doi:10.4319/lo.2005.50.1.0036

Battin, T. J., S. Luyssaert, L. A. Kaplan, A. K. Aufdenkampe, A. Richter, and L. J. Tranvik. 2009. The boundless carbon cycle. Nat. Geosci. **2**: 598–600. doi:10.1038/ngeo618

Bauer J. E., and T. S. Bianchi 2011. Particulate organic carbon cycling and transformation, p. 69–117. *In* J. Middelburg and R. Laane [eds.], Treatise on Coastal and Estuarine Science, v. **5**: Biogeochemistry. Academic Press.

Bellamy, A. R., J. E. Bauer, and A. G. Grottoli. 2017. Influence of land use and lithology on sources and ages of nutritional subsidies to stream macroinvertebrates: A multi-isotopic approach. Aquat. Sci. doi:10.1007/s00027-017-0542-3

Berggren, M., L. Ström, H. Laudon, J. Karlsson, A. Jonsson, R. Giesler, A. K. Bergström, and M. Jansson. 2010. Lake secondary production fueled by rapid transfer of low molecular weight organic carbon from terrestrial sources to aquatic consumers. Ecol. Lett. **13**: 870–880. doi:10.1111/j.1461-0248.2010.01483.x

Blair, N. E., E. L. Leithold, H. Brackley, N. Trustrum, M. Page, and L. Childress. 2010. Terrestrial sources and export of particulate organic carbon in the Waipaoa sedimentary system: Problems, progress and processes. Mar. Geol. **270**: 108–118. doi:10.1016/j.margeo.2009.10.016

Blair, N. E., and R. C. Aller. 2012. The fate of terrestrial organic carbon in the marine environment. Ann. Rev. Mar. Sci. **4**: 401–423. doi:10.1146/annurev-marine-120709-142717

Boix-Fayos, C., J. de Vente, J. Albaladejo, and M. Martínez-Mena. 2009. Soil carbon erosion and stock as affected by land use changes at the catchment scale in Mediterranean ecosystems. Agric. Ecosyst. Environ. **133**: 75–85. doi:10.1016/j.agee.2009.05.013

Brady, A. L., G. Slater, B. Laval, and D. S. Lim. 2009. Constraining carbon sources and growth rates of freshwater microbialites in Pavilion Lake using ^{14}C analysis. Geobiology **7**: 544–555. doi:10.1111/j.1472-4669.2009.00215.x

Broecker, W. S., and A. Walton. 1959. The geochemistry of ^{14}C in fresh-water systems. Geochim. Cosmochim. Acta **16**: 15–38. doi:10.1016/0016-7037(59)90044-4

Butman, D., and P. A. Raymond. 2011. Significant efflux of carbon dioxide from streams and rivers in the United States. Nat. Geosci. **4**: 839–842. doi:10.1038/ngeo1294

Butman, D., P. A. Raymond, K. Butler, and G. Aiken. 2012. Relationships between Δ^{14}C and the molecular quality of dissolved organic carbon in rivers draining to the coast from the conterminous United States. Global Biogeochem. Cycles **26**: 1–15. doi:10.1029/2012GB004361

Butman, D. E., H. F. Wilson, R. T. Barnes, M. A. Xenopoulos, and P. A. Raymond. 2015. Increased mobilization of aged carbon to rivers by human disturbance. Nat. Geosci. **8**: 112–116. doi:10.1038/NGEO2322

Caraco, N., and J. Cole. 2004. When terrestrial organic matter is sent down the river: The importance of allochthonous carbon inputs to the metabolism of lakes and rivers, p. 301–316. *In* G. A. Polis, M. E. Power, and G. R. Huxel [eds.], Food webs at the landscape level. Univ. of Chicago Press.

Caraco, N., J. E. Bauer, J. J. Cole, S. Petsch, and P. Raymond. 2010. Millennial-aged organic carbon subsidies to a modern river food web. Ecology **91**: 2385–2393. doi:10.1890/09-0330.1

Carbone, M. S., C. I. Czimczik, T. F. Keenan, P. F. Murakami, N. Pederson, P. G. Schaberg, X. Xu, and A. D. Richardson. 2013. Age, allocation and availability of nonstructural carbon in mature red maple trees. New Phytol. **200**: 1145–1155. doi:10.1111/nph.12448

Chanton, J. P., J. E. Bauer, P. A. Glaser, D. I. Siegel, C. A. Kelley, S. C. Tyler, E. H. Romanowicz, and A. Lazrus. 1995. Radiocarbon evidence for the substrates supporting methane formation within northern Minnesota peatlands. Geochim. Cosmochim. Acta **59**: 3663–3668. doi:10.1016/0016-7037(95)00240-Z

Cherkinsky, A. E. 1986. ^{14}C dating and soil organic matter dynamics in Arctic and subarctic ecosystems. Radiocarbon **38**: 241–245. doi:10.1017/S0033822200017616

Cherrier, J., J. E. Bauer, E. R. Druffel, R. B. Coffin, and J. P. Chanton. 1999. Radiocarbon in marine bacteria: Evidence for the ages of assimilated carbon. Limnol. Oceanogr. **44**: 730–736. doi:10.4319/lo.1999.44.3.0730

Ciais, P., and others. 2013. Carbon and other biogeochemical cycles, p. 465–570. *In* T. F. Stocker and others [eds.], Climate change 2013: The physical science basis. Contribution of Working Group I to the Fifth Assessment Report of the Intergovernmental Panel on Climate Change. Cambridge Univ. Press.

Clark, I. D., and P. Fritz. 1997. Environmental isotopes in hydrogeology. CRC press.

Cole, J. J., S. R. Carpenter, J. F. Kitchell, and M. L. Pace. 2002. Pathways of organic carbon utilization in small lakes: Results from a whole-lake 13 C addition and coupled model. Limnol. Oceanogr. **47**: 1664–1675. doi:10.4319/lo.2002.47.6.1664

Cole, J. J., and others. 2007. Plumbing the global carbon cycle: Integrating inland waters into the terrestrial carbon budget. Ecosystems **10**: 172–185. doi:10.1007/s10021-006-9013-8

Cole, J. J., S. R. Carpenter, J. Kitchell, M. L. Pace, C. T. Solomon, and B. Weidel. 2011. Strong evidence for terrestrial support of zooplankton in small lakes based on stable isotopes of carbon, nitrogen, and hydrogen. Proc. Natl. Acad. Sci. USA **108**: 1975–1980. doi:10.1073/pnas.1012807108

Copard, Y., P. Amiotte-Suchet, and C. Di-Giovanni. 2007. Storage and release of fossil organic carbon related to weathering of sedimentary rocks. Earth Planet. Sci. Lett. **258**: 345–357. doi:10.1016/j.epsl.2007.03.048

Currie, L. A., T. I. Eglinton, B. A. Benner, and A. Pearson. 1997. Radiocarbon "dating" of individual chemical compounds in atmospheric aerosol: First results comparing direct isotopic and multivariate statistical apportionment of specific polycyclic aromatic hydrocarbons. Nucl. Instrum. Methods Phys. Res. B **123**: 475–486. doi:10.1016/S0168-583X(96)00783-5

Deines, P., and J. Grey. 2006. Site-specific methane production and subsequent midge mediation within Esthwaite Water, UK. Arch. Hydrobiol. **167**: 317–334. doi:10.1127/0003-9136/2006/0167-0317

Deines, P., M. J. Wooller, and J. Grey. 2009. Unravelling complexities in benthic food webs using a dual stable isotope (hydrogen and carbon) approach. Freshw. Biol. **54**: 2243–2251. doi:10.1111/j.1365-2427.2009.02259.x

DelVecchia, A. G., J. A. Stanford, and X. Xu. 2016. Ancient and methane-derived carbon subsidizes contemporary food webs. Nat. Commun. **7**: 1–9. doi:10.1038/ncomms13163

Dodds, W. K., F. Tromboni, W. Aparecido Saltarelli, F. Cunha, and D. Gasparini. 2017. The root of the problem: Direct influence of riparian vegetation on estimation of stream ecosystem metabolic rates. Limnol. Oceanogr. Lett. **2**: 9–17. doi:10.1002/lol2.10032

Doucett, R. R., J. C. Marks, D. W. Blinn, M. Caron, and B. A. Hungate. 2007. Measuring terrestrial subsidies to aquatic food webs using stable isotopes of hydrogen. Ecology **88**: 1587–1592. doi:10.1890/06-1184

Evans, T. M. 2012. Assessing food and nutritional resources of native and invasive lamprey larvae using natural abundance isotopes. M.S. thesis. The Ohio State Univ.

Fellman, J. B., R. G. Spencer, P. A. Raymond, N. E. Pettit, G. Skrzypek, P. J. Hernes, and P. F. Grierson. 2014. Dissolved organic carbon biolability decreases along with its modernization in fluvial networks in an ancient landscape. Ecology **95**: 2622–2632. doi:10.1890/13-1360.1

Fellman, J. B., E. Hood, P. A. Raymond, J. Hudson, M. Bozeman, and M. Arimitsu. 2015. Evidence for the assimilation of ancient glacier organic carbon in a proglacial stream food web. Limnol. Oceanogr. **60**: 1118–1128. doi:10.1002/lno.10088

Fernandes, R., S. Bergemann, S. Hartz, P. M. Grootes, M.-J. Nadeau, F. Melzner, A. Rakowski, and M. Hüls. 2012. Mussels with meat: Bivalve tissue-shell radiocarbon age differences and archaeological implications. Radiocarbon **54**: 953–965. doi:10.1017/S0033822200047597

Fernandes, R., A. Dreves, M. J. Nadeau, and P. Grootes. 2013. A freshwater lake saga: Carbon routing within the aquatic food web of Lake Schwerin. Radiocarbon **55**: 1102–1113. doi:10.1017/S0033822200048013

Fernandes, R., C. Rinne, M.-J. Nadeau, and P. Grootes. 2016. Towards the use of radiocarbon as a dietary proxy: Establishing a first wide-ranging radiocarbon reservoir effects baseline for Germany. Environ. Archaeol. **21**: 285–294. doi:10.1179/1749631414Y.0000000034

Fernández, P., and others. 2003. Factors governing the atmospheric deposition of polycyclic aromatic hydrocarbons to remote areas. Environ. Sci. Technol. **37**: 3261–3267. doi:10.1021/es020137k

Finlay, J. C. 2001. Stable-carbon-isotope ratios of river biota: Implications for energy flow in lotic food webs. Ecology **82**: 1052–1064. doi:10.1890/0012-9658(2001)082[1052:SCIROR]2.0.CO;2]

Finlay, J. C. 2003. Controls of streamwater dissolved inorganic carbon dynamics in a forested watershed. Biogeochemistry **62**: 231–252. doi:10.1023/A:1021183023963

Finlay, J. C., R. R. Doucett, and C. McNeely. 2010. Tracing energy flow in stream food webs using stable isotopes of hydrogen. Freshw. Biol. **55**: 941–951. doi:10.1111/j.1365-2427.2009.02327.x

Fry, B. 2007. Stable isotope ecology. Springer-Verlag.

Garnett, M. H., and M. F. Billett. 2007. Do riparian plants fix CO_2 lost by evasion from surface waters? An investigation using carbon isotopes. Radiocarbon **49**: 993–1002. doi:10.1017/S0033822200042855

Gaudinski, J. B., S. E. Trumbore, E. A. Davidson, and S. Zheng. 2000. Soil carbon cycling in a temperate forest: Radiocarbon-based estimates of residence times, sequestration rates and partitioning of fluxes. Biogeochemistry **51**: 33–69. doi:10.1023/A:1006301010014

Gaudinski, J. B., and others. 2009. Use of stored carbon reserves in growth of temperate tree roots and leaf buds: Analyses using radiocarbon measurements and modeling. Glob. Chang. Biol. **15**: 992–1014. doi:10.1111/j.1365-2486.2008.01736.x

Goldewijk, K. K. 2001. Estimating global land use change over the past 300 years: The HYDE database. Global Biogeochem. Cycles **15**: 417–433. doi:10.1029/1999GB001232

Grey, J. 2016. The incredible lightness of being methane-fuelled: Stable isotopes reveal alternative energy pathways in aquatic ecosystems and beyond. Front. Ecol. Evol. **4**: 1–14. doi:10.3389/fevo.2016.00008

Griffith, D. R., R. T. Barnes, and P. A. Raymond. 2009. Inputs of fossil carbon from wastewater treatment plants to U.S. rivers and oceans. Environ. Sci. Technol. **43**: 5647–5651. doi:10.1021/es9004043

Guillemette, F., S. L. McCallister, and P. A. del Giorgio. 2016. Selective consumption and metabolic allocation of terrestrial and algal carbon determine allochthony in lake bacteria. ISME J. **10**: 1373–1382. doi:10.1038/ismej.2015.215

Guillemette, F., T. S. Bianchi, and R. G. Spencer. 2017. Old before your time: Ancient carbon incorporation in contemporary aquatic foodwebs. Limnol. Oceanogr. doi:10.1002/lno.10525

Guo, L., and R. W. Macdonald. 2006. Source and transport of terrigenous organic matter in the upper Yukon River: Evidence from isotope (δ^{13}C, Δ^{14}C, and δ^{15}N) composition of dissolved, colloidal, and particulate phases. Global Biogeochem. Cycles **20**: 1–12. doi:10.1029/2005GB002593

Hagvar, S., and M. Ohlson. 2013. Ancient carbon from a melting glacier gives high ^{14}C age in living pioneer invertebrates. Sci. Rep. **3**: 1–4. doi:10.1038/srep02820

Hagvar, S., M. Ohlson, and J. E. Brittain. 2016. A melting glacier feeds aquatic and terrestrial invertebrates with ancient carbon and supports early succession. Arct. Antarct. Alp. Res. **48**: 551–562. doi:10.1038/srep02820

Hall, R. O., J. B. Wallace, and S. L. Eggert. 2000. Organic matter flow in stream food webs with reduced detrital resource base. Ecology **81**: 3445–3463. doi:10.2307/177506

Hedges, J. I. 1992. Global biogeochemical cycles: Progress and problems. Mar. Chem. **39**: 67–93. doi:10.1016/0304-4203(92)90096-S

Hedges, J. I., and others. 1986. Organic carbon-14 in the Amazon River system. Science **231**: 1129–1131. doi:10.1126/science.231.4742.1129

Hedges, J. I., R. G. Keil, and R. Benner. 1997. What happens to terrestrial organic matter in the ocean? Org. Geochem. **27**: 195–212. doi:10.1016/S0146-6380(97)00066-1

Heitkamp, M. A., and B. T. Johnson. 1984. Impact of an oil field effluent on microbial activities in a Wyoming river. Can. J. Microbiol. **30**: 786–792. doi:10.1139/m84-120

Hilton, R. G., A. Galy, N. Hovius, M.-C. Chen, M. J. Horng, and H. Chen. 2008. Tropical-cyclone-driven erosion of the terrestrial biosphere from mountains. Nat. Geosci. **1**: 759–762. doi:10.1038/ngeo333

Hood, E., J. Fellman, R. G. Spencer, P. J. Hernes, R. Edwards, D. D'Amore, and D. Scott. 2009. Glaciers as a source of ancient and labile organic matter to the marine environment. Nature **462**: 1044–1047. doi:10.1038/nature08580

Hood, E., T. J. Battin, J. Fellman, S. O'Neel, and R. G. Spencer. 2015. Storage and release of organic carbon from glaciers and ice sheets. Nat. Geosci. **8**: 91–96. doi:10.1038/ngeo2331

Hope, D., M. F. Billett, and M. S. Cresser. 1997. Exports of organic carbon in two river systems in NE Scotland. J. Hydrol. **193**: 61–82. doi:10.1016/S0022-1694(96)03150-2

Hossler, K., and J. E. Bauer. 2012. Estimation of riverine carbon and organic matter source contributions using time-based isotope mixing models. J. Geophys. Res. **117**: 1–15. doi:10.1029/2012JG001988

Hossler, K., and J. E. Bauer. 2013. Amounts, isotopic character and ages of organic and inorganic carbon exported from rivers to ocean margins: 1. Estimates of terrestrial losses and inputs to the Middle Atlantic Bight. Global Biogeochem. Cycles **27**: 331–346. doi:10.1002/gbc.20033

Hsueh, D. Y., N. Y. Krakauer, J. T. Randerson, X. Xu, S. E. Trumbore, and J. R. Southon. 2007. Regional patterns of radiocarbon and fossil fuel-derived CO2 in surface air across North America. Geophys. Res. Lett. **34**: 1–6. doi: 10.1029/2006GL027032

Iavorivska, L., E. W. Boyer, and J. W. Grimm. 2017. Wet atmospheric deposition of organic carbon: An underreported source of carbon to watersheds in the northeastern United States. J. Geophys. Res. Atmos. doi:10.1002/2016JD026027

Ishikawa, N. F., M. Uchida, Y. Shibata, and I. Tayasu. 2010. A new application of radiocarbon ^{14}C concentrations to stream food web analysis. Nucl. Instrum. Methods Phys. Res. Sect. B Beam Interact. Mater. At. **268**: 1175–1178. doi: 10.1016/j.nimb.2009.10.127

Ishikawa, N. F., F. Hyodo, and I. Tayasu. 2013. Use of carbon-13 and carbon-14 natural abundances for stream food web studies. Ecol. Res. **28**: 759–769. doi:10.1007/s11284-012-1003-z

Ishikawa, N. F., M. Uchida, Y. Shibata, and I. Tayasu. 2014. Carbon storage reservoirs in watersheds support stream food webs via periphyton production. Ecology **95**: 1264–1271. doi:10.1890/13-0976.1

Ishikawa, N. F., M. Yamane, H. Suga, N. O. Ogawa, Y. Yokoyama, and N. Ohkouchi. 2015. Chlorophyll a-specific $\Delta^{14}C$, $\delta^{13}C$ and $\delta^{15}N$ values in stream periphyton: Implications for aquatic food web studies. Biogeosciences 12: 6781–6789. doi:10.5194/bg-12-6781-2015

Ishikawa, N. F., and others. 2016. Terrestrial–aquatic linkage in stream food webs along a forest chronosequence: Multi-isotopic evidence. Ecology **97**: 1146–1158. doi: 10.1890/15-1133.1

Jenkinson, D. S., P. R. Poulton, and C. Bryant. 2008. The turnover of organic carbon in subsoils. Part 1. Natural and bomb radiocarbon in soil profiles from the Rothamsted long-term field experiments. Eur. J. Soil Sci. **59**: 391–399. doi:10.1111/j.1365-2389.2008.01025.x

Junger, M., and D. Planas. 1994. Quantitative use of stable carbon isotope analysis to determine the trophic base of invertebrate communities in a boreal forest lotic system. Can. J. Fish. Aquat. Sci. **51**: 52–61. doi:10.1139/f94-007

Kao, S. J., and K. K. Liu. 1996. Particulate organic carbon export from a subtropical mountainous river (Lanyang Hsi) in Taiwan. Limnol. Oceanogr. **41**: 1749–1757. doi: 10.4319/lo.1996.41.8.1749

Kao, S. J., and others. 2014. Preservation of terrestrial organic carbon in marine sediments offshore Taiwan: Mountain building and atmospheric carbon dioxide sequestration. Earth Surf. Dyn. **2**: 127–139. doi:10.5194/esurf-2-127-2014

Keaveney, E. M., and P. J. Reimer. 2012. Understanding the variability in freshwater radiocarbon reservoir offsets: A cautionary tale. J. Archaeol. Sci. **39**: 1306–1316. doi: 10.1016/j.jas.2011.12.025

Keaveney, E. M., P. J. Reimer, and R. H. Foy 2015*a*. Young, old, and weathered carbon- part 2: Using radiocarbon and stable isotopes to identify terrestrial carbon support of the food web in an alkaline, humic lake. Radiocarbon **57**: 425–438. doi:10.2458/azu_rc.57.18355

Keaveney, E. M., P. J. Reimer, and R. H. Foy 2015*b*. Young, old and weathered carbon- part 1: Using radiocarbon and stable isotopes to identify carbon sources in an alkaline, humic lake. Radiocarbon **57**: 407–423. doi:10.2458/azu_rc.57.18354

Kerner, M., H. Hohenberg, S. Ertl, M. Reckermann, and A. Spitzy. 2003. Self-organization of dissolved organic matter to micelle-like microparticles in river water. Nature **422**: 150–154. doi:10.1038/nature01469

Kleber, M. 2010. What is recalcitrant soil organic matter? Environ. Chem. **7**: 320–332. doi:10.1071/EN10006

Kleber, M., and M. G. Johnson. 2010. Advances in understanding the molecular structure of soil organic matter: Implications for interactions in the environment. Adv. Agron. **106**: 77–142. doi:10.1016/S0065-2113(10)06003-7

Komada, T., E. R. Druffel, and S. E. Trumbore. 2004. Oceanic export of relict carbon by small mountainous rivers. Geophys. Res. Lett. **31**: 1–4. doi:10.1029/2004GL019512

Kruger, B. R., J. P. Werne, D. K. Branstrator, T. R. Hrabik, Y. Chikaraishi, N. Ohkouchi, and E. C. Minor. 2016. Organic matter transfer in Lake Superior's food web: Insights from bulk and molecular stable isotope and radiocarbon analyses. Limnol. Oceanogr. **61**: 149–164. doi:10.1002/lno.10205

Lambin, E. F., and P. Meyfroidt. 2011. Global land use change, economic globalization, and the looming land scarcity. Proc. Natl. Acad. Sci. USA **108**: 3465–3472. doi: 10.1073/pnas.1100480108

Layman, C. A., and others. 2012. Applying stable isotopes to examine food-web structure: An overview of analytical tools. Biol. Rev. **87**: 545–562. doi:10.1111/j.1469-185X.2011.00208.x

Leberfinger, K., I. Bohman, and J. Herrmann. 2011. The importance of terrestrial resource subsidies for shredders in open-canopy streams revealed by stable isotope analysis. Freshw. Biol. **56**: 470–480. doi:10.1111/j.1365-2427.2010.02512.x

Leithold, E. L., N. E. Blair, and D. W. Perkey. 2006. Geomorphologic controls on the age of particulate organic carbon from small mountainous and upland rivers. Global Biogeochem. Cycles **20**: 1–11. doi:10.1029/2005GB002677

Li, H., E. C. Minor, and P. K. Zigah. 2013. Diagenetic changes in Lake Superior sediments as seen from FTIR and 2D correlation spectroscopy. Org. Geochem. **58**: 125–136. doi:10.1016/j.orggeochem.2013.03.002

Longworth, B. E., S. T. Petsch, P. A. Raymond, and J. E. Bauer. 2007. Linking lithology and land use to sources of dissolved and particulate organic matter in headwaters of a temperate, passive-margin river system. Geochim. Cosmochim. Acta **71**: 4233–4250. doi:10.1016/j.gca.2007.06.056

Lu, Y. H., E. A. Canuel, J. E. Bauer, and R. M. Chambers. 2014. Effects of watershed land use on sources and nutritional value of particulate organic matter in temperate headwater streams. Aquat. Sci. **76**: 419–436. doi:10.1007/s00027-014-0344-9

Lützow, M. v., I. Kögel-Knabner, K. Ekschmitt, E. Matzner, G. Guggenberger, B. Marschner, and H. Flessa. 2006. Stabilization of organic matter in temperate soils: Mechanisms and their relevance under different soil conditions–a review. Eur. J. Soil Sci. **57**: 426–445. doi:10.1029/2005GB002677

Mahowald, N. M., R. Scanza, J. Brahney, C. L. Goodale, P. G. Hess, J. K. Moore, and J. Neff. 2017. Aerosol deposition impacts on land and ocean carbon cycles. Curr. Clim. Change Rep. **3**: 16–31. doi:10.1007/s40641-017-0056-z

Marín-Spiotta, E., K. E. Gruley, J. Crawford, E. E. Atkinson, J. R. Miesel, S. Greene, and R. G. M. Spencer. 2014. Paradigm shifts in soil organic matter research affect interpretations of aquatic carbon cycling: Transcending disciplinary and ecosystem boundaries. Biogeochemistry **117**: 279–297. doi:10.1007/s10533-013-9949-7

Marwick, T. R., F. Tamooh, C. R. Teodoru, A. V. Borges, F. Darchambeau, and S. Bouillon. 2015. The age of river-transported carbon: A global perspective. Global Biogeochem. Cycles **29**: 122–137. doi:10.1002/2014GB004911

Masiello, C. A., and E. R. Druffel. 2001. Carbon isotope geochemistry of the Santa Clara River. Global Biogeochem. Cycles **15**: 407–416. doi:10.1029/2000GB001290

Mayorga, E., A. K. Aufdenkampe, C. A. Masiello, A. V. Krusche, J. I. Hedges, P. D. Quay, J. E. Richey, and T. A. Brown. 2005. Young organic matter as a source of carbon dioxide outgassing from Amazonian rivers. Nature **436**: 538–541. doi:10.1038/nature03880

McCallister, S. L., J. E. Bauer, J. E. Cherrier, and H. W. Ducklow. 2004. Assessing sources and ages of organic matter supporting river and estuarine bacterial production: A multiple-isotope (Δ^{14}C, δ^{13}C, and δ^{15}N) approach. Limnol. Oceanogr. **49**: 1687–1702. doi:10.4319/lo.2004.49.5.1687

McCallister, S. L., and P. A. del Giorgio. 2008. Direct measurement of the δ^{13}C signature of carbon respired by bacteria in lakes: Linkages to potential carbon sources, ecosystem baseline metabolism, and CO_2 fluxes. Limnol. Oceanogr. **53**: 1204–1216. doi:10.1029/2005GB002677

McCallister, S. L., and P. A. del Giorgio. 2012. Evidence for the respiration of ancient terrestrial organic C in northern temperate lakes and streams. Proc. Natl. Acad. Sci. USA **109**: 16963–16968. doi:10.1073/pnas.1207305109

McNichol, A. P., and L. I. Aluwihare. 2007. The power of radiocarbon in biogeochemical studies of the marine carbon cycle: Insights from studies of dissolved and particulate organic carbon (DOC and POC). Chem. Rev. **107**: 443–466. doi:10.1021/cr050374g

Melymuk, L., and others. 2014. From the city to the lake: Loadings of PCBs, PBDEs, PAHs and PCMs from Toronto to Lake Ontario. Environ. Sci. Technol. **48**: 3732–3741. doi:10.1021/es403209z

Moore, J. W., and B. X. Semmens. 2008. Incorporating uncertainty and prior information into stable isotope mixing models. Ecol. Lett. **11**: 470–480. doi:10.1111/j.1461-0248.2008.01163.x

Pace, M. L., and others. 2004. Whole-lake carbon-13 additions reveal terrestrial support of aquatic food webs. Nature **427**: 240–243. doi:10.1038/nature02227

Pessenda, L. C. R., E. P. E. Valencia, P. B. Camargo, E. C. Telles, L. A. Martinelli, C. C. Cerri, R. Aravena, and K. Rozanski. 1996. Natural radiocarbon measurements in Brazilian soils developed on basic rocks. Radiocarbon **38**: 203–208. doi:10.1017/S0033822200017574

Peterson, B. J., and B. Fry. 1987. Stable isotopes in ecosystem studies. Annu. Rev. Ecol. Syst. **18**: 293–320. doi:10.1146/annurev.es.18.110187.001453

Petsch, S. T., T. I. Eglinton, and K. J. Edwards. 2001. ^{14}C-dead living biomass: Evidence for microbial assimilation of ancient organic carbon during shale weathering. Science **292**: 1127–1131. doi:10.1029/2005GB002677

Philipsen, B., and J. Heinemeier. 2013. Freshwater reservoir effect variability in Northern Germany. Radiocarbon **55**: 1085–1101. doi:10.1017/S0033822200048001

Phillips, D. L., and J. W. Gregg. 2001. Uncertainty in source partitioning using stable isotopes. Oecologia **127**: 171–179. doi:10.1007/s004420000578

Phillips, D. L., and J. W. Gregg. 2003. Source partitioning using stable isotopes: Coping with too many sources. Oecologia **136**: 261–269. doi:10.1007/s00442-003-1218-3

Phillips, D. L., R. Inger, S. Bearhop, A. L. Jackson, J. W. Moore, A. C. Parnell, B. X. Semmens, and E. J. Ward. 2014. Best practices for use of stable isotope mixing models in food-web studies. Can. J. Zool. **92**: 823–835. doi:10.1139/cjz-2014-0127

Pohlman, J. W., J. E. Bauer, E. A. Canuel, K. S. Grabowski, D. L. Knies, C. S. Mitchell, M. J. Whiticar, and R. B. Coffin. 2009. Methane sources in gas hydrate-bearing cold seeps: Evidence from radiocarbon and stable isotopes. Mar. Chem. **115**: 102–109. doi:10.1016/j.marchem.2009.07.001

Raymond, P. A., and J. E. Bauer. 2001. Riverine export of aged terrestrial organic matter to the North Atlantic Ocean. Nature **409**: 497–500. doi:10.1038/35054034

Raymond, P. A., J. E. Bauer, N. F. Caraco, J. J. Cole, B. Longworth, and S. T. Petsch. 2004. Controls on the variability of organic matter and dissolved inorganic carbon ages in northeast US rivers. Mar. Chem. **92**: 353–366. doi:10.1016/j.marchem.2004.06.036

Raymond, P. A., and others. 2013. Global carbon dioxide emissions from inland waters. Nature **503**: 355–359. doi:10.1038/nature12760

Regnier, P., and others. 2013. Anthropogenic perturbation of the carbon fluxes from land to ocean. Nat. Geosci. **6**: 597–607. doi:10.1038/ngeo1830

Restrepo, J. D., A. J. Kettner, and J. P. M. Syvitski. 2015. Recent deforestation causes rapid increase in river sediment load in the Colombian Andes. Anthropocene **10**: 13–28. doi:10.1016/j.ancene.2015.09.001

Roditi, H. A., N. S. Fisher, and S. A. Sañudo-Wilhelmy. 2000. Uptake of dissolved organic carbon and trace elements by zebra mussels. Nature **407**: 78–80. doi:10.1038/35024069

Salmon, V., S. Derenne, E. Lallier-Verges, C. Largeau, and B. Beaudoin. 2000. Protection of organic matter by mineral matrix in a Cenomanian black shale. Org. Geochem. **31**: 463–474. doi:10.1016/S0146-6380(00)00013-9

Schell, D. M. 1983. Carbon-13 and carbon-14 abundances in Alaskan aquatic organisms: Delayed production from peat in arctic food webs. Science **219**: 1068–1071. doi:10.1126/science.219.4588.1068

Schillawski, S., and S. Petsch. 2008. Release of biodegradable dissolved organic matter from ancient sedimentary rocks. Global Biogeochem. Cycles **22**: 1–8. doi:10.1029/2007GB002980

Schmidt, M. W., and others. 2011. Persistence of soil organic matter as an ecosystem property. Nature **478**: 49–56. doi:10.1038/nature10386

Schrumpf, M., K. Kaiser, G. Guggenberger, T. Persson, I. Kögel-Knabner, and E. D. Schulze. 2013. Storage and stability of organic carbon in soils as related to depth, occlusion within aggregates, and attachment to minerals. Biogeosciences **10**: 1675–1691. doi:10.5194/bg-10-1675-2013

Sobek, S., L. J. Tranvik, Y. T. Prairie, P. Kortelainen, and J. J. Cole. 2007. Patterns and regulation of dissolved organic carbon: An analysis of 7,500 widely distributed lakes. Limnol. Oceanogr. **52**: 1208–1219. doi:10.4319/lo.2007.52.3.1208

Solomon, C. T., S. R. Carpenter, M. K. Clayton, J. J. Cole, J. J. Coloso, M. L. Pace, M. Jake Vander Zanden, and B. C. Weidel. 2011. Terrestrial, benthic, and pelagic resource use in lakes: Results from a three-isotope Bayesian mixing model. Ecology **92**: 1115–1125. doi:10.1890/i0012-9658-92-5-1115

Spencer, R. G., and others. 2012. An initial investigation into the organic matter biogeochemistry of the Congo River. Geochim. Cosmochim. Acta **84**: 614–627. doi:10.1016/j.gca.2012.01.013

Spencer, R. G., W. Guo, P. A. Raymond, T. Dittmar, E. Hood, J. Fellman, and A. Stubbins. 2014*a*. Source and biolability of ancient dissolved organic matter in glacier and lake ecosystems on the Tibetan Plateau. Geochim. Cosmochim. Acta **142**: 64–74. doi:10.1016/j.gca.2014.08.006

Spencer, R. G., A. Vermilyea, J. Fellman, P. Raymond, A. Stubbins, D. Scott, and E. Hood. 2014*b*. Seasonal variability of organic matter composition in an Alaskan glacier outflow: Insights into glacier carbon sources. Environ. Res. Lett. **9**: 1–7. doi:10.1088/1748-9326/9/5/055005

Spencer, R. G., P. J. Mann, T. Dittmar, T. I. Eglinton, C. McIntyre, R. Max Holmes, N. Zimov, and A. Stubbins. 2015. Detecting the signature of permafrost thaw in Arctic rivers. Geophys. Res. Lett. **42**: 2830–2835. doi:10.1002/2015GL063498

Stuiver, M., and H. A. Polach. 1977. Discussion reporting of ^{14}C data Radiocarbon **19**: 355–363. doi:10.1017/S0033822200003672

Taipale, S., P. Kankaala, and R. I. Jones. 2007. Contributions of different organic carbon sources to Daphnia in the pelagic foodweb of a small polyhumic lake: Results from mesocosm $DI^{13}C$-additions. Ecosystems **10**: 757–772. doi:10.1007/s10021-007-9056-5

Tanentzap, A. J., and others. 2017. Terrestrial support of lake food webs: Synthesis reveals controls over cross-ecosystem resource use. Sci. Adv. **3**: e1601765. doi:10.1126/sciadv.1601765

Taylor, R. E. 2016. Radiocarbon dating: Development of a nobel method, p. 21–44. *In* E. A. G. Schurr, E. R. M. Druffel, and S. E. Trumbore [eds.], Radiocarbon and climate change. Springer International.

Tittel, J., O. Büttner, K. Freier, A. Heiser, R. Sudbrack, and G. Ollesch. 2013. The age of terrestrial carbon export and rainfall intensity in a temperate river headwater system. Biogeochemistry **115**: 53–63. doi:10.1007/s10533-013-9896-3

Tranvik, L. J., and others. 2009. Lakes and reservoirs as regulators of carbon cycling and climate. Limnol. Oceanogr. **54**: 2298–2314. doi:10.4319/lo.2009.54.6_part_2.2298

Trumbore, S. 2000. Age of soil organic matter and soil respiration: Radiocarbon constraints on belowground C dynamics. Ecol. Appl. **10**: 399–411. doi:10.2307/2641102

Trumbore, S. 2009. Radiocarbon and soil carbon dynamics. Annu. Rev. Earth Planet. Sci. **37**: 47–66. doi:10.1146/annurev.earth.36.031207.124300

Turnbull, J. C., H. Graven, and N. Y. Krakauer. 2016. Radiocarbon in the atmosphere, p. 83–137. *In* E. A. G. Schurr, E. R. M. Druffel, and S. E. Trumbore [eds.], Radiocarbon and climate change. Springer International.

Vannote, R. L., G. W. Minshall, K. W. Cummins, J. R. Sedell, and C. E. Cushing. 1980. The river continuum concept. Can. J. Fish. Aquat. Sci. **37**: 130–137. doi:10.1139/f80-017

Vonk, J. E., and others. 2013. High biolability of ancient permafrost carbon upon thaw. Geophys. Res. Lett. **40**: 2689–2693. doi:10.1002/grl.50348

Vörösmarty, C. J., and others. 2010. Global threats to human water security and river biodiversity. Nature **467**: 555–561. doi:10.1038/nature09440

Wallace, J. B., S. L. Eggert, J. L. Meyer, and J. R. Webster. 1997. Multiple trophic levels of a forest stream linked to terrestrial litter inputs. Science **277**: 102–104. doi:10.1126/science.277.5322.102

Wallace, J. B., S. L. Eggert, J. L. Meyer, and J. R. Webster. 2015. Stream invertebrate productivity linked to forest subsidies: 37 stream-years of reference and experimental data. Ecology **96**: 1213–1228. doi:10.1890/14-1589.1

Wang, X., H. Ma, R. Li, Z. Song, and J. Wu. 2012. Seasonal fluxes and source variation of organic carbon transported by two major Chinese Rivers: The Yellow River and Changjiang (Yangtze) River. Global Biogeochem. Cycles **26**: 1–10. doi:10.1029/2011GB004130

Wang, Y., B. Gu, M.-K. Lee, S. Jiang, and Y. Xu. 2014. Isotopic evidence for anthropogenic impacts on aquatic food web dynamics and mercury cycling in a subtropical wetland ecosystem in the US. Sci. Total Environ. **487**: 557–564. doi:10.1016/j.scitotenv.2014.04.060

Watanabe, K., and T. Kuwae. 2015. Radiocarbon isotopic evidence for assimilation of atmospheric CO_2 by the seagrass *Zostera marina*. Biogeosciences **12**: 6251–6258. doi:10.5194/bg-12-6251-2015

Weber, A. E., J. E. Bauer, and G. T. Watters. 2017. Assessment of nutritional subsidies to freshwater mussels using a multiple natural abundance isotope approach. Freshw. Biol. **62**: 615–629. doi:10.1111/fwb.12890

Wilkinson, G. M., M. L. Pace, and J. J. Cole. 2013. Terrestrial dominance of organic matter in north temperate lakes. Global Biogeochem. Cycles **27**: 43–51. doi:10.1029/2012GB004453

Wozniak, A. S., J. E. Bauer, and R. M. Dickhut. 2011. Fossil and contemporary aerosol particulate organic carbon in the eastern United States: Implications for deposition and inputs to watersheds. Global Biogeochem. Cycles **25**: 1–14. doi:10.1029/2010GB003855

Wozniak, A. S., J. E. Bauer, R. M. Dickhut, L. Xu, and A. P. McNichol. 2012*a*. Isotopic characterization of aerosol organic carbon components over the eastern United States. J. Geophys. Res. Atmos. **117**: 1–14. doi:10.1029/2011JD017153

Wozniak, A. S., J. E. Bauer, and R. M. Dickhut. 2012*b*. Characteristics of water-soluble organic carbon associated with aerosol particles in the eastern United States. Atmos. Environ. **46**: 181–188. doi:10.1016/j.atmosenv.2011.10.001

Xu, B., and others. 2009. Black soot and the survival of Tibetan glaciers. Proc. Natl. Acad. Sci. USA **106**: 22114–22118. doi:10.1073/pnas.0910444106

Zeng, F.-W., C. A. Masiello, and W. C. Hockaday. 2011. Controls on the origin and cycling of riverine dissolved inorganic carbon in the Brazos River, Texas. Biogeochemistry **104**: 275–291. doi:10.1007/s10533-010-9501-y

Zigah, P. K., E. C. Minor, J. P. Werne, and S. L. McCallister. 2011. Radiocarbon and stable carbon isotopic insights into provenance and cycling of carbon in Lake Superior. Limnol. Oceanogr. **56**: 867–886. doi:10.4319/lo.2011.56.3.0867

Zigah, P. K., E. C. Minor, J. P. Werne, and S. L. Leigh McCallister. 2012*a*. An isotopic ($\Delta^{14}C$, $\delta^{13}C$, and $\delta^{15}N$) investigation of the composition of particulate organic matter and zooplankton food sources in Lake Superior and across a size-gradient of aquatic systems. Biogeosciences **9**: 3663–3678. doi:10.5194/bg-9-3663-2012

Zigah, P. K., E. C. Minor, and J. P. Werne. 2012*b*. Radiocarbon and stable-isotope geochemistry of organic and inorganic carbon in Lake Superior. Global Biogeochem. Cycles **26**: 1–20. doi:10.1029/2011GB004132

Zigah, P. K., E. C. Minor, H. A. Abdulla, J. P. Werne, and P. G. Hatcher. 2014. An investigation of size-fractionated organic matter from Lake Superior and a tributary stream using radiocarbon, stable isotopes and NMR. Geochim. Cosmochim. Acta **127**: 264–284. doi:10.1016/j.gca.2013.11.037

Acknowledgments

We thank Dr. Jonathan Cole for his encouragement for undertaking this review and synthesis and for providing constructive feedback and suggestions on earlier drafts of the manuscript. Comments and suggestions made by two outside reviewers, and by the associate editor and editor-in-chief of this journal, greatly improved the final manuscript. This work was partially supported by National Science Foundation awards DEB-0234533, EAR-0403949 and OCE-0961860 to J.E.B., the Hudson River Foundation to J.E.B. and A.R.B, and funding from The Ohio State University.

19

How have recent temperature changes affected the efficiency of ocean biological carbon export?

B. B. Cael,[1,2]* *Kelsey Bisson*,[3] *Michael J. Follows*[1]

[1]Department of Earth, Atmosphere, and Planetary Sciences, Massachusetts Institute of Technology, Cambridge, Massachusetts; [2]Department of Physical Oceanography, Woods Hole Oceanographic Institution, Woods Hole, Massachusetts; [3]Earth Research Institute, University of California, Santa Barbara, California

Scientific Significance Statement

The ocean's biologically mediated carbon reservoirs are an integral component of the global carbon cycle, and may feed back on climate if the total carbon exported out of the surface ocean is affected by surface temperatures. However, the magnitude of this feedback is difficult to quantify. Using long-term temperature records and a simple metabolic model, we show a 1.5% ± 0.4% decline in the fraction of primary production removed from the surface ocean over the past three decades of climate change, suggesting increased global temperatures have reduced the efficiency with which the ocean exports carbon into the deep. The larger temperature changes in Earth's history may have reduced this efficiency much more so.

Abstract

The ocean's large, microbially mediated reservoirs of carbon are intimately connected with atmospheric CO_2 and climate, yet quantifying the feedbacks between them remains an unresolved challenge. Through an idealized mechanistic model, we consider the impact of documented climate change during the past few decades on the efficiency of biological carbon export out of the surface ocean. This model is grounded in universal metabolic phenomena, describing export efficiency's temperature dependence in terms of the differential temperature sensitivity of phototrophic and heterotrophic metabolism. Temperature changes are suggested to have caused a statistically significant decrease in export efficiency of 1.5% ± 0.4% over the past 33 yr. Larger changes are suggested in the midlatitudes and Arctic. This interpretation is robust across multiple sea surface temperature and net primary production data products. The same metabolic mechanism may have resulted in much larger changes e.g., in response to the large temperature shifts between glacial and interglacial time periods.

The ocean's "biological pumps" sustain large reservoirs of carbon, mediated by microbial activity, with significant leverage on atmospheric CO_2 and climate (Volk and Hoffert 1985). While the general significance of the biological pumps for the carbon cycle and climate is clearly demonstrated (Volk and Hoffert 1985; Cox et al. 2000) the details

*Correspondence: snail@mit.edu

Author Contribution Statement: BBC conceived the study, performed the analyses, and wrote the paper. KB performed the analyses and wrote the paper. MJF conceived the study, performed the analyses, and wrote the paper.

of their relationship remain elusive (Boyd 2015). Understanding the current functioning of the biological pumps is limited because in situ data is sparse (Boyd and Trull 2007), data collection is difficult and expensive, and the system is extremely complex and variable (Buesseler and Boyd 2009).

Even so, there exists a clear and quantifiable imprint of metabolic sensitivity to temperature in the ocean system (e.g., Eppley 1972), which can be exploited to understand global changes in carbon export with climate change. In particular, differential sensitivities of phototrophic and heterotrophic metabolisms to environmental temperature are documented (Eppley 1972; Huntley and Lopez 1992), and a model of their effect on export efficiency (*ef*, the ratio of the flux of organic matter exported across the base of the euphotic zone to the integrated primary production within that layer) explains the observed dependence of *ef* on temperature (Cael and Follows 2016).

Here we ask: how has the documented trend in global ocean temperatures over the past few decades impacted the efficiency of this export flux? We use temperature records with the above model to infer temperature's contribution to global change in *ef* through time. We focus on multidecadal changes (after Henson et al. 2010) over the past 33 yr (the duration over which suitable data products are available) using multiple data products to examine the sensitivity of calculated changes to inputs used.

A metabolic model of export efficiency

Export efficiency is a combination of growth, respiration, sinking, remineralization, and other processes. It has primarily been considered as a function of temperature (T), primary production ($\mathcal{P}$), and community structure (i.e., the size distribution of plankton, who is eating whom, and so forth) (Michaels and Silver 1988; Eppley 1989; Laws et al. 2000). While all three are important, community structure variables and their influence on *ef* are challenging to assess, quantify, and measure, making an estimation of how recent climatic shifts have produced global shifts in *ef* via community structure challenging. Previous studies disagree substantially on both global trends in $\mathcal{P}$ (Siegel et al. 2013; Behrenfeld et al. 2016; and references therein) and the relationship between *ef* and $\mathcal{P}$ (Buesseler 1998; Laws et al. 2011; Maiti et al. 2013), making an estimation of how climatic shifts have affected *ef* via $\mathcal{P}$ similarly intractable.

In contrast, global trends in sea surface temperature (SST) over the past few decades are well-characterized (IPCC 2014). SST is commonly used as a proxy for upper-ocean temperature, is anticorrelated with *ef* (e.g., Laws et al. 2000; Henson et al. 2011), and is also one of the only variables for which long-term, global observational records exist (Ishii et al. 2005; Reynolds et al. 2007; Dee et al. 2011). Do SST observations, suggest a shift in *ef*?

Recently, a simple model was proposed (Cael and Follows 2016; herein the model will be referred to as CF16) to explain the *ef*-*T* relationship seen in observations. Heterotrophic and phototrophic growth rates increase with temperature, but the former increase more so (Eppley 1972; Huntley and Lopez 1992); metabolic ecological theory relates this to the different activation energies of respiration and photosynthesis (Lopez-Urrutia et al. 2006). As originally posited in Laws et al. (2000), this differential dependence suggests that increasing temperatures should increase community respiration relative to production and therefore decrease *ef*. Rather than absorbing these dependencies into a numerical food-web model as in Laws et al. (2000), CF16 considers *ef* as a random variable scaled by temperature according to these dependencies, and this description is shown to be consistent with observations. We refer the reader to Cael and Follows (2016) for a full description and discussion of CF16, but describe it briefly below.

Within a basic differential equation for plankton biomass p in the euphotic layer,

$$\dot{p} = \mu p - \lambda p - \lambda' p - wp \quad (1)$$

where μ is the growth rate, λ is the grazing rate, λ' is the loss rate due to factors other than grazing, and w is the sinking rate, in steady state $ef = \frac{wp}{\mu p}$ can be written as

$$ef = 1 - \frac{\lambda + \lambda'}{\mu} \quad (2)$$

Then one can find the maximum efficiency by neglecting λ' and incorporating the temperature dependencies of phototrophy and heterotrophy as $\mu \propto e^{0.063T}$, $\lambda \propto e^{0.11T}$. This yields a curve of maximum export efficiency as a function of temperature:

$$ef_{\max}(T) = 1 - \alpha e^{\beta T} \quad (3)$$

where the parameter* α is one minus the maximum efficiency at $T = 0$, which is estimated empirically to be α=0.24, and the parameter β=0.11−0.063=0.047. *ef* values can be rescaled by $ef_{\max}(T)$ to extract this temperature dependence; that is, $ef/ef_{\max}(T)$ is variable but independent of temperature, suggesting that this rescaling captures all of the temperature dependency of *ef* (Cael and Follows 2016).

Here we then use this temperature scaling to derive an average $\langle ef \rangle$ as a function of temperature†:

---

*Figure S1 (*see* Supporting Information) shows the sensitivity of CF16 to changes in the parameters α and β.

†$\langle \tilde{ef} \rangle$=0.37 for the observations used in Cael and Follows (2016) so we use that value here nominally, though we note the value of this factor is largely irrelevant for the analyses of this paper because we focus on percent changes.

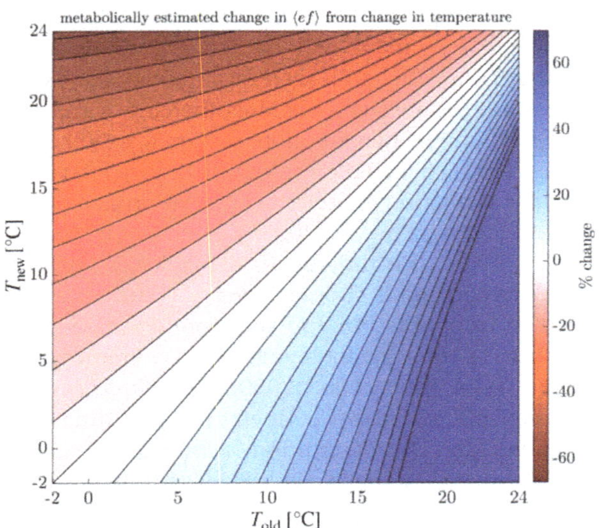

Fig. 1. Percent change in $\langle ef \rangle$ after a temperature change, as estimated by CF16 (see Eq. 3), as a function of the initial and final temperatures (T_{old} and T_{new}). Contours are spaced at 5%.

Table 1. Percent change in $\langle\langle ef \rangle\rangle$, during the period 1982–2014, as estimated by CF16, for different data products. All changes suggest a decrease in $\langle\langle ef \rangle\rangle$ and are statistically significant ($p < 0.001$; *see* Supporting Information). Mean and standard deviation of percent change across each SST-$\mathcal{P}$ pair are $-1.5\% \pm 0.4\%$.

SST	$\mathcal{P}$	% Change
ERA	CbPMv2	−0.83
ERA	VGPM	−1.12
COBE	CbPMv2	−1.55
COBE	VGPM	−1.98
OISST	CbPMv2	−1.52
OISST	VGPM	−1.91

$$\langle ef \rangle (T) = \langle \tilde{ef} \rangle \times ef_{max}(T). \tag{4}$$

While empirical models have been proposed to relate temperature to *ef* (Laws et al. 2000, 2011; Dunne et al. 2005; Henson et al. 2011), we focus on CF16 because it isolates a single, understood metabolic mechanism—the hypothesized and observed differential response of temperature on phototrophic and heterotrophic metabolisms. We emphasize that CF16 does not seek to be a complete model for or explain all the variability in *ef*; it isolates the variability in *ef* due to the differential temperature effect on metabolism. Because the differential temperature response is assumed to arise from chemical kinetics, namely the activation energies of respiration and photosynthesis, it is assumed to be constant over time.

Figure 1 shows the estimated percent change of $\langle ef \rangle$ resulting from a temperature change from T_{old} to T_{new} as predicted by CF16. Percent changes are a function of both and as Eq. 4 varies by a factor of three from low to high temperatures, percent changes in response to large temperature differences can be very large. Note that temperature differences shown in Fig. 1 are larger than those that have been observed over the past 33 yr, which are <1°C (IPCC 2014).

Global estimates of multidecadal change in ef

CF16 can be used to calculate $\langle ef \rangle$ from SST,[‡] so it can be used to infer trends in globally averaged *ef* (for which we will use the symbol[§] $\langle\langle ef \rangle\rangle$) from SST reanalyses. Does CF16 indicate a trend? Is this trend consistent between data products?

To test these questions, we generate three time series of $\langle ef \rangle$ from different SST reanalyses and Eq. 4. We use the ECMWF ERA-Interim SST (Dee et al. 2011), the NOAA OISST (Reynolds et al. 2007), and the ESRL COBE-SST (Ishii et al. 2005) products. Description and discussion of how these reanalyses are developed can be found in the above references. For consistent comparison, we use a common

- spatial resolution of 1°, the resolution of the coarsest product
- temporal resolution of 1 month, i.e., monthly averages available for each product
- start time of January 1982, the January of the earliest year common to all three products
- end time of December 2014, the December of the latest year common to all three products

From each SST product we compute global time series of $\langle ef \rangle$ for each 1° bin.

Because we define $\langle\langle ef \rangle\rangle$ as the ratio of globally integrated export flux to globally integrated production, to compute it $\langle ef \rangle$ for each 1° bin must be weighted by both area and $\mathcal{P}$. We use climatologies[¶] from the two most common algorithms to estimate $\mathcal{P}$: the Carbon-based Productivity Model (CbPMv2) (Westberry et al. 2008), and the Vertically

[‡]As a proxy for upper-ocean temperature.

[§]All export efficiency notation: $ef :=$ export flux divided by primary production. $\mathcal{P} :=$ primary production. $T :=$ temperature. $ef_{max}(T) :=$ theoretical maximum *ef* for a given temperature. $\alpha = 0.24$: one minus the maximum efficiency at $T = 0$, $\beta = 0.047$: the differential temperature sensitivity of phototrophy and heterotrophy. $\tilde{ef} := ef / ef_{max}(T)$. $\langle ef \rangle :=$ mean *ef* averaged over a spatial region, e.g., a 1° box or a latitudinal band. $\langle\langle ef \rangle\rangle :=$ globally averaged *ef*.

[¶]Both available at http://www.science.oregonstate.edu/ocean.productivity/standard.product.php. We use climatologies rather than time series because of the lack of $\mathcal{P}$ time series over the duration of the SST time series. This is justifiable in light of the disagreement on global trends in $\mathcal{P}$ (Siegel et al. 2013; Behrenfeld et al. 2016).

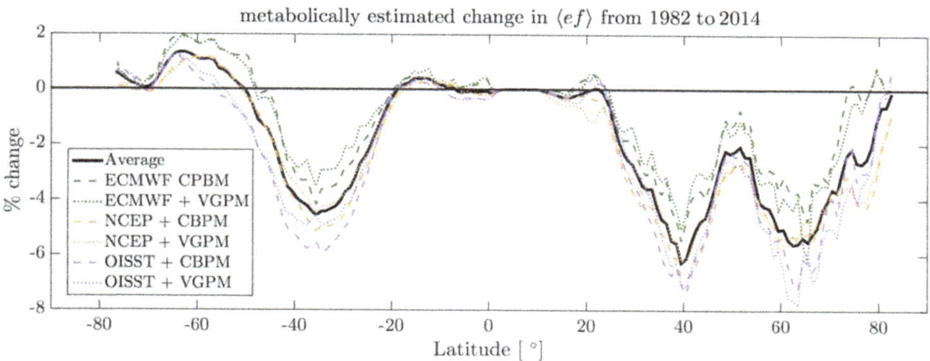

Fig. 2. Percent change in $\langle ef \rangle$ from 1982 to 2014, as estimated by CF16, as a function of latitude, for different data products. Color corresponds to SST product and line type to $\mathcal{P}$ product. Black curve is the average across the SST-$\mathcal{P}$ pairs.

Generalized Productivity Model (VGPM) (Behrenfeld and Falkowski 1997). In total, the time series $\langle\langle ef \rangle\rangle(t)$ is calculated by

$$\langle\langle ef \rangle\rangle(t) := \frac{\sum_{x,y} A(x,y)\mathcal{P}(x,y,t \bmod 12)\langle ef \rangle(x,y,t)}{\sum_{x,y} A(x,y)\mathcal{P}(x,y,t \bmod 12)} \tag{5}$$

where (x, y) are latitude and longitude, $A(x, y)$ is the area of the 1°×1° box at (x, y), and t mod 12 is the month of the $\mathcal{P}$-climatology.

The above three SST and two $\mathcal{P}$ products yield six time series of $\langle\langle ef \rangle\rangle(t)$. We regress each against time using the simplest statistical model that resolves a seasonal cycle and a linear trend:

$$\langle\langle ef \rangle\rangle(t) = mt + b(\text{month}) + \text{error} \tag{6}$$

regressing $\langle\langle ef \rangle\rangle(t)$ against a time variable that runs $t=1, 2, \ldots$ along with an indicator variable for each month (*see* Supporting Information). To estimate statistical significance of the trends, we use both the standard method and two resampling methods, one of which accounts for autocorrelation in the time series (*see* Supporting Information). The estimated rates of change m can be multiplied by the duration of the time series to estimate a percent change in $\langle\langle ef \rangle\rangle(t)$ from 1982 to 2014 for each SST-$\mathcal{P}$ pair; *see* Table 1.

Independent of SST and $\mathcal{P}$ product, a global decline in $\langle\langle ef \rangle\rangle$ is observed of 1.5% ± 0.4%, where the uncertainty is the standard deviation across SST-$\mathcal{P}$ pairs. All declines are found to be significant ($p < 0.001$) by all three significance estimation procedures. Over the entire timeseries, CF16 predicts a $\langle\langle ef \rangle\rangle$ value of 0.128 ± 0.016.

How consequential is a 1.5% ± 0.4% decrease in $\langle\langle ef \rangle\rangle$? A simple box model of the carbon cycle (Ito and Follows 2005; Williams and Follows 2011; *see* Supporting Information) suggests a 1.5% ± 0.4% decline in $\langle\langle ef \rangle\rangle$ would result in a 1.2% ± 0.3% increase in the mixing ratio of atmospheric CO_2 on millennial timescales (when the solubility pump has equilibrated but prior to carbonate compensation). Generalizing this result, the sensitivity of the soft-tissue carbon pump to changes in mean sea surface temperature is predicted to be ~ 7 ppm K^{-1}, comparable to the ~ 10 ppm K^{-1} sensitivity of the solubility pump to global mean ocean temperature suggested by theory and models (Williams and Follows 2011, their Fig. 13.9b). Interestingly, the combined sensitivity of solubility and metabolic effects predicts a ~ 70 ppm drawdown of atmospheric CO_2 for a 4K global cooling of the ocean associated with the last Glacial Maximum (Adkins et al. 2002).

We note that substantial latitudinal variation in SST trends have occurred during the record of these SST reanalyses. Thus, we employ the same procedure as above, but only averaging zonally, yielding an estimated percent change in $\langle ef \rangle$ (here the ratio of longitudinally integrated export flux to longitudinally integrated production) at each degree of latitude from 1982 to 2014. See Fig. 2; substantial latitudinal variation exists, with percent changes ranging between +2% and −8%. $\langle ef \rangle$ in the Southern Ocean increases slightly, corresponding to cooling, while $\langle ef \rangle$ changes little in the tropics where SST is high. Thus, it appears the global decrease of 1–2% is driven by decreases in the Arctic and at midlatitudes, where $\langle ef \rangle$ decreases on the order of 5%. While variation exists between each of the six SST-$\mathcal{P}$ pairs, their latitudinal dependence is similar.

Conclusion

Neither the existence, magnitude, nor driving mechanisms of a change in the biological pump over the past few decades of recent climate change can be established definitively. This is an unavoidable result of searching for small shifts in a system exhibiting substantial variability on all time scales that is challenging to measure adequately.

A simple metabolic perspective accounting for the differential temperature dependencies of autotrophy and heterotrophy underpins a model of export efficiency, which explains its observed dependence on temperature. Driving that model with observed changes in SST leads to a suggestion that global export efficiency has decreased 1.5% ± 0.4% over the past few decades, with larger decreases in midlatitudes and Arctic. This decrease is robust across SST and primary production data products. Larger temperature differences than those observed over the past few decades are predicted to cause larger changes in global export efficiency.

References

Adkins, J. F., K. McIntyre, and D. P. Schrag. 2002. The salinity, temperature and $\delta^{18}O$ of the Glacial deep ocean. Science **298**: 1769–1773. doi:10.1126/science.1076252

Behrenfeld, M. J., and P. G. Falkowski. 1997. Photosynthetic rates derived from satellite-based chlorophyll concentration. Limnol. Oceanogr. **42**: 1–20. doi:10.4319/lo.1997.42.1.0001

Behrenfeld, M. J., and others. 2016. Revaluating ocean warming impacts on global phytoplankton. Nat. Clim. Chang. **6**: 323–330. doi:10.1038/nclimate2838

Boyd, P. W. 2015. Toward quantifying the response of the oceans' biological pump to climate change. Front. Mar. Sci. **2**: 77. doi:10.3389/fmars.2015.00077

Boyd, P. W., and T. W. Trull. 2007. Understanding the export of biogenic particles in oceanic waters: Is there consensus? Prog. Oceanogr. **72**: 276–312. doi:10.1016/j.pocean.2006.10.007

Buesseler, K. O. 1998. The decoupling of production and particular export in the surface ocean. Global Biogeochem. Cycles **12**: 297–310. doi:10.1029/97GB03366

Buesseler, K. O., and P. W. Boyd. 2009. Shedding light on processes that control particle export and flux attenuation in the twilight zone of the open ocean. Limnol. Oceanogr. **54**: 1210–1232. doi:10.4319/lo.2009.54.4.1210

Cael, B. B., and M. J. Follows. 2016. On the temperature dependence of oceanic export efficiency. Geophys. Res. Lett. **43**: 5170–5175. doi:10.1002/2016GL068877

Cox, P. M., R. A. Betts, C. D. Jones, S. A. Spall, and I. J. Totterdell. 2000. Acceleration of global warming due to carbon-cycle feedbacks in a coupled climate model. Nature **408**: 184–187. doi:10.1038/35041539

Dee, D. P., and others. 2011. The ERA-Interim reanalysis: Configuration and performance of the data assimilation system. Q. J. R. Meteorol. Soc. **137**: 553–597. doi:10.1002/qj.828

Dunne, J. P., R. A. Armstrong, A. Gnanadesikan, and J. L. Sarmiento. 2005. Empirical and mechanistic models for the particle export ratio. Global Biogeochem. Cycles **19**: GB4026. doi:10.1029/2004GB002390

Eppley, R. W. 1972. Temperature and phytoplankton growth in the sea. Fish. Bull. **70**: 1063–1085.

Eppley, R. W. 1989. New production: History, methods, problem, p. 85–97. *In* W. H. Berger, V. S. Smetacek, and G. Wefer [eds.], Productivity of the ocean. John Wiley.

Henson, S. A., J. L. Sarmiento, J. P. Dunne, L. Bopp, I. Lima, S. C. Doney, J. John, and C. Beaulieu. 2010. Detection of anthropogenic climate change in satellite records of ocean chlorophyll and productivity. Biogeosciences **7**: 621–640. doi:10.5194/bg-7-621-2010

Henson, S. A., R. Sanders, E. Madsen, P. J. Morris, F. Le Moigne, and G. D. Quartly. 2011. A reduced estimate of the strength of the ocean's biological carbon pump. Geophys. Res. Lett. **38**: L04606. doi:10.1029/2011GL046735

Huntley, M. E., and M. D. Lopez. 1992. Temperature-dependent production of marine copepods: A global synthesis. Am. Nat. **140**: 201–242. doi:10.1086/285410

IPCC. 2014. Climate Change 2014: Synthesis Report. *In* Core Writing Team, R. K. Pachauri, and L. A. Meyer [eds.], Contribution of Working Groups I, II and III to the Fifth Assessment Report of the Intergovernmental Panel on Climate Change, 151 pp. IPCC, Geneva, Switzerland.

Ishii, M., A. Shouji, S. Sugimoto, and T. Matsumoto. 2005. Objective analyses of sea-surface temperature and marine meteorological variables for the 20th century using ICOADS and the Kobe collection. Int. J. Climatol. **25**: 865–879. doi:10.1002/joc.1169

Ito, T., and M. J. Follows. 2005. Preformed phosphate, soft tissue pump and atmospheric CO_2. J. Mar. Res. **63**: 813–839. doi:10.1357/0022240054663231

Laws, E. A., P. G. Falkowski, W. O. Smith, H. Ducklow, and J. J. McCarthy. 2000. Temperature effects on export production in the open ocean. Global Biogeochem. Cycles **14**: 1231–1246. doi:10.1029/1999GB001229

Laws, E. A., E. D'Sa, and P. Naik. 2011. Simple equations to estimate ratios of new or export production to total production from satellite-derived estimates of sea surface temperature and primary production. Limnol. Oceanogr.: Methods **9**: 593–601. doi:10.4319/lom.2011.9.593

López-Urrutia Á., E. San Martin, R. P. Harris, and X. Irigoien. 2006. Scaling the metabolic balance of the oceans. Proc. Nat. Acad. Sci. **103**: 8739–8744.

Maiti, K., M. A. Charette, K. O. Buesseler, and M. Kahru. 2013. An inverse relationship between production and export efficiency in the Southern Ocean. Geophys. Res. Lett. **40**: 1557–1561. doi:10.1002/grl.50219

Michaels, A. F., and M. W. Silver. 1988. Primary production, sinking fluxes and the microbial food web. Deep-Sea Res. Part A **35**: 473–490. doi:10.1016/0198-0149(88)90126-4

Reynolds, R. W., T. M. Smith, C. Liu, D. B. Chelton, K. S. Casey, and M. G. Schlax. 2007. Daily high-resolution-blended analyses for sea surface temperature. J. Clim. **20**: 5473–5496. doi:10.1175/2007JCLI1824.1

Siegel, D. A., and others. 2013. Regional to global assessments of phytoplankton dynamics from the SeaWiFS mission. Remote Sens. Environ. **135**: 77–91. doi:10.1016/j.rse.2013.03.025

Volk, T., and M. I. Hoffert. 1985. Ocean carbon pumps: Analysis of relative strengths and efficiencies in ocean-driven atmospheric CO_2 changes, p. 99–110. *In* E. T. Sundquist and W. S. Broecker [eds.], The carbon cycle and atmospheric CO_2: Natural variations archean to present. Washington, DC: American Geophysical Union.

Westberry, T., M. J. Behrenfeld, D. A. Siegel, and E. Boss. 2008. Carbon-based primary productivity modeling with vertically resolved photoacclimation. Global Biogeochem. Cycles **22**: GB2024. doi:10.1029/2007GB003078

Williams, R. G., and M. J. Follows. 2011. Ocean dynamics and the carbon cycle. Cambridge Univ. Press.

Acknowledgments

This material is based upon work supported by the National Science Foundation Graduate Research Fellowship Program, and the National Science Foundation Award Nos. OCE-1315201 and OCE-1259388. MJF is grateful for support from the Simons Foundation (SCOPE Award, No. 329108, Follows) and the Gordon and Betty Moore Foundation Marine Microbiology Initiative grant #3778.

20

Ordination obscures the influence of environment on plankton metacommunity structure

Tad A. Dallas,[1,2] *Andrew M. Kramer,*[1] *Marcus Zokan,*[3] *John M. Drake*[1]

[1]University of Georgia, Odum School of Ecology, Athens, Georgia; [2]University of California, Environmental Science and Policy, Davis, California; [3]US Fisheries and Wildlife, SE, Social Circle, Georgia

Scientific Significance Statement

Freshwater systems are often thought to operate as metacommunities, or discrete communities connected through dispersal. There are different ways to analyze metacommunities to understand the underlying factors controlling structure, most of which use some form of ordination. Here, we modify one analysis, showing where the effect of environmental controls on metacommunity structure in plankton metacommunties can be obscured by traditional ordination and we propose a novel way to better visualize metacommunity structure.

Abstract

The composition of plankton communities in individual habitats is often influenced by environmental conditions like pH or hydroperiod. At larger scales, environmental gradients can influence community structure across interconnected local communities. Detecting the role of environmental and spatial factors on metacommunity structure depends on the ordering of sites and species prior to analysis. We investigated this ordination in two wetland metacommunities; a well-sampled, hyper-diverse zooplankton metacommunity, and a Central American phytoplankton metacommunity. We calculated coherence, turnover, and boundary clumping to classify the structure of the metacommunity, and we propose a statistic that responds to variation in both coherence and turnover. Traditional ordination approaches failed to discern metacommunity structure, while significant structure existed along abiotic gradients in both zooplankton and phytoplankton systems. This shows that abiotic controls on community composition may not be detectable with traditional analyses, and suggests an alternative of ordering sites by known abiotic gradients.

Metacommunity ecology is a rapidly developing research area that seeks to explain local community composition by reference to landscape-level (e.g., spatial orientation of communities) and local (e.g., site-level environmental variables) factors (Logue et al. 2011). Local factors can be separated from the regional effects of dispersal through statistical modeling. However, spatial autocorrelation in environmental conditions, and the use of dispersal proxies, such as spatial distance, make the task of disentangling environmental and spatial influences difficult (Smith and Lundholm 2010; Tuomisto et al. 2012). Two common approaches to the analysis of metacommunities are variance partitioning on ordinated community data (Cottenie 2005; Tuomisto et al. 2012), and metacommunity classification using statistics of the site-species matrix (Leibold and Mikkelson 2002). Differentiating local patch effects and regional dispersal effects is a core goal

*Correspondence: tdallas@ucdavis.edu

Author Contribution Statement: AMK, MZ and JMD designed the zooplankton sampling study. MZ conducted the zooplankton sampling. TAD performed the statistical analyses. All authors contributed to the writing of the manuscript.

of metacommunity ecology. This combination of theory from community ecology and biogeography has been applied to a range of complex ecosystems, including zooplankton communities in ponds and lakes (Cottenie et al. 2003), parasite communities of host species (Dallas and Presley 2014), and microbial communities of pitcher plants (Miller and Kneitel 2005). While conceptually compelling, there is not agreement about what quantitative framework best captures the factors influencing metacommunity structure (i.e., quantifiable patterns of nestedness or clustering of patches in environmental or geographic space) (Ulrich and Gotelli 2013).

One approach to quantifying metacommunity structure is the pattern-based framework proposed by Leibold and Mikkelson (2002). This framework characterizes metacommunity structure according to three statistics: coherence, turnover, and boundary clumping. Coherence measures the continuity of a species range, quantified as the number of absences within the species range boundaries (i.e., embedded absences). Significantly positive coherence (i.e., low number of embedded absences) is a necessary condition for the other two statistics to be interpretable in the Leibold and Mikkelson (2002) framework. The second statistic, turnover, measures the tendency of species to replace one another at their range boundaries, and is closely related to measures of species co-occurrence [i.e., checkerboard score; Stone and Roberts (1990)]. The final statistic of the Leibold and Mikkelson (2002) framework is boundary clumping, which measures the tendency of species range boundaries to occur together. Inferences may be made by comparing each of these statistics (apart from boundary clumping) to a null distribution generated by permutation of species occurrences subject to metacommunity-level constraints. The metacommunity statistics (coherence, turnover, and boundary clumping) are calculated on the empirical interaction matrix, compared to the null distributions of statistics from permutation analysis, and the metacommunity is classified as one of six (Leibold and Mikkelson 2002) [or ten, Presley et al. (2010)] possible structures.

This framework has been criticized, however, due to variation in performance on simulated data and the known sensitivity of estimated statistics to matrix ordering (Ulrich and Gotelli 2013). This critique could be addressed by using statistics that are invariant to matrix order [e.g., Barber's modularity *Q*; Barber (2007)], or by ordering sites and species along known environmental gradients (Gotelli and Ulrich 2012). Despite these potential shortcomings, we believe the Leibold and Mikkelson (2002) framework may nevertheless be of considerable value. Particularly, it has the virtue of providing quantities that map clearly to basic metacommunity concepts and may be modified and extended as needed to represent new ideas (Presley et al. 2010).

We used the Leibold and Mikkelson (2002) framework, and addressed the critique of Ulrich and Gotelli (2013) in two novel ways. First, we compared metacommunity classifications based on the reciprocal averaging approach of Leibold and Mikkelson (2002) to an approach that orders sites and species along known environmental gradients. Reciprocal averaging (also known in community ecology as "correspondence analysis") places sites with similar species, and species with similar distributions, closer together in the site-by-species interaction matrix. Commonly, these ordination scores are related to environmental or spatial covariates to infer the relative influence of covariates on metacommunity structure (Willig et al. 2011; López-González et al. 2012; Dallas and Presley 2014). However, correlations between covariates and ordination scores may be misleading, while ordering sites by known environmental gradients may lead to different conclusions. Specifically, assessing metacommunity structure along multiple environmental gradients could result in different classifications (e.g., *random, nested*), which could suggest conflicting influences of environmental covariates on metacommunity structure. We demonstrate the sensitivity of the analysis to matrix ordering, and recommend ways to address this issue. Second, we extend the Leibold and Mikkelson (2002) framework by advocating the use of a continuous measures of the magnitude of coherence and turnover to quantify variation in metacommunity classifications within a continuous space [see Heino et al. (2015)]. In our analysis, a metacommunity occupies a single point within a continuous space of coherence and turnover. Successional changes, species extinctions, immigration events, and species invasions may not alter metacommunity classification, but will affect the location of the metacommunity in the coherence-turnover phase space. Previous studies have visualized different metacommunities in this phase space (Heino et al. 2015). We propose instead to study the location in this space of a single metacommunity, ordered according different environmental variables. The distance between points in coherence-turnover space offers a novel way to examine the influence of time or null model choice on metacommunity structure.

We examined two aquatic metacommunity systems (one zooplankton and one phytoplankton) to investigate our extension to the Leibold and Mikkelson (2002) framework. Plankton metacommunities are well-studied, although the relative importance of spatial (Shurin 2000; Rojo et al. 2016), and environmental (Cottenie et al. 2001, 2003) covariates remains unclear (Soininen et al. 2007; Dallas and Drake 2014). Both phytoplankton and zooplankton community data were obtained through repeated sampling of two series of wetlands. Zooplankton were sampled from 14 Carolina Bays, and phytoplankton from 30 wetlands in Costa Rica and Nicaragua (Rojo et al. 2016). For both data sources, we analyzed metacommunities ordered by traditional ordination analysis, and compared the outcome when sites were ordered along known environmental gradients. We demonstrated that (1) classification of plankton metacommunity structure is sensitive to the order of sites and species, and (2) plankton

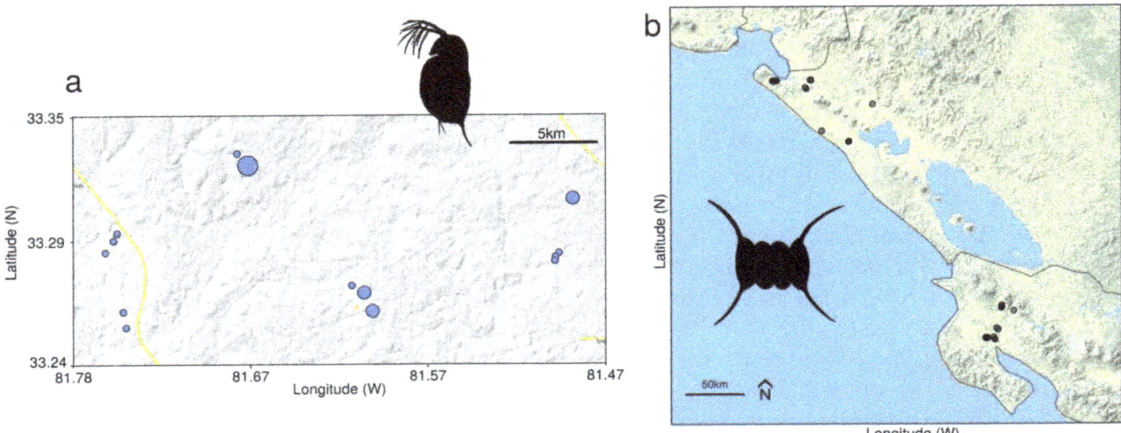

Fig. 1. The spatial distribution of (**a**) the 14 Carolina Bays sampled at the Savannah River Site, where points indicate bay position and point size is proportional to bay area, and (**b**) the 30 Central American wetlands sampled for phytoplankton.

assemblages were structured by environmental covariates that were only weakly related to ordination scores. Taken together, these findings provide evidence of environmental structure in zooplankton and phytoplankton communities, reveal the pitfalls of the traditional ordination approach, and demonstrate a useful extension to the Leibold and Mikkelson (2002) framework for characterizing metacommunity structure.

Methods

Zooplankton communities

Zooplankton communities were sampled every other week between January 2009 and February 2011 from 14 ephemeral water bodies within the Savannah River Site (SRS), a nuclear reserve owned by the United States Department of Energy located in the upper coastal plain of South Carolina (Fig. 1). These ephemeral wetlands, also referred to as Carolina Bays, are an ideal system to examine metacommunity structure, as communities are extremely diverse, well-sampled, and occupy discrete habitats. These ephemeral wetlands usually have no natural surface drainage and depend largely on precipitation for filling, resulting in typically low conductivity and acidic conditions (Newman and Schalles 1990). Sampled wetlands ($n = 14$) are typically inundated in the winter and dry in summer [although some stay inundated year round except in drought years; Sharitz (2003)]. These wetlands are strikingly species-rich, as single wetlands can have as many as 60 zooplankton species (Zokan 2016), and over 100 zooplankton species recorded throughout the system of ephemeral wetlands (DeBiase and Taylor 2005). A total of 84 zooplankton species were recorded over the course of 49 potential sampling events over the time period examined here. Further information related to water quality and sampling protocols can be found in the Supporting Information.

Phytoplankton communities

Phytoplankton communities were sampled by Rojo et al. (2016). Sites were sampled at three times between September 2010 and June 2011, corresponding to periods of inundation and dessication of wetlands. Here, we combine data from two sampling periods (January and June 2011), excluding the September 2010 sampling as some of the wetlands sampled later were not sampled in September. This resulted in a total of 295 phytoplankton species across the 30 sampled wetlands. See Rojo et al. (2016) for more detailed information on sampled sites and methodology, as well as experimental data [provided as Supporting Information in Rojo et al. (2016)].

Metacommunity analyses

We used the Leibold and Mikkelson (2002) framework to analyze zooplankton species assemblages among ephemeral bays, and phytoplankton assemblages among wetlands, both represented as site-species matrices of wetlands (rows) and species (columns). The ordering of sites and species can influence statistical values of the Leibold and Mikkelson (2002) framework, and may alter metacommunity classification. Sites were ordered by either reciprocal averaging scores or observed environmental gradients. Reciprocal averaging concentrates species occurrences along the matrix diagonal to group species with similar ranges together, and sites with similar species assemblages together. Column ordination was based on reciprocal averaging in all cases, while row ordering was based on central tendencies of measured environmental variables. For zooplankton communities, environmental variables included mean values of conductivity, water depth, hydroperiod (length of time the bay contains water), pH, and temperature. For phytoplankton communities, sites were ordered based on mean values of conductivity, water depth, pH, temperature, dissolved oxygen, chlorophyll *a*, bicarbonate, nitrate, and phosphate.

Table 1. Coherence, measured as the number of embedded absences (c), for sites are ordered along different environmental gradients ("Covariate").

Covariate	c	z	p	$\bar{C}$	σ_C
Conductivity	448	0.43	0.66	459	26
Depth	417	1.13	0.26	447	26
Hydroperiod	453	−0.42	0.67	443	25
pH	415	2.02	0.04	465	25
Temperature	450	0.02	0.98	451	25
Recip. avg.	375	1.35	0.18	412	27
Bicarbonate	1424	0.89	0.37	1516	103
Chlorophyll *a*	1408	1.28	0.20	1551	112
Conductivity	1374	1.82	0.07	1564	105
Depth	1348	1.12	0.26	1493	129
Dissolved oxygen	1272	2.17	0.03	1552	129
Nitrate	1355	2.00	0.05	1590	117
pH	1303	2.09	0.04	1560	123
Phosphate	1318	1.83	0.07	1533	117
Temperature	1344	1.49	0.14	1509	111
Turbidity	1385	1.73	0.08	1558	100
Recip. avg.	1178	1.59	0.11	1348	107

Null values of c (denoted with an upper-case C) were calculated based on the swap null model, which fixes row and column totals. These null values (mean = $\bar{C}$, variance = σ_C) were compared to the empirical c to assess significance and calculate divergence from null expectation (z). "Recip. avg." refers to the metacommunity ordered based on reciprocal averaging ordination scores.

Table 2. Species turnover (β) values for sites ordered along different environmental gradients ("Covariate").

Covariate	β	z	p	$\bar{B}$	σ_B
Conductivity	938	−0.38	0.70	674	695
Depth	116	0.79	0.43	658	688
Hydroperiod	74	1.82	0.07	2164	1145
pH	3444	−6.51	<0.001	384	47
Temperature	230	0.54	0.59	599	680
Recip. avg.	8938	−2.24	0.03	5633	1475
Bicarbonate	17361	0.73	0.47	28642	15459
Chlorophyll *a*	38849	−0.65	0.51	28575	15750
Conductivity	8299	1.32	0.19	25613	13141
Depth	42187	−0.38	0.70	36029	16217
Dissolved oxygen	73712	−3.55	<0.001	34112	11165
Nitrate	27032	0.03	0.98	27372	13433
pH	58133	−2.70	0.01	25026	12282
Phosphate	4811	0.30	0.76	6327	5010
Temperature	22436	0.56	0.58	29769	13204
Turbidity	10220	0.41	0.68	14535	10544
Recip. avg.	97533	−0.21	0.83	91419	29130

Null values for β (denoted with a B) were calculated based on the swap null model, which fixes row and column totals. These null values (mean = $\bar{B}$, variance = σ_B) were compared to the empirical β to assess significance and calculate divergence from null expectation (z). "Recip. avg." refers to the metacommunity ordered based on reciprocal averaging ordination scores.

Following Leibold and Mikkelson (2002), three statistics were calculated on each ordered site-species matrix; coherence, turnover, and boundary clumping. Coherence is quantified as the number of absences between the extremes of species ranges, and measures the tendency for species ranges to be constrained to a subset of similar sites (Leibold and Mikkelson 2002). Turnover is a measure of overlap at the extremes of species ranges, and is calculated after species ranges are made completely coherent (embedded absences are removed). Boundary clumping measures the tendency of the extremes of species ranges to coincide with the range extremes of other species, and is quantified using Morisita's index (Leibold and Mikkelson 2002). This measure is conceptually similar to the field of community detection and measures of modularity in graph theory, although measures of modularity are typically invariant to site-species matrix ordering (Barber 2007). Here, we use Morisita's index since we aimed to examine the sensitivity of the Leibold and Mikkelson (2002) framework to matrix ordering.

Statistical significance was determined relative to the distribution of the statistic calculated on randomized matrices (i.e., a null distribution) for coherence and turnover. Significance of boundary clumping (Morisita's index) was assessed relative to a chi-squared distribution. The randomization of species occurrences, much like the ordering of the interaction matrix, can influence statistical results and metacommunity classification. We chose a conservative null model [sequential swap algorithm; Gotelli and Entsminger (2003)] that maintains the number of sites occupied for each species and the number of species in each local community (also called a *fixed-fixed* null model). However, we relax this null model in the Supporting Information, and instead assign species occurrences proportional to the number of observed occurrences [called a *fixed-proportional* null model; Gotelli (2000)].

Significance was determined by calculating each statistic for a set of 1000 null matrices, and comparing the statistic calculated on the empirical matrix using a z-test. However, significance does not address the magnitude of divergence from the null expectation. To do so, we used the z statistic as a measure of the divergence of our empirical statistic from the statistic's null distribution. We computed the z statistic as the mean simulated statistic minus the observed statistic divided by the standard deviation of the simulated statistic.

Relationship of environmental gradients to reciprocal averaging gradient

Values for each environmental gradient were compared to ordination scores (first axis of reciprocal averaging ordination) to determine the strength of association between

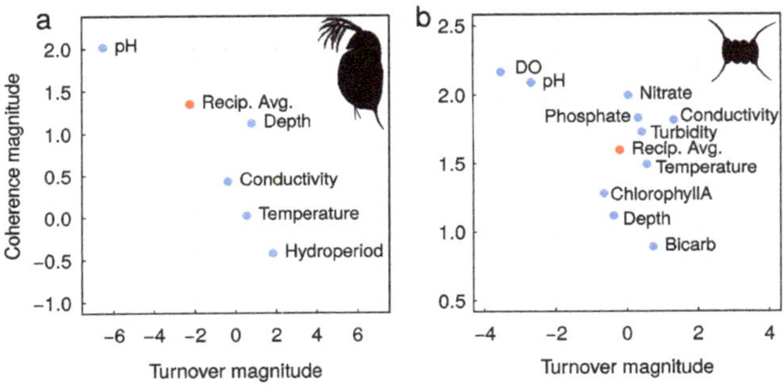

Fig. 2. Zooplankton (**a**) and phytoplankton (**b**) metacommunities were structured differently when sites were ordered by environmental variables or reciprocal averaging gradient (red point; "Recip. avg."). The magnitude of the difference between the null expectation and the empirical statistics for coherence (*y*-axis), and turnover (*x*-axis) determines the location of the metacommunity in the phase space. Gray lines separate regions based on statistical significance ($\alpha < 0.05$) relative to a null expectation, where larger positive values of coherence magnitude indicate fewer embedded absences relative to the null model, and larger negative turnover magnitude indicates more species replacements than expected under the null model.

environmental gradient and the latent ordination gradient. The underlying assumption made by previous studies (Willig et al. 2011; López-González et al. 2012) is that environmental factors important for metacommunity structure would be strongly correlated with the ordination scores. To test this, we examined correlations between environmental gradients and the first ordination axis, hypothesizing that environmental gradients along which the metacommunity was strongly structured would be more strongly associated with the structuring gradient obtained from the reciprocal averaging ordination.

Results

Metacommunity structure along environmental gradients

When the site-species matrix was ordered by reciprocal averaging ordination scores, the zooplankton metacommunity did not differ from the expectation under the *fixed-fixed* (sequential swap) null model with respect to the number of embedded absences in species ranges (i.e., coherence; Table 1) or species turnover (Table 2). Based on the framework of Leibold and Mikkelson (2002), this metacommunity would be considered *random.* However, when sites were ordered based on pH, the metacommunity contained fewer embedded absences than expected by chance, indicating non-random metacommunity structure (Table 1).

Apart from being non-random, the metacommunity ordered based on pH had significantly positive turnover (more species turnover than expected under the fixed row and column null model; Table 2) and boundary clumping (pH $M = 10.98$, $df = 84$, $p < 0.0001$) suggesting that species replaced one another in discrete groups across the pH gradient (i.e., a *Clementsian* metacommunity). Results for the

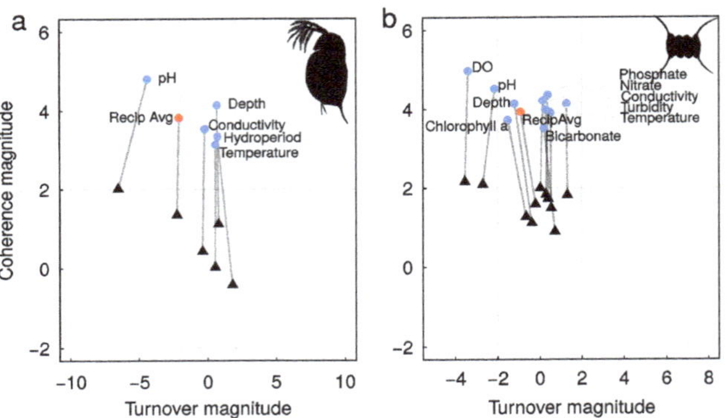

Fig. 3. The choice of null model influenced the results of both zooplankton (**a**) and phytoplankton (**b**) metacommunities. Results from both null models are plotted in the coherence and turnover phase space, with dark gray lines connecting the same environmental gradients for both *fixed-fixed* (gray triangles) and *fixed-proportional* (blue circles) null models. Gray horizontal and vertical lines correspond to significance levels at $\alpha = 0.05$.

fixed-proportional null model were similar, although the metacommunity ordered along the reciprocal averaging gradient and conductivity become non-random (Fig. 3 and Supporting Information).

Similar results were found for the phytoplankton metacommunity, where the number of embedded absences (i.e., coherence) in the traditionally ordered metacommunity was not significantly different than expected under the *fixed-fixed* null model, while the same metacommunity ordered along gradients of dissolved oxygen and pH yielded significantly non-random metacommunities (Table 1). Moreso, metacommunities ordered based on dissolved oxygen and pH had significantly greater turnover than expected (Table 2) and significantly clumped range boundaries (dissolved oxygen $M = 9.98$, $p < 0.001$; pH $M = 12.09$, $p < 0.001$). When sites and species were ordered using the traditional reciprocal averaging ordination, the metacommunity did not differ from the null expectation with respect to coherence, and would be classified as *random.* Thus, the same metacommunity ordered along the traditional reciprocal averaging gradient and two environmental gradients (pH and dissolved oxygen), produce two starkly different results.

Magnitude measures

The two metacommunity statistics that were compared relative to null models (i.e., coherence and turnover) allow for the creation of quantitative measures for the magnitude of effect, as described above. These measures create a phase space, where metacommunities ordered by different factors occupy different regions (Fig. 2). This continuous space adds to the simple classification of metacommunities by providing information on the distance between classification states. For instance, the zooplankton metacommunity was *Clementsian* when ordered by pH, suggesting that species replaced one another in discrete communities across the pH gradient,

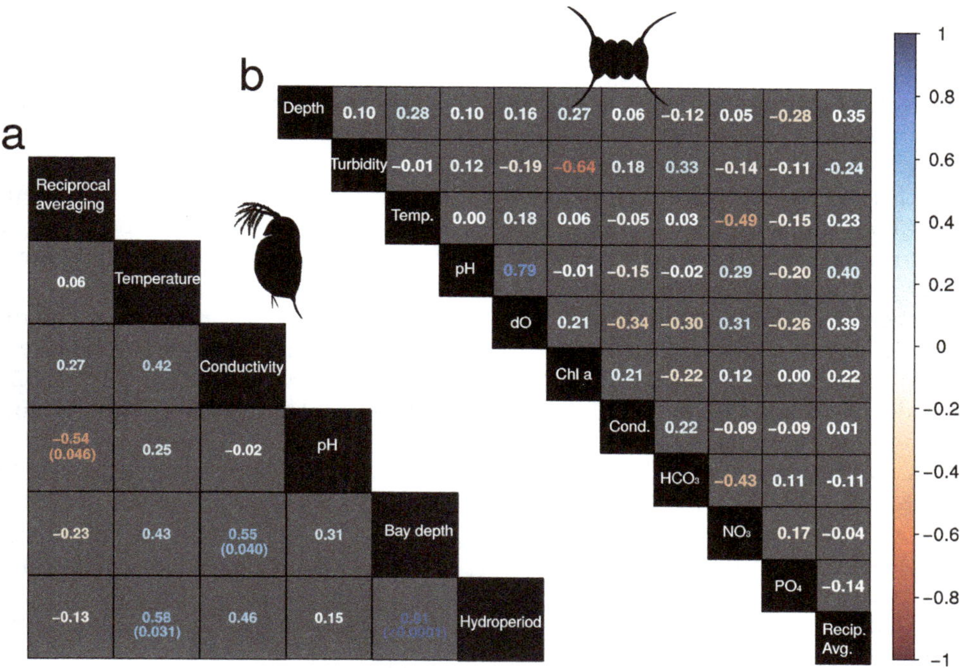

Fig. 4. Correlation matrix of relationships between mean environmental conditions and site ordination scores for zooplankton (**a**) and phytoplankton (**b**) metacommunities. Pearson's correlation coefficients are provided in the upper triangle (*p*-values less than 0.05 provided below correlation coefficients). Scaled circles represent the size of correlation, and color (red to blue) indicates the direction of the correlation (negative to positive). Reciprocal averaging gradients were not strongly correlated with environmental covariates. Some covariate names are abbreviated for clarity (Temp. = temperature, dO = dissolved Oxygen, Chl a = chlorophyll a).

which can be visualized in phase space (Fig. 2). The position of a metacommunity in this phase space changes when using a different null model (Fig. 3), providing a means to evaluate the robustness of metacommunity classifications.

Associations between environmental covariates and the reciprocal averaging gradient

The environmental gradient responsible for structuring the zooplankton metacommunity was pH, as discussed above (Tables 1, 2). However, the metacommunity was classified as *random* when ordered using the traditional analytical approach (i.e., ordering by reciprocal averaging scores). The gradient obtained from reciprocal averaging was unrelated to the conductivity gradient (Fig. 4; $t = 0.96$, $df = 12$, $p = 0.356$), and only weakly correlated with the pH gradient (Fig. 4; $t = -2.23$, $df = 12$, $p = 0.046$). Hydroperiod did not influence zooplankton assemblages, but was strongly related to other environmental covariates (i.e., mean bay depth and temperature).

Similarly for the phytoplankton metacommunity, the reciprocal averaging gradient did not result in the same metacommunity classification compared to ordering sites along known environmental gradients. The phytoplankton metacommunity would be considered *random* when ordered along the traditional ordination gradient. Meanwhile, environmental gradients of dissolved oxygen and pH resulted in significantly structured metacommunites. Both dissolved oxygen (*rho* $= 0.07$, $df = 28$, $p = 0.03$) and pH (*rho* $= 0.40$, $df = 28$, $p = 0.03$) were weakly related to the reciprocal averaging ordination gradient (Fig. 4).

Discussion

The zooplankton metacommunity was classified as *Clementsian* along a gradient of pH, suggesting that discrete subsets of the zooplankton community have overlapping pH tolerances, resulting in a modular interaction pattern. However, the metacommunity was classified as *random* when sites were ordered by their reciprocal averaging scores [i.e., the traditional Leibold and Mikkelson (2002) approach]. Further, the phytoplankton metacommunity was classified as *Clementsian* when sites were ordered along gradients of dissolved oxygen and pH, while the metacommunity was classified as *random* when sites were ordered along the reciprocal averaging gradient. Regardless of metacommunity classification, the structuring environmental gradients for both zooplankton and phytoplankton were weakly related or unrelated to the reciprocal averaging gradient, and correlations often differed in the direction of the relationship. This suggests that the reciprocal averaging gradient was able to capture some environmental variation, but still masked the influence of individual environmental gradients. Perhaps more importantly, this suggests that the reciprocal averaging gradient may not capture all of the relevant biological information, leading to incorrect conclusions about metacommunity structure.

The Leibold and Mikkelson (2002) framework has been criticized as too simplistic, and potentially inaccurate, based on a simulation study examining several commonly used measures of species aggregation and segregation (Ulrich and Gotelli 2013). However, Ulrich and Gotelli (2013) found that the Leibold and Mikkelson (2002) statistics performed comparably to other measures of aggregation and segregation, despite being taken out of the context of the analytical chain proposed in Leibold and Mikkelson (2002). Specifically, matrices generated randomly, including those that would be classified as *random*, were used to assess the statistical properties of the other statistics in the Leibold and Mikkelson (2002) framework. Here, we alleviate some of the concerns raised in Ulrich and Gotelli (2013) by extending the Leibold and Mikkelson (2002) framework; introducing a continuous measure of coherence and turnover, calculating statistics relative to a number of realistic null models (see Supporting Information), and ordering sites along known biological gradients (Gotelli and Ulrich 2012).

Using our Leibold and Mikkelson (2002) extension, we found pH is important to the structure of ephemeral pond zooplankton communities. Previous work in lakes in the northeast United States (Dallas and Presley 2014), boreal shield lakes (Derry et al. 2009), and interconnected ponds (Cottenie et al. 2001) has suggested the importance of pH as a determinant of zooplankton community structure. This is arguably because zooplankton species have relatively narrow regions of pH conditions at which they can exist (Holt et al. 2003). The Carolina bays studied are rain-driven, and are typically more acidic than more permanent water bodies that may have a groundwater source. Hydroperiod can influence zooplankton diversity (Serrano and Fahd 2005; Zokan and Drake 2015), resilience (Angeler and Moreno 2007), and colonization dynamics (Chaparro et al. 2016) in ephemeral waterbodies. We failed to find an effect of hydroperiod on zooplankton community structure in our system. This may be because we didn't explicitly examine temporal patterns in zooplankton community composition (i.e., succession). To date, few studies of metacommunities have considered how metacommunity structure may change over time (but see Keith et al. 2011; Newton et al. 2012). However, the Leibold and Mikkelson (2002) framework may be extended to explain successional metacommunities by examining metacommunities as multi-layer networks (Pilosof et al. 2015), or examining changes in the coherence-turnover phase space we suggest here as a measure of metacommunity change through time.

References

Angeler, D. G., and J. M. Moreno. 2007. Zooplankton community resilience after press-type anthropogenic stress in temporary ponds. Ecol. Appl. **17**: 1105–1115. doi:10.1890/06-1040

Barber, M. J. 2007. Modularity and community detection in bipartite networks. Phys. Revi. E **76**: 066102. doi:10.1103/PhysRevE.76.066102

Chaparro, G., M. S. Fontanarrosa, and I. OFarrell. 2016. Colonization and succession of zooplankton after a drought: Influence of hydrology and free-floating plant dynamics in a floodplain lake. Wetlands **36**: 85–100. doi:10.1007/s13157-015-0718-3

Cottenie, K. 2005. Integrating environmental and spatial processes in ecological community dynamics. Ecol. Lett. **8**: 1175–1182. doi:10.1111/j.1461-0248.2005.00820.x

Cottenie, K., N. Nuytten, E. Michels, and L. De Meester. 2001. Zooplankton community structure and environmental conditions in a set of interconnected ponds. Hydrobiologia **442**: 339–350. doi:10.1023/A:1017505619088

Cottenie, K., E. Michels, N. Nuytten, and L. De Meester. 2003. Zooplankton metacommunity structure: Regional vs. local processes in highly interconnected ponds. Ecology **84**: 991–1000. doi:10.1890/0012-9658(2003)084[0991:ZMSRVL]2.0.CO;2

Dallas, T., and J. M. Drake. 2014. Relative importance of environmental, geographic, and spatial variables on zooplankton metacommunities. Ecosphere **5**: art104. doi:10.1890/ES14-00071.1

Dallas, T., and S. J. Presley. 2014. Relative importance of host environment, transmission potential and host phylogeny to the structure of parasite metacommunities. Oikos **123**: 866–874. doi:10.1111/oik.00707

DeBiase, A. E., and B. E. Taylor. 2005. Microcrustaceans (Branchiopoda and Copepoda) of wetland ponds and impoundments on the Savannah river site, Aiken, South Carolina. Tech. rep., Savannah River Ecology Laboratory (SREL), Aiken, SC.

Derry, A. M., S. E. Arnott, J. A. Shead, P. D. Hebert, and P. T. Boag. 2009. Ecological linkages between community and genetic diversity in zooplankton among boreal shield lakes. Ecology **90**: 2275–2286. doi:10.1890/07-1852.1

Gotelli, N. J. 2000. Null model analysis of species co-occurrence patterns. Ecology **81**: 2606–2621. doi:10.1890/0012-9658(2000)081[2606:NMAOSC]2.0.CO;2

Gotelli, N. J., and G. L. Entsminger. 2003. Swap algorithms in null model analysis. Ecology **84**: 532–535. doi:10.1890/0012-9658(2003)084[0532:SAINMA]2.0.CO;2

Gotelli, N. J., and W. Ulrich. 2012. Statistical challenges in null model analysis. Oikos **121**: 171–180. doi:10.1111/j.1600-0706.2011.20301.x

Heino, J., J. Soininen, J. Alahuhta, J. Lappalainen, and R. Virtanen. 2015. A comparative analysis of metacommunity types in the freshwater realm. Ecol. Evol. **5**: 1525–1537. doi:10.1002/ece3.1460

Holt, C., N. Yan, and K. Somers. 2003. pH 6 as the threshold to use in critical load modeling for zooplankton community change with acidification in lakes of south-central Ontario: Accounting for morphometry and geography. Can. J. Fish. Aquat. Sc. **60**: 151–158. doi:10.1139/f03-008

Keith, S. A., A. C. Newton, M. D. Morecroft, D. J. Golicher, and J. M. Bullock. 2011. Plant metacommunity structure remains unchanged during biodiversity loss in English woodlands. Oikos **120**: 302–310. doi:10.1111/j.1600-0706.2010.18775.x

Leibold, M. A., and G. M. Mikkelson. 2002. Coherence, species turnover, and boundary clumping: Elements of metacommunity structure. Oikos **97**: 237–250. doi:10.1034/j.1600-0706.2002.970210.x

Logue, J. B., N. Mouquet, H. Peter, H. Hillebrand, Metacommunity Working Group. 2011. Empirical approaches to metacommunities: A review and comparison with theory. Trends Ecol. Evol. **26**: 482–491. doi:10.1016/j.tree.2011.04.009

López-González, C., S. J. Presley, A. Lozano, R. D. Stevens, and C. L. Higgins. 2012. Metacommunity analysis of Mexican bats: Environmentally mediated structure in an area of high geographic and environmental complexity. J. Biogeogr. **39**: 177–192. doi:10.1111/j.1365-2699.2011.02590.x

Miller, T. E., and J. M. Kneitel. 2005. Inquiline communities in pitcher plants as a prototypical metacommunity, p. 122–145. *In* M. Holyoak, M. A. Leibold, and R. D. Holt [eds.], Metacommunities: spatial dynamics and ecological communities. University of Chicago Press.

Newman, M. C., and J. F. Schalles. 1990. The water chemistry of Carolina bays: A regional survey. Arch. Hydrobiol. **118**: 147–168. doi:0003-9136/90/0118-0147

Newton, A. C., R. M. Walls, D. Golicher, S. A. Keith, A. Diaz, and J. M. Bullock. 2012. Structure, composition and dynamics of a calcareous grassland metacommunity over a 70-year interval. J. Ecol. **100**: 196–209. doi:10.1111/j.1365-2745.2011.01923.x

Pilosof, S., M. A. Porter, and S. Kéfi. 2015. Ecological multilayer networks: A new frontier for network ecology. arXiv Preprint arXiv:1511.04453.

Presley, S. J., C. L. Higgins, and M. R. Willig. 2010. A comprehensive framework for the evaluation of metacommunity structure. Oikos **119**: 908–917. doi:10.1111/j.1600-0706.2010.18544.x

Rojo, C., F. Mesquita-Joanes, J. S. Monrós, J. Armengol, M. Sasa, F. Bonilla, R. Rueda, J. Benavent-Corai, R. Piculo, and M. M. Segura. 2016. Hydrology affects environmental and spatial structuring of microalgal metacommunities in tropical pacific coast wetlands. PLoS One **11**: e0149505. doi:10.1371/journal.pone.0149505

Serrano, L., and K. Fahd. 2005. Zooplankton communities across a hydroperiod gradient of temporary ponds in the Donana National Park (SW Spain). Wetlands **25**: 101–111. doi:10.1672/0277-5212(2005)025[0101:ZCAAHG]2.0.CO;2

Sharitz, R. R. 2003. Carolina bay wetlands: Unique habitats of the southeastern United States. Wetlands **23**: 550–562. doi:10.1672/0277-5212(2003)023[0550:CBWUHO]2.0.CO;2

Shurin, J. B. 2000. Dispersal limitation, invasion resistance, and the structure of pond zooplankton communities. Ecology **81**: 3074–3086. doi:10.1890/0012-9658(2000)081[3074:DLIRAT]2.0.CO;2

Smith, T. W., and J. T. Lundholm. 2010. Variation partitioning as a tool to distinguish between niche and neutral processes. Ecography **33**: 648–655. doi:10.1111/j.1600-0587.2009.06105.x

Soininen, J., M. Kokocinski, S. Estlander, J. Kotanen, and J. Heino. 2007. Neutrality, niches, and determinants of plankton metacommunity structure across boreal wetland ponds. Ecoscience **14**: 146–154. doi:10.2980/1195-6860(2007)14[146:NNADOP]2.0.CO;2

Stone, L., and A. Roberts. 1990. The checkerboard score and species distributions. Oecologia **85**: 74–79. doi:10.1007/BF00317345

Tuomisto, H., L. Ruokolainen, and K. Ruokolainen. 2012. Modelling niche and neutral dynamics: On the ecological interpretation of variation partitioning results. Ecography **35**: 961–971. doi:10.1111/j.1600-0587.2012.07339.x

Ulrich, W., and N. J. Gotelli. 2013. Pattern detection in null model analysis. Oikos **122**: 2–18. doi:10.1111/j.1600-0706.2012.20325.x

Willig, M. R., S. J. Presley, C. P. Bloch, I. Castro-Arellano, L. M. Cisneros, C. L. Higgins, and B. T. Klingbeil. 2011. Tropical metacommunities along elevational gradients: Effects of forest type and other environmental factors. Oikos **120**: 1497–1508. doi:10.1111/j.1600-0706.2011.19218.x

Zokan, M. 2016. Zooplankton species diversity in the temporary wetland system of the Savannah River Site, South Carolina, USA. Univ. of Georgia. Dissertation. p. 152.

Zokan, M., and J. M. Drake. 2015. The effect of hydroperiod and predation on the diversity of temporary pond zooplankton communities. Ecol. Evol. **5**: 3066–3074. doi:10.1002/ece3.1593

Acknowledgments

We wish to thank Rebecca Sharitz (SREL/UGA) and Linda Lee (SREL) for the use of laboratory space, equipment, and logistical support for the Savannah River Site zooplankton sampling. We also thank Chris Dibble for providing comments on earlier drafts. Funding was provided from the Savannah River Ecology Laboratory, the University of Georgia and the Odum School of Ecology.

Permissions

All chapters in this book were first published in L&O LETTERS, by John Wiley & Sons Ltd.; hereby published with permission under the Creative Commons Attribution License or equivalent. Every chapter published in this book has been scrutinized by our experts. Their significance has been extensively debated. The topics covered herein carry significant findings which will fuel the growth of the discipline. They may even be implemented as practical applications or may be referred to as a beginning point for another development.

The contributors of this book come from diverse backgrounds, making this book a truly international effort. This book will bring forth new frontiers with its revolutionizing research information and detailed analysis of the nascent developments around the world.

We would like to thank all the contributing authors for lending their expertise to make the book truly unique. They have played a crucial role in the development of this book. Without their invaluable contributions this book wouldn't have been possible. They have made vital efforts to compile up to date information on the varied aspects of this subject to make this book a valuable addition to the collection of many professionals and students.

This book was conceptualized with the vision of imparting up-to-date information and advanced data in this field. To ensure the same, a matchless editorial board was set up. Every individual on the board went through rigorous rounds of assessment to prove their worth. After which they invested a large part of their time researching and compiling the most relevant data for our readers.

The editorial board has been involved in producing this book since its inception. They have spent rigorous hours researching and exploring the diverse topics which have resulted in the successful publishing of this book. They have passed on their knowledge of decades through this book. To expedite this challenging task, the publisher supported the team at every step. A small team of assistant editors was also appointed to further simplify the editing procedure and attain best results for the readers.

Apart from the editorial board, the designing team has also invested a significant amount of their time in understanding the subject and creating the most relevant covers. They scrutinized every image to scout for the most suitable representation of the subject and create an appropriate cover for the book.

The publishing team has been an ardent support to the editorial, designing and production team. Their endless efforts to recruit the best for this project, has resulted in the accomplishment of this book. They are a veteran in the field of academics and their pool of knowledge is as vast as their experience in printing. Their expertise and guidance has proved useful at every step. Their uncompromising quality standards have made this book an exceptional effort. Their encouragement from time to time has been an inspiration for everyone.

The publisher and the editorial board hope that this book will prove to be a valuable piece of knowledge for researchers, students, practitioners and scholars across the globe.

List of Contributors

Ryan D. Batt and Stephen R. Carpenter
Center for Limnology, University of Wisconsin - Madison, Madison, Wisconsin

Anthony R. Ives
Department of Zoology, University of Wisconsin - Madison, Madison, Wisconsin

Walter K. Dodds
Kansas State University, Manhattan, Kansas

Flavia Tromboni
Departamento de Ecologia, IBRAG, Universidade do Estado do Rio de Janeiro, Rio de Janerio, Brazil

Wesley Aparecido Saltarelli and Davi Gasparini Fernandes Cunha
Departamento de Hidráulica e Saneamento, Escola de Engenharia de São Carlos, Universidade de São Paulo, São Paulo, Brazil

Hilary A. Dugan and John J. Magnuson
Center for Limnology, University of Wisconsin - Madison, Madison, Wisconsin

Greta Helmueller
Center for Limnology, University of Wisconsin - Madison, Madison, Wisconsin
University of South Florida, College of Marine Science, St. Petersburg, Florida

Tamar Guy-Haim
Israel Oceanographic and Limnological Research, National Institute of Oceanography, Haifa, Israel
Marine Biology Department, The Leon H. Charney School of Marine Sciences, University of Haifa, Mt. Carmel, Haifa, Israel

Erez Yeruham
Israel Oceanographic and Limnological Research, National Institute of Oceanography, Haifa, Israel

Orit Hyams-Kaphzan and Ahuva Almogi-Labin
Geological Survey of Israel, Jerusalem, Israel

James T. Carlton
Williams College - Mystic Seaport, Maritime Studies Program, Mystic, Connecticut

Nicole M. Hayes
Department of Biology, University of Regina, Regina, Saskatchewan, Canada

Bridget R. Deemer
School of the Environment, Washington State University-Vancouver, Vancouver, Washington

Jessica R. Corman
Center for Limnology, University of Wisconsin-Madison, Madison, Wisconsin

N. Roxanna Razavi
Finger Lakes Institute, Hobart and William Smith Colleges, Geneva, New York

Kristin E. Strock
Environmental Science Department, Dickinson College, Carlisle, Pennsylvania

Shane L. Hogle, Randelle M. Bundy and Katherine A. Barbeau
Geosciences Research Division, Scripps Institution of Oceanography, La Jolla, California

Jessica M. Blanton and Eric E. Allen
Marine Biology Research Division, Scripps Institution of Oceanography, La Jolla, California

Emma S. Kritzberg
Department of Biology/Aquatic Ecology, Lund University, Lund, Sweden

Elizabeth B. Kujawinski, Krista Longnecker and Cara L. Fiore
Department of Marine Chemistry and Geochemistry, Woods Hole Oceanographic Institution, Woods Hole, Massachusetts

Sheean T. Haley and Sonya T. Dyhrman
Department of Earth and Environmental Science and the Lamont-Doherty Earth Observatory, Columbia University, Palisades, New York

Winifred M. Johnson
MIT-WHOI Joint Program in Oceanography/ Applied Ocean Science and Engineering, Cambridge, Massachusetts and Woods Hole, Massachusetts

Harriet Alexander
Department of Earth and Environmental Science and the Lamont-Doherty Earth Observatory, Columbia University, Palisades, New York
MIT-WHOI Joint Program in Oceanography/ Applied Ocean Science and Engineering, Cambridge, Massachusetts and Woods Hole, Massachusetts

Gastón L. Miño
Ralph M. Parsons Laboratory, Department of Civil and Environmental Engineering, Massachusetts Institute of Technology, Cambridge, Massachusetts
Laboratorio de Microscopía Aplicada a Estudios Moleculares y Celulares (LAMAE), Centro de Investigaciones y Transferencia de Entre Ríos (CITER), Facultad de Ingeniería, Universidad Nacional de Entre Ríos (FIUNER), Oro Verde, Argentina

M. A. R. Koehl
Department of Integrative Biology, University of California, Berkeley, California

Nicole King
Howard Hughes Medical Institute and the Department of Molecular and Cell Biology, University of California, Berkeley, California

Roman Stocker
Ralph M. Parsons Laboratory, Department of Civil and Environmental Engineering, Massachusetts Institute of Technology, Cambridge, Massachusetts
Institute of Environmental Engineering, Department of Civil, Environmental, and Geomatic Engineering, ETH Zurich, Zurich, Switzerland

S. M. Powers, S. G. Labou and S. E. Hampton
Center for Environmental Research, Education and Outreach, Washington State University, Pullman, Washington

H. M. Baulch
School of Environment and Sustainability, and Global Institute for Water Security, University of Saskatchewan, Saskatoon, Saskatchewan, Canada

R. J. Hunt
US Geological Survey, Wisconsin Water Science Center, Middleton, Wisconsin

N. R. Lottig
Trout Lake Station, Center for Limnology, University of Wisconsin–Madison, Boulder Junction, Wisconsin

E. H. Stanley
Center for Limnology, University of Wisconsin-Madison, Madison, Wisconsin

Antonietta Quigg
Department of Marine Biology, Texas A & M University at Galveston, Galveston, Texas
Department of Oceanography, Texas A & M University, College Station, Texas

Uta Passow
Marine Science Institute, University of California Santa Barbara, Santa Barbara, California

Wei-Chun Chin
School of Engineering, University of California - Merced, Merced, California

Chen Xu
Department of Marine Science, Texas A & M University at Galveston, Galveston, Texas

Shawn Doyle, Jason B. Sylvan, Anthony H. Knap and Terry L. Wade
Department of Oceanography, Texas A & M University, College Station, Texas

Laura Bretherton, Manoj Kamalanathan and Alicia K. Williams
Department of Marine Biology, Texas A & M University at Galveston, Galveston, Texas

Zoe V. Finkel
Environmental Science, Mount Allison University, New Brunswick, Sackville, Canada

Kathleen A. Schwehr, Saijin Zhang and Luni Sun
Department of Marine Science, Texas A & M University at Galveston, Galveston, Texas

Wassim Obeid and Patrick G. Hatcher
Department of Chemistry and Biochemistry, Old Dominion University, Norfolk, Virginia

Peter H. Santschi
Department of Oceanography, Texas A & M University, College Station, Texas
Department of Marine Science, Texas A & M University at Galveston, Galveston, Texas

Kevin C. Rose
Department of Biological Sciences, Rensselaer Polytechnic Institute, Troy, New York

Luke A. Winslow and Jordan S. Read
U.S. Geological Survey, Office of Water Information, Middleton, Wisconsin

Gretchen J. A. Hansen
Bureau of Science Services, Wisconsin Department of Natural Resources, Madison, Wisconsin

Luciana M. Sanders
Southern Cross Geoscience, Southern Cross University, Lismore, New South Wales, Australia
National Marine Science Centre, School of Environment, Science and Engineering, Southern Cross University, Coffs Harbour, New South Wales, Australia

Kathryn H. Taffs
Southern Cross Geoscience, Southern Cross University, Lismore, New South Wales, Australia

Debra J. Stokes
Marine Ecology Research Centre, Southern Cross University, Lismore, New South Wales, Australia

Christian J. Sanders, Paul A. Macklin and Isaac R. Santos
National Marine Science Centre, School of Environment, Science and Engineering, Southern Cross University, Coffs Harbour, New South Wales, Australia

Joseph M. Smoak
Department of Environmental Science, University of South Florida, St. Petersburg, Florida, USA

Alex Enrich-Prast
Department of Environmental Change, Linköping University, Linköping, Sweden

Humberto Marotta
Ecosystems and Global Change Laboratory (LEMG-UFF) / International Laboratory of Global Change (LINCGlobal), Biomass and Water Management Research Center (NAB-UFF), Graduated Program in Geosciences (Environmental Geochemistry), Universidade Federal Fluminense (UFF), Niterói, Rio de Janeiro, Brazil
Sedimentary and Environmental Processes Laboratory (LAPSA-UFF), Department of Geography, Graduated Program in Geography, Universidade Federal Fluminense (UFF), Niterói, Rio de Janeiro, Brazil

Jasmine E. Saros, Robert M. Northington and Dennis S. Anderson
Climate Change Institute & School of Biology and Ecology, University of Maine, Orono, Maine

Nicholas John Anderson
Department of Geography, Loughborough University, Leicestershire, LE11 3TU, United Kingdom

S. Simoncelli and D. J. Wain
Department of Architecture and Civil Engineering, University of Bath, Claverton Down, Bath, United Kingdom

S. J. Thackeray
Centre for Ecology & Hydrology, Lancaster Environment Centre, Bailrigg, Lancaster, United Kingdom

Rachel E. Sipler, Donglai Gong, Marta P. Sanderson, Quinn N. Roberts and Deborah A. Bronk
The Virginia Institute of Marine Science, College of William & Mary, Gloucester Point, Virginia

Steven E. Baer
The Virginia Institute of Marine Science, College of William & Mary, Gloucester Point, Virginia
Bigelow Laboratory for Ocean Sciences, East Boothbay, Maine

Margaret R. Mulholland
Department of Ocean, Earth and Atmospheric Sciences, Old Dominion University, Norfolk, Virginia

James Z. Sippo, Damien T. Maher, Douglas R. Tait, Christian J. Sanders and Isaac R. Santos
School of Environment, Science and Engineering, Southern Cross University, Lismore, 2480 Australia
National Marine Science Centre, Southern Cross University, Coffs Harbour, New South Wales 2450, Australia

Sergio Ruiz-Halpern
School of Environment, Science and Engineering, Southern Cross University, Lismore, 2480 Australia

Amber R. Bellamy and James E. Bauer
Aquatic Biogeochemistry Laboratory, Department of Evolution, Ecology and Organismal Biology, Ohio State University, Columbus, Ohio

B. B. Cael
Department of Earth, Atmosphere, and Planetary Sciences, Massachusetts Institute of Technology, Cambridge, Massachusetts
Department of Physical Oceanography, Woods Hole Oceanographic Institution, Woods Hole, Massachusetts

Michael J. Follows
Department of Earth, Atmosphere, and Planetary Sciences, Massachusetts Institute of Technology, Cambridge, Massachusetts

Kelsey Bisson
Earth Research Institute, University of California, Santa Barbara, California

Tad A. Dallas
University of Georgia, Odum School of Ecology, Athens, Georgia
University of California, Environmental Science and Policy, Davis, California

Andrew M. Kramer and John M. Drake
University of Georgia, Odum School of Ecology, Athens, Georgia

Marcus Zokan
US Fisheries and Wildlife, SE, Social Circle, Georgia

Index

Q

R

S

T

V

W

Z

www.ingramcontent.com/pod-product-compliance
Lightning Source LLC
Chambersburg PA
CBHW080643200326
41458CB00013B/4723

* 9 7 8 1 6 4 1 1 6 1 1 4 5 *